Materials in Eighteenth-Century Science

Transformations: Studies in the History of Science and Technology
Jed Buchwald, general editor

Mordechai Feingold, editor, *Jesuit Science and the Republic of Letters*

Larrie D. Ferreiro, *Ships and Science: The Birth of Naval Architecture in the Scientific Revolution, 1600–1800*

Sungook Hong, *Wireless: From Marconi's Black-Box to the Audion*

Jeff Horn, *The Path Not Taken: French Industrialization in the Age of Revolution, 1750–1830*

Myles W. Jackson, *Harmonious Triads: Physicists, Musicians, and Instrument Makers in Nineteenth-Century Germany*

Myles W. Jackson, *Spectrum of Belief: Joseph von Fraunhofer and the Craft of Precision Optics*

Mi Gyung Kim, *Affinity, That Elusive Dream: A Genealogy of the Chemical Revolution*

Ursula Klein and Wolfgang Lefèvre, *Materials in Eighteenth-Century Science: A Historical Ontology*

John Krige, *American Hegemony and the Postwar Reconstruction of Science in Europe*

Janis Langins, *Conserving the Enlightenment: French Military Engineering from Vauban to the Revolution*

Wolfgang Lefèvre, editor, *Picturing Machines 1400–1700*

Staffan Müller-Wille and Hans-Jörg Rheinberger, editors, *Heredity Produced: At the Crossroads of Biology, Politics, and Culture, 1500–1870*

William R. Newman and Anthony Grafton, editors, *Secrets of Nature: Astrology and Alchemy in Early Modern Europe*

Gianna Pomata and Nancy G. Siraisi, editors, *Historia: Empiricism and Erudition in Early Modern Europe*

Alan J. Rocke, *Nationalizing Science: Adolphe Wurtz and the Battle for French Chemistry*

George Saliba, *Islamic Science and the Making of the European Renaissance*

Materials in Eighteenth-Century Science

A Historical Ontology

Ursula Klein and Wolfgang Lefèvre

The MIT Press
Cambridge, Massachusetts
London, England

For information about special quantity discounts, please email special_sales@mitpress.mit.edu.

This book was set in Times by Heinz Reddner, MPI for the History of Science, Berlin, and was printed and bound in the United States of America.

Printed on recycled paper.

Library of Congress Cataloging-in-Publication Data

Klein, Ursula.
Materials in eighteenth-century science : a historical ontology / Ursula Klein and Wolfgang Lefèvre.
   p. cm. — (Transformations)
 Includes bibliographical references and index.
 ISBN 978-0-262-11306-9 (hardcover : alk. paper)
 1. Chemistry—History—18th century. 2. Ontology—History—18th century.
 3. Classification of sciences. I. Lefèvre, Wolfgang, 1941– II. Title.
QD15.K54 2007
540.9'033—dc22
                      2006039115

10 9 8 7 6 5 4 3 2 1

Contents

Conclusion: Multidimensional Objects and Materiality 295

List of Figures

Introduction

This book presents a new history of material objects in general, and of eighteenth-century chemistry in particular. It interweaves three historical and philosophical themes: ontologies of materials, practices of making, identifying and classifying materials, and the science of materials, from the late seventeenth century until the early nineteenth century.

In the eighteenth century the science of materials was chemistry. Though learned inquiries into materials also took place in mineralogy, botany, pharmacy, architecture, engineering, and a few other areas, chemistry was the only scientific culture in the eighteenth century where materials were studied persistently, comprehensively, and from multiple perspectives. Eighteenth-century chemists treated materials as useful commodities, perceptible objects of nature, and entities carrying imperceptible features. They invested commodities with new meaning when they ordered them according to natural origin, analyzed their invisible components, and explored their affinities in chemical transformations. In so doing, chemists constituted objects of inquiry that reached out to cultures of natural history and experimental philosophy. Chemical substances were multidimensional objects of inquiry that could be investigated in practical and theoretical contexts and that amalgamated perceptible and imperceptible, useful and philosophical, technological and scientific, social and natural features. The many faces of eighteenth-century chemical substances challenge our current understanding of objects of inquiry in the history of science and distinctions between scientific and quotidian technological objects, and objects of natural history and natural philosophy.

Our main approach to eighteenth-century chemists' ontology of materials is the scrutiny of practices of identification and classification. Identifying and classifying things are human activities that structure the world by ordering single things into kinds of things and by establishing relationships between the different kinds. Studies of classifications inform historians what objects were handled in the past and how the historical actors understood these objects; that is, they lay bare the rough ontological structures of the past. In the course of the eighteenth century, chemists' ontology of materials shifted in various ways, in keeping with changes in their classificatory practices. For example, until the middle of the eighteenth century chemists regarded plant materials primarily as remedies; in keeping with pharmaceutical classification they ordered plant materials into two different classes, namely, the pharmaceutical simples purchased from merchants, such as sugar, camphor, natural balsams, wax, gums, and resins, and the chemical remedies prepared in their laboratories, such as distilled oils, distilled waters, extracts, and essential salts. Around the middle of the eighteenth century they epistemically elevated plant materials as compound components or proximate principles of plants, while grouping together the pharmaceutical simples and the chemical remedies into the unified class of proximate principles of plants. Some four decades later they began to highlight "organic substances" created by processes of life, grouping together proximate principles of plants and of animals. Yet beginning in the late 1820s many of these early organic substances, such as balsams, distilled vegetable oils and animal fats, disappeared from chemists' agenda, and were replaced by a new type of scientific objects, namely the stoichiometrically pure carbon compounds that were classified according to chemical composition and constitution represented by chemical formulae.

Eighteenth-century chemistry has often been studied as a science of atoms, corpuscles, and Newtonian forces. Instead, our approach depicts the chemistry of that period as a science of materials. The material substances studied by eighteenth-century chemists were for the most part commodities procured, sold, or tested in apothecary's shops, foundries, assaying laboratories, arsenals, dye manufactories, distilleries, coffee shops, and so on. Even the few substances that were genuine inventions or discoveries of the academic laboratory were soon transferred to the mundane world, where they found application as remedies and other goods. But chemists also studied substances as natural objects that carry perceptible and imperceptible features. We argue that chemically processed substances and natural raw materials played such a central role in eighteenth-century chemistry because they lent themselves to multifarious ways of inquiry: descriptive (in the *historia* tradition), technological, and philosophical. Our focus on substances allows us to grasp issues traditionally highlighted as characteristic of the science of chemistry—composition, affinities, and similar entities akin to the imperceptible objects of experimental philosophy—alongside themes traditionally treated as centerpieces of chemical technology. We thus obtain a larger picture of chemistry from the late seventeenth century until the early nineteenth century, outlined with broad strokes but extending from its mundane artisanal practices to experimental and natural histories and all the way to conceptual or philosophical inquiry.

Speaking of eighteenth-century chemistry almost inevitably brings up the theme of the Chemical Revolution. The Lavoisierian Chemical Revolution of the last third of the eighteenth century has certainly been the most debated topic in the history of chemistry. It also spurred controversies in the history and philosophy of science more broadly. If we are right to claim that eighteenth-century chemical substances were multidimensional objects of inquiry, which invited chemists to switch from perceptible properties and commercial uses to imperceptible features, our approach should also provide new insight into the Chemical Revolution. This is indeed the case. The two studies presented in parts II and III of this book—the classification of pure chemical substances in the *Méthode de nomenclature chimique* of 1787, which has always been regarded as a central achievement of the Chemical Revolution, as well as chemists' classification of plant materials before and after roughly 1790—challenge the current understanding of the Chemical Revolution. Seen from our new perspective, Lavoisier and his collaborators reaped the rewards of a century. In so doing, they introduced reforms of concepts, theories, analytical methods, and language. Yet they neither initiated an ontological rupture nor overthrew the existing taxonomic structure of their main objects of inquiry, the chemical substances.

Chemists continued to live in largely the same world of objects of inquiry before and after the Lavoisierian reforms. Well into the nineteenth century, in large areas of chemistry, especially plant and animal chemistry, the modes of individuating, identifying and classifying substances were similar to artisans' and naturalists' classificatory practices. This began to change in the 1830s with the emergence of the new experimental culture of organic or carbon chemistry. The pure stoichiometric substances produced, individuated, identified, and classified in carbon chemistry were tangled in a web of new types of experiments and work on paper with chemical formulae that did not exist outside academic chemistry at the time. As material things, however, these novel types of substances remained potential commodities—a potential that began to be realized some twenty years later with the rise of the synthetic dye industry. There was no move away from perceptible, applicable substances and towards the study of imperceptible chemical composition or molecular structure in nineteenth-century chemistry. The significance of perceptible substances was not transformed into that of mere targets that allowed experimental investigations of molecular structure. Rather, substances remained chemists' predominant objects of inquiry well into the twentieth century, and studies of composition, constitution, and molecular structure remained closely tied to studies of the perceptible and applicable dimension of these things.

A striking discovery in our research was just how diverse eighteenth-century chemists' classificatory practices were. Eighteenth-century chemists did not order substances under a single conceptual umbrella or paradigm, and they did not create one comprehensive taxonomic system. Apart from the many different ways of classifying materials in contexts of technological inquiry or "applied chemistry," we found two main differences in their ways of classification in the context of conceptual investigations: classification according to chemical composition, and classification according to provenance and perceptible properties. This striking difference in eighteenth-century chemists' modes of classification informs the organization of the book and its division into two main empirical parts—part II analyzing the domain of substances classified according to chemical composition, and part III studying plant materials classified according to provenance, or ways of chemical preparation, and perceptible properties. A long introductory part I tackles historical and philosophical questions concerning the kinds of materials studied by eighteenth-century chemists, chemists' collective practices of studying these objects, and the uses of studies of classification for historians and philosophers of science. In a final conclusion we examine the role played by materiality in the existence and maintenance of the major difference in chemists' order of materials.

Acknowledgments

For helpful comments on the manuscript of this book, we would like to thank our colleagues at the Max Planck Institute for the History of Science (MPIWG), Berlin, the two anonymous referees, and, last but not least, Jed Z. Buchwald, who included the book in his *Transformations* series.

We also want to thank Sara Meirowitz and her team at The MIT Press, particularly Mel Goldsipe and Krista Magnuson. Special thanks go to Heinz Reddner (MPIWG), who set the book.

In part III we have drawn on material that was previously published in the following articles by Ursula Klein: "Shifting Ontologies, Changing Classifications: Plant Materials from 1700 to 1830" (*Studies in History and Philosophy of Science* 36, 2005, pp. 261–329); "Contexts and Limits of Lavoisier's Analytical Plant Chemistry: Plant Materials and their Classification" (*Ambix* 52, 2005, pp. 107–157).

Part I
Materials in Eighteenth-Century Science
Contexts and Practices

Introduction to Part I

Materials are essential to our life. It is hard to imagine how human civilization would have developed without stone, wood, clay, bronze, iron, brass, copper, and other kinds of stuff our tools and instruments are made of. Olive oil and other vegetable oils, animal fats, wine and beer, honey and beeswax, juices and syrups, sea salt and rock salt, milk, and water have been used as food and medicines since prehistoric times. Materials are part and parcel of the material culture of human societies. Hence it is not surprising that they have served to demarcate large historical periods in human civilization, such as the Stone Age, Bronze Age, or Iron Age. Materials are even indispensable for the most accomplished forms of human life. There is no modern painting without canvas and pigments, no poem without paper and ink, no classical symphony without wood and metals to build the violin and the piano.

The extension and diversification of materials, extracted from natural objects and produced in workshops, factories, and laboratories, are among the most impressive manifestations of technological and scientific productivity. Today the databases of chemists and material scientists list more than ten million different materials and chemical substances.[1] Science has been deeply involved in these developments, long before "materials science" was established in the second half of the twentieth century.[2] Our essay turns to an early period of the entanglement of science with the making of materials and material knowledge: the eighteenth century. In the eighteenth century different groups of learned men and (very few) women were concerned with materials, among them architects, engineers, physicians, apothecaries, botanists, mineralogists, metallurgists, assayers, cameralist officials, educated manufacturers, members of academies, and experimental philosophers.[3] But the most outstanding group of learned men and women, those who possessed the broadest range of skills and knowledge on materials and had the most authoritative voice in this respect, were the chemists. In almost all European countries, governments commissioned chemists to analyze raw minerals; control the quality of dyestuffs; explore the secrets of making porcelain and steel; find native substitutes for precious imported goods such as sugar, dyestuffs, and balsams; and survey the production processes of materials in state manufactories.[4] In their laboratories, eighteenth-century chemists experimented with materials, studied possibilities of improving their production, examined their physical and chemical properties, explored their invisible reactions, and analyzed

---

1   See Schummer [1997]. In the following we use the term "materials" for all kinds of stuff that have been applied in industry and everyday consumption, and "substances" when we wish to highlight the scientific circumscription of these objects. For a philosophical history of materials from antiquity until the twentieth century see Bensaude-Vincent [1998].
2   For the establishment of materials science in the U.S., see Bensaude-Vincent [2001b].
3   Here, we use the term "learned" in a broad sense, including maker's knowledge that relied on forms of learning other than apprenticeship.
4   On the relationship between eighteenth-century chemical science and technology see Bensaude-Vincent and Nieto-Galan [1999]; Clow and Clow [1952]; Fester [1923]; Fors [2003]; Frängsmyr [1974]; Gee [1989]; Gillispie [1957], [1980]; Guerlac [1977]; Gustin [1975]; Hickel [1978]; Homburg [1993], [1999a]; Hufbauer [1982]; Klein [2003b], [2005b]; Lindquist [1984]; Meinel [1983]; Multhauf [1965], [1966], [1972], [1984], [1996]; Nieto-Galan [2001]; Porter [1981]; Schneider [1968–1975], [1972]; Simon [2005]; Smith [1979]; Teich [1975]; and Tomic [2003].

Figure I.1: Substances from the "Chemical Museum" at Leeds, founded in 1874 (courtesy of the University Collection, University of Leeds).

their composition. In the eighteenth century, writing about and teaching chemistry meant, first of all, the compilation and teaching of experimental histories of substances. The students attending public or private lectures of chemistry at the time — most of whom were students of medicine, physicians, apothecaries, mineralogists, botanists, assayers, dyers, colorists, cameralist officials, and other practitioners — were keen to improve their connoisseurship of materials and their analytical knowledge about their composition. Even the concepts and theories that characterized eighteenth-century chemistry — "chemical compound," "composition," "affinity," and the theories of salts, calcination, fermentation, composition of plants, combustion, and so on — evolved not primarily around "atoms" or "corpuscles," but chemical substances.

Our book focuses on eighteenth-century chemistry as the most authoritative science of materials at the time. We will study this early science of materials from a broad, comparative view that provides insight into the types of materials handled and studied in eighteenth-century European chemistry as well as the apothecary trade, metallurgy, and other closely related arts and crafts. What types of materials did eighteenth-century chemists study in their experiments, lectures, and textbooks? Where did these materials come from? How did chemists transform quotidian materials when they studied them as scientific objects? How did chemical substances change in history?

Our goal is ambitious: we want to write a history of the most significant scientific objects of classical chemistry—chemical substances—covering European chemistry in the time period from the beginning of the eighteenth century until the early nineteenth century.[5] And we want to philosophically analyze conditions of the constitution of these scientific objects and changes in their materiality and meaning. How can we do this? It is all too obvious that it would be impossible to write such a history and philosophy by following the traces of each single material and analyzing hundreds (or even thousands) of experimental reports written by French, German, British, Swedish, and other European chemists over the course of the eighteenth century. It would be impossible not only because of the almost insurmountable amount of work required, but also because the result would be an overwhelming number of details rather than a broader picture of chemists' ontology of materials. Therefore we have chosen an approach starting not from the single materials and the single experiments performed with them, but from a more structured level of historical objects and practices: eighteenth-century chemists' modes of identifying and classifying materials.

Identifying, classifying, and naming things are ubiquitous human activities, which structure the world by defining what kinds of things exist and how they relate to each other. Whereas in everyday life and in knowledge traditions outside the sciences classificatory structures are powerful but often inarticulated schemes, carried only by ordinary language, in the sciences they range from inarticulated schemes to explicitly formulated partial taxonomic orders to technically elaborated comprehensive taxonomic systems. In whatever form classifications occur in the history of science, they tell historians of science what kinds of objects were handled in scientific practices of the past, how historical actors conceived of these objects, and how they selected and highlighted those of their manifold features they considered significant. Historical studies of modes of classification provide insight not only into the development of exact scientific methodologies, what has been called the "quantifying spirit,"[6] but also into historical actors' ontologies and historical shifts of ontologies.[7]

Philosophers have often treated classification as an abstract intellectual achievement, next to theory and independent of practice and social contexts.[8] This understanding of classification has long been questioned by anthropologists, sociologists, historians, and several philosophers, too, who have studied classification as a historically situated activity, entrenched in social and cultural institutions and informed not only by perception and cognition, but also by goals and interests, meaning, and col-

---

5 For an argument that eighteenth-century chemistry was a European science, see Holmes [1995]; Kragh [1998]; and Nieto-Galan [2001] pp. 123–136. For a more general discussion of the interaction of European scholars in the Enlightenment and afterwards, see Daston [1991].
6 See Frängsmyr et al. [1990].
7 For the notion of "ontology," or "historical ontology," used here, see also Hacking [2002] p. 1. A related concept is Daston's "applied metaphysics," which "studies the dynamic world of what emerges and disappears from the horizon of working scientists" Daston [2000] p. 1; see also Daston [2004a]. For differences between Hacking's concept of historical ontology and Daston's concept of "applied metaphysics," see Hacking [2002] pp. 10–11.
8 This has been the case especially in the metaphysical discussion on "natural kinds;" see the classical papers by Kripke [1980]; Putnam [1975a]; and Quine [1969]. For a recent discussion on natural kinds in chemistry, see van Brakel [2000], and the papers included in Harré [2005a].

lective ways of working with and handling objects.[9] Our study builds on this recent line of studies of classification. If classification is studied as a practice linked with other practices, it allows us to bring to the foreground both the types of material objects historical actors were dealing with and their epistemic approach to these objects. Studies of classification thus contribute to a historical ontology of the past that takes into account the actors' material culture and their ways of making and knowing.

---

9   See, in particular, Barnes et al. [1996]; Bowker and Star [1999]; Buchwald [1992]; Daston [2004b]; Douglas and Hull [1992]; Dupré [1993], [2001]; Foucault [1970]; Hacking [1991], [1993], [2002]; Kuhn [1989], [1993]; McOuat [1996]; and Roth [2005].

1

Commodities and Natural Objects

Eighteenth-century chemists studied an astonishingly rich arsenal of materials, ranging from entire plants, roots, leaves, flowers, bones, hair, nails, and other organized
vegetable and animal parts to balsams, resins, gums, oils, fats, and blood extracted
from plants and animals, to composite materials such as ceramics, porcelain, and
glass, all the way to processed chemical substances such as metals, mineral acids,
alkalis, and salts. When we include all kinds of raw materials and processed substances eighteenth-century chemists studied in their laboratories, described in their
experimental histories, and classified at their writing desks, their number amounts to
thousands. And even if we omitted the many composite and organized materials
which eighteenth-century chemists reproduced, extracted, analyzed, and further
explored in their laboratories, focusing only on those processed substances which
look more like the typical "chemical substances," the number of those selected substances would still be astonishing. The famous table of chemical nomenclature and
the adjoined chemical lexicon, published in 1787 by Antoine-Laurent Lavoisier and
his collaborators, listed—apart from an impressive number of metals, acidifiable
bases, alkalis, earths, metal oxides, compounds of metal oxides, alloys, and compounds of acidifiable bases—hundreds of salts made from twenty-six different acids.

1.1 Origin from the three natural kingdoms

Where did these materials come from? Almost all eighteenth-century chemists classified materials according to their origin from the three natural kingdoms. Especially in
the teaching of chemistry and chemistry textbooks, they ordered materials in this naturalistic fashion, thus representing them as objects of nature. For example, Herman
Boerhaave's (1668–1738) *Elementa Chemiae* (1732),[1] one of the most influential
early eighteenth-century chemical textbooks, presents chapters on the history of minerals, vegetables, and animals in its "theoretical part." It retains this division in its
"practical part," which first describes 88 "chemical operations upon vegetables," followed by descriptions of 39 "chemical operations upon animals" and 100 "chemical
operations upon minerals." In France, Nicolas Lemery's (1645–1715) famous *Cours
de Chymie* (1675), the last French edition of which appeared in 1757, was divided
into three main parts with the headings "Of Minerals," "Of Vegetables," and "Of Animals."[2] Lemery's *Cours* belonged to a tradition of seventeenth-century French textbooks that organized their practical parts along the naturalistic tripartite distinction.[3]
Guillaume François Rouelle (1703–1770), whose teaching at the Parisian *Jardin du
Roi* between 1742 and 1768 made chemistry a prominent Enlightenment subject, also
structured his experimental lectures according to the three natural kingdoms.[4] Two
other famous French chemical textbooks of the mid-eighteenth century, Pierre Joseph

---

1   Boerhaave [1732]. For life and work of Boerhaave, see Gillispie [1970–1980] vol. II p. 224ff.
2   See the contemporary English translation of the *Cours:* Lemery [1677]. For life and work of Lemery,
    see Gillispie [1970–1980] vol. VIII p. 172ff.

Macquer's (1718–1784) *Elemens de Chymie-Pratique* (1751) and Antoine Baumé's (1728–1804) *Manuel de Chymie* (1763) continued this tradition.[5] In the same vein, Gabriel François Venel (1723–1775) wrote in his article *Chymie*, presented in Diderot's *Encyclopédie*, that the three kingdoms of nature provide "three large divisions according to which we have distributed chemical subjects; the minerals, vegetables and animals fill out these divisions."[6] Many additional examples could be mentioned, including the period immediately after the Chemical Revolution in the last third of the eighteenth century.

In the chemistry lectures and textbooks of the eighteenth century, the materials grouped together under headings such as plants or vegetable substances, animals or animal substances, and minerals were extremely diverse. The class of vegetable substances, for example, comprised entire plants, roots, leaves, fruits, seeds, and other organs of plants; materials extracted from plants such as resins, gums, essential and fatty oils, balsams, and salts; vinegar, wine, beer, and spirit of wine obtained from fermented plants; as well as composite pharmaceuticals and other artificial chemical preparations made from natural plant materials (figure 1.1). Chemists used a set of different criteria to order these materials further, such as natural origin from a species of plant or provenance, the mode of chemical extraction, and perceptible properties, both physical and chemical. A striking feature of this mode of classification is the combination of apparently conflicting criteria of classification. The classification of materials as "plant materials," for example, identified them as natural objects. But on the rank of genus and species, the same material was often identified by the way it was prepared chemically; that is, as a product of chemical art. The class of distilled waters, for example, was ordered into the higher taxon of plant materials and at the same time identified by its mode of chemical preparation, namely distillation. Elixirs and other composite preparations made from many different ingredients, ardent spirits like spirit of wine and pure alcohol, and the various kinds of ethers are additional examples of classes of materials that were ordered into the class of plant substances despite the fact that chemists considered them to be not natural, but chemically altered materials.

From a broad comparative view, the eighteenth-century chemical order according to the three natural kingdoms is indeed remarkable. In the sixteenth century and early seventeenth century, most alchemists and chemists divided substances into natural raw materials on the one hand and chemical preparations on the other.[7] From the per-

3 The French chemical textbook tradition began with the *Elemens de Chymie* by Jean Beguin (1550–1620), the Latin edition of which appeared in 1610 (Beguin [1624]; for life and work of Beguin, see Gillispie [1970–1980] vol. I p. 571f.). The naturalistic tripartite division was introduced in the chemical textbook (1633–1635) by William Davison (1593–c. 1669), the first professor of chemistry at the Parisian *Jardin du Roi.* (For life and work of Davison, see ibid. vol. III p. 596f.) When introducing the naturalistic tripartite division in the last, practical part of his textbook, Davison referred to the Arabic physician Rhazes (c. 860–925) (see Partington [1961–1970] vol. III p. 6: *Rasis in libro Diuinitatis*). The tripartite naturalistic classification was adopted in the practical parts of the chemical textbooks by de Clave, Le Febvre, Glaser and Lemery; see de Clave [1646]; Le Febvre [1664]; Glaser [1676]; and Lemery [1677].
4 See Rouelle [n.d.]. For life and work of G. F. Rouelle, see Gillispie [1970–1980] vol. XI p. 562ff.
5 See Macquer [1751]; Baumé [1763].
6 Diderot and d'Alembert [1966] vol. III p. 418. All translations are our own, except where stated.

spective of the second half of the nineteenth century and afterward, the ordering of all kinds of raw materials and processed substances according to their origin from the three natural kingdoms is equally curious. In the inorganic and organic chemistry of that later period chemists classified chemical substances according to composition, constitution, and molecular structure. For this mode of classification the origin of substances, natural or experimental, was irrelevant.

Why did eighteenth-century chemists classify materials according to their origin from the three natural kingdoms? At first glance it seems obvious to interpret chemists' acceptance of the naturalistic tripartite distinction as a mere convenience allowing them to order a plethora of materials. But even if we consider this kind of classification to be a convenience, it must be admitted that it was a convenience in the absence of compelling alternatives. Chemists' classification of materials according to the three natural kingdoms is significant since it informs us about the absence of, or rather chemists' collective reluctance toward, an alternative already developing in the early eighteenth century; namely, classification based on knowledge about the composition of chemical substances and a chemical theory of composition. Chemists did not consider knowledge of composition to be a reliable resource for ordering all of the kinds of materials they were dealing with, especially plant and animal materials.[8] Analytical knowledge and the theory of composition were not the organizing grid for the *entire* culture of eighteenth-century chemistry. But this still does not explain sufficiently why the majority of European chemists accepted the division of substances according to the three natural kingdoms. If chemists did not consider chemical analysis to be reliable in all areas of chemistry, other alternatives of classifying materials

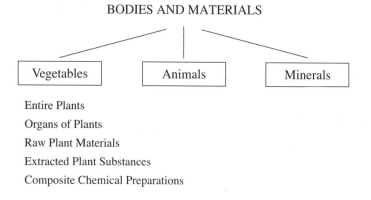

BODIES AND MATERIALS

| Vegetables | Animals | Minerals |

Entire Plants

Organs of Plants

Raw Plant Materials

Extracted Plant Substances

Composite Chemical Preparations

Figure 1.1: Classification according to the three natural kingdoms.

---

7   See below, chapter 2. As we will show in part III, this classification still had an impact in the early eighteenth century, especially in chemical-pharmaceutical contexts where chemists divided the natural *simplicia* from the artificial preparations and *composita* (see chapter 11.3).

8   Eighteenth-century chemists' cautionary remarks about the inadequacy of their analytical means in plant and animal chemistry can hardly be overlooked. They are repeated over and over again in practically all of the eighteenth-century publications dealing with the analysis of plant and animal substances. See also chapters 12 and 13.

existed as well. Classification according to types of preparation or according to chemical properties would have been possible alternatives and, indeed, were preferred by a few chemists.[9]

It is well known that the distinction between the three natural kingdoms was not an invention of chemists, but had been widely accepted by Enlightenment naturalists and philosophers. Therefore the fact that chemists shared this classification with other savants is also telling about the relationship between eighteenth-century chemistry and other learned cultures. It provides information about chemists' self-representation in broader eighteenth-century culture. Whereas classification according to types of chemical operations and chemical properties would have highlighted chemistry as a manipulative art, the naturalistic mode of classifying emphasized chemists' inquiry into nature. It rendered chemists as naturalists and natural philosophers, and connected chemistry with the grand theme of Enlightenment: nature. Unlike early seventeenth-century chemists, who justified chemistry, chemical pharmacy, and chemical medicine by emphasizing the power of chemical art and the superiority of chemical preparations over natural materials, and unlike late nineteenth-century chemists who envisioned chemistry as a productive enterprise spurring industrialization and creating a new world of synthetic artifacts, eighteenth-century chemists presented their art and science to the Enlightenment public in terms of nature and natural objects and processes.

1.2 Commodities

Like naturalists, eighteenth-century chemists presented their objects of inquiry as natural things and processes. Yet, when we "go out into nature" we do not find pure metals (with the exception of gold), acids, salts, alkalis, essential oils, distilled waters, and so on. What we do find in mountains and fields are roots, fruits, herbs, beetles and worms, ores, stones, spring water, and many other raw materials. But those substances that strike us as typically "chemical" are absent in nature. Rethinking the common order of representing chemical substances, the French chemist Pierre Joseph Macquer remarked in 1751 that salts, metals, half-metals, and other typical—that is, processed—chemical substances were "far from being presented to us by nature in the state of perfection and degree of purity we normally assume when we introduce them in chemical textbooks." He then proposed that an elementary chemical textbook should not start with metals, acids, salts, and other processed substances well-known to chemists, but rather with the natural raw materials from which these substances were separated.[10]

Many of the substances that eighteenth-century chemists studied were indeed not natural objects; that is, not substances "found in nature." This is quite obvious for

---

9 Among the chemists who classified materials on the highest taxonomic rank according to types of operations were Cartheuser [1753]; Vogel [1755]; Erxleben [1784]; and Guyton de Morveau [1777–1778].
10 See Macquer [1751] vol. I p. vi.

composite materials, such as glass, ceramics, porcelain, and composite remedies, which were manufactured in workshops and pharmaceutical laboratories. But how did chemists get access to the more typical chemical substances? We tend to suppose that mercury, vitriolic acid, nitric acid, copper sulfate, silver nitrate, lime, potash, and so on were chemical substances procured in the chemical or alchemical laboratory. And there is no question that these substances were ubiquitous in eighteenth-century chemical laboratories and often used to perform chemical experiments. As the French chemist Etienne François Geoffroy (1672–1731) emphasized, they were "the principal materials with which one usually works in chemistry."[11] In 1718, when Geoffroy made this utterance, metals, acids, salts, earths, and alkalis, which were the bulk of substances represented in his famous table of affinities (or *rapports*), already had a long history as chemical laboratory substances. Many eighteenth-century chemical novices first encountered these substances when they entered a chemical laboratory. But a great number of eighteenth-century chemists first met with these substances at quite different sites, namely in apothecary's shops, foundries and salt works, or when traveling to mines and taking courses in assaying.

What do we mean when we say that mercury, vitriolic acid, nitric acid, copper sulfate, potash, and so on were typical chemical substances used and procured in the eighteenth-century chemical laboratory? Do we mean that they were discovered by learned men in the course of their attempts to unravel the secrets of nature? And do we assume that the seventeenth- and eighteenth-century alchemical or chemical laboratory was mainly, or even exclusively, a site for the acquisition of natural knowledge? In the words of F. L. Holmes, were the processed chemical substances the results of "thriving investigative activity" by eighteenth-century chemists or the previous alchemists?[12] Or do we mean, alternatively, that they were reproduced in academic laboratories, but originally produced at different social sites? And is it not possible that the early modern laboratory was also a place for artisanal production and inquiry—as the Latin *laborare*, "to work," indicates? These are questions that concern a central aspect of the history of science: the coming into being of the objects themselves that are at stake in scientific investigation.

Nonetheless, historians of chemistry have seldom raised questions like these. Based on the historical actors' presentation of chemical substances as natural objects stemming from the three kingdoms of nature, they have paid attention to experiments performed with substances, and to theories about salts, calces, gases, and other classes of substances. But they have not seriously examined the question of the provenance of the processed chemical substances that late seventeenth-century and eighteenth-century academic chemists studied in their laboratories and wrote about in their chemical histories. Neither did they wonder why chemists in the second half of the seventeenth century, when chemistry became accepted in academic institutions such as the Royal Society, the Paris Academy of Sciences, the Paris Royal Botanical Garden, and German medical faculties, intensified their studies of processed sub-

---

11 Geoffroy [1718] p. 203.
12 Holmes [1989] p. 1.

stances. Nor did they ask how processed substances came into their horizon of interests and practice.

The bulk of the processed substances studied by eighteenth-century chemists—in particular in the first half of the century—were neither found in nature nor were they the outcome of disinterested academic investigations. Rather, they were products long known in mining and metallurgy, in distilleries and apothecary's shops, or newer commodities, imported by merchants from foreign countries and prepared in native pharmaceutical and alchemical laboratories.[13] Gold, silver, copper, bronze, brass, tin, lead, and iron had been well-known metals since ancient times, widely applied in the manufacture of tools, vessels, coins, ornaments, mirrors, pigments, and remedies. Mercury, antimony, and arsenic were equally quotidian materials, which became prominent in sixteenth-century Paracelsian chemical pharmacy and medicine. Also of ancient origin were the alkalis potash and soda, used as remedies and in the making of soap and glass. The various kinds of earth analyzed by eighteenth-century chemists were materials applied in the manufacture of ceramics and porcelain. Likewise, the most important kinds of mineral acids used in eighteenth-century chemical laboratories—nitric acid, *oleum sulfuris* or vitriolic acid, and muriatic acid—were also applied as commodities. In the fifteenth century, nitric acid became widely applied in metallurgy for the separation of gold from silver; vitriolic acid, which was first used only as a reagent in late medieval alchemical laboratories, became applied pharmaceutically in the sixteenth century; and muriatic acid was introduced by Rudolph Glauber in the seventeenth century and thereafter used as a remedy or an ingredient for the preparation of remedies. Almost all mineral salts analyzed and re-synthesized by eighteenth-century chemists were also applied as medicines, as well as for dyeing (in the cases of copper and iron vitriol, alum, and sal-ammoniac), the making of pigments for painting (vitriols), the soldering of metals (borax and sal-ammoniac), and cooking (rock salt and sea salt).[14] Furthermore, almost all of the chemically extracted vegetable and animal materials examined by eighteenth-century chemists—vegetable extracts, animal fats, essential oils, distilled waters, ardent spirits, and so on—were commodities applied as remedies, cosmetics, and food. Even the apparently most natural objects with which eighteenth-century chemists experimented—entire plants and the roots, leaves, fruits, seeds, and other organs of plants—were for the most part herbal drugs obtained from medical and botanical gardens or purchased from merchants and druggists.

Trade and the arts and crafts constituted the space in which most of the objects of inquiry of eighteenth-century chemists came into being. Early eighteenth-century academic chemists purchased materials from merchants, apothecaries, miners, and other practitioners and reproduced them in their laboratories. They further purified these useful materials, studied their perceptible properties, and analyzed their composition. Performing analyses and re-syntheses, in the second half of the eighteenth

---

13 One of the best overviews on the history of chemical substances is still the one contained in Hermann Kopp's history of chemistry (see Kopp [1966a] vols. 3 and 4); see also Multhauf [1966]. For an overview of the history of the practical uses of materials, see also Fester [1923].
14 See the tables in chapter 8.

century in particular, chemists also introduced new kinds of substances and new experimental techniques that went beyond the existing artisanal types of materials, instruments, and operations, as was the case with the different "kinds of air," or gases. But even such apparently exotic substances like gases did not unambiguously belong to a world of scientific objects. Rather, many of these new chemical substances left the site of their discovery to become applied in medicine, the apothecary trade, and the mundane social world more broadly.[15] Technology constituted a space of possibilities for eighteenth-century chemists' experiments as well as their ontology of materials.[16] Materials went back and forth from academic institutions to sites of production and consumption. We will come back to this important issue in connection with our historical analyses in parts II and III.

Two possible objections against our general argument concerning the origin of the materials studied by eighteenth-century chemists shall be addressed briefly at this point. The first objection concerns alchemy. It may be accepted that a certain number of eighteenth-century chemical substances—such as gold, silver, copper and mercury, metal alloys, calcareous earth, soda and potash, common salt, vegetable balsams, resins, wax, honey, milk, and so on—were ancient artisanal products. But, in a nutshell, it may also be argued that many of the eighteenth-century chemical substances had first been introduced in medieval and early modern alchemy. We have no problems with this view as long as "alchemy" is not defined unambiguously as an occult tradition, whose practical part aspired primarily to the preparation of the philosopher's stone and the transmutation of innoble metals into gold.[17] Most of the acids, salts, and other chemical preparations invented by sixteenth- and seventeenth-century alchemists—such as Paracelsus (1493–1541),[18] Oswald Croll (1560–1609),[19] Otto Tachenius (1610–1680),[20] Johann Rudolph Glauber (1604–1670), Johann Kunckel (c. 1630–1703),[21] and Johann Joachim Becher (1635–1682)—were sold and applied as chemical remedies. In the course of the second half of the seventeenth century these chemical remedies were included in official pharmacopoeias and other apothecary books. Indeed, all of the new acids and salts displayed in the first chemical table of affinity introduced by E. F. Geoffroy in 1718 were also used as chemical remedies.[22] As to alchemists like those mentioned above, it makes little sense to sort out their practical, commercial goals and interests from their philosoph-

---

15 For the application of gases in eighteenth-century medicine, see Levere [1994].

16 In this book, we use the term technology—which was coined only in the late eighteenth century—not as an actors' term, but as an analytical category to denote the ensemble of sites, techniques, instruments, materials, and products in the eighteenth-century arts and crafts (including medicine and pharmacy), as well as the forms of knowledge tied to that ensemble.

17 William Newman pointed out recently that "before the end of the seventeenth century, the word [alchemy] was widely used by early modern writers as a synonym for 'chymistry,' a discipline that included iatrochemistry and a host of technologies such as the refining of salts and metals, the production of acids, alcoholic libations, and pigments, and finally, the transmutation of base metals into noble ones." Newman [2004] p. xiii.

18 For life and work of Paracelsus, see Gillispie [1970–1980] vol. X p. 304ff.

19 For life and work of Croll, see ibid. vol. III p. 471f.

20 For life and work of Tachenius, see ibid. vol. XIII p. 234f.

21 For life and work of Kunckel, see ibid. vol. VII p. 524ff.

22 See chapter 8.

ical and religious objectives. They were craftsmen and entrepreneurs and at the same time learned men and authors of books. The mixed commercial and epistemic objectives of these educated artisans become obvious when we study their careers and their writings.[23] Many of their publications were collections of recipes addressing the reader as a potential maker-experimenter in the second person singular, giving him or her detailed instructions about instruments, techniques, substances, and the practical, mostly pharmaceutical, uses of chemical preparations.[24]

Second, one may object that the question of the origin of chemical substances, and of chemists' access to them, is merely a question of supply. This is a serious objection, in particular from a comparative point of view. Imagine a nineteenth-century merchant or gardener who delivered plants to a botanist, or a breeder who sold sheep to a geneticist. It may have been the case that the merchant, gardener, and breeder had interesting things to say to the botanist and geneticist about the origin, travel, storage, and ways of growing and breeding of the plant or animal specimen. But depending on the kind of scientific investigation, the information provided by these practitioners may have been totally uninteresting to the scientists. It may have been treated as a technical issue, which was left to the practitioner and could be black-boxed by the scientist.[25] There may have been cases—for example when obstacles arose—where the scientist went back to the practitioner to make inquiries that contributed to the solution of problems. But such cases were rare, and merchants, gardeners, and breeders were social groups distinct from academic botanists and geneticists.

Yet we assert that the relationship between eighteenth-century academic chemists and practitioners directly involved in the trade and commercial production was different from the example just discussed. There was close and stable interaction between eighteenth-century academic chemists and certain groups of educated practitioners, especially apothecaries, assayers, mining officials, and, especially in France, commissioners of state manufactories. This relationship was not merely the technical, commercial relationship of purchasers and providers. Although these social groups were not identical, they shared many interests and goals as well as materials, instruments, techniques, laboratories, and other items of material culture. Moreover, eighteenth-century chemistry itself was not restricted to academies, medical faculties, places of public and private teaching, botanical gardens, museums, and other institutions of learned inquiry. Academically educated chemists performed chemical operations and gathered experiential knowledge at non-academic sites as well, such as pharmaceutical laboratories, laboratories for assaying, arsenals, and manufactories for dyestuffs, ceramics, and sugar; and a significant of practitioners, especially apothecaries, were acknowledged as chemists. Eighteenth-century chemists were far from

---

23 For the alchemical laboratory practice and the mixed careers of seventeenth-century alchemists, see Newman [2000]; Newman and Principe [2002]; Nummedal [2002]; Smith [1994], and [2004]; and Moran [2005].

24 This implies that sixteenth- and seventeenth-century alchemy was far from being a secretive practice throughout.

25 For the distinction between technical and epistemic objects at stake here, see Rheinberger [1997].

being unambiguous savants and "experimental philosophers." Many of them were hybrid savant-technologists who were actively involved in the arts and crafts as apothecaries, mining and metallurgical officials, state commissioners in manufactories, and project makers sponsored by the state.[26] We will argue further in the next chapter that eighteenth-century chemists often pursued questions and objectives related to the arts and crafts, especially in the technological strands of their experiments and their experimental histories.

1.3 Learned inquiry into materials

The bulk of the substances studied by eighteenth-century chemists originated in the artisanal world, and many of the new chemical substances that were discovered in academic chemists' laboratories soon became transformed into remedies and other commodities. In the eighteenth century substances traveled from pharmaceutical laboratories and workshops to academic chemical laboratories, and vice versa, to become reproduced, analyzed, technically improved, and applied as useful materials. But were the materials studied, produced, and applied by apprenticed artisans and craftsmen and investigated by academic chemists actually identical kinds of objects of inquiry? This is a key question tackled in detail in parts II and III of our book.

   In this introductory part we discuss more general outlines of eighteenth-century chemical practices that contribute to answering this question. Materials were indeed transformed when they became objects of inquiry for academic chemists. Chemists invested them with new meaning, and sometimes even transformed their boundaries by splitting them into different kinds of substances. New individuations and identifications of substances—such as the division of air into different kinds of air—went hand in hand with material transformations. We have argued above that technology constituted a space of possibilities for eighteenth-century chemists' ontology of materials. Now we wish to add that chemists' ontology of materials was also conditioned by their distinctive chemical concepts, theories, analytical methods, and types of experiments. In the eighteenth century, the space of production and representation of chemical substances was constituted both by technology and by an ensemble of practices and concepts developed in scientific inquiry.

   We distinguish three main styles and practices of studying materials in eighteenth-century academic chemistry that we will discuss later: experimental history, technological improvement, and experimental philosophy. Eighteenth-century chemists studied substances not only as applied materials, but also as perceptible objects of nature and as objects carrying imperceptible features. Their epistemic transformation of materials produced and applied in the arts and crafts into natural objects becomes manifest in their classification according to the three natural kingdoms. A second epistemic transformation, built on the first one, took place when chemists studied imperceptible features of chemical substances. Eighteenth-century chemists ascribed

---

26 For further arguments concerning the hybrid persona of the eighteenth-century chemist, see Klein [2005b].

new philosophical or theoretical meaning to substances when they studied their composition, reactions, and chemical affinities. The vast majority of materials studied by eighteenth-century chemists were commodities and at the same time objects of learned inquiry or scientific objects.

It is of central importance to the historical and philosophical questions tackled in this book that eighteenth-century chemists' most important objects of inquiry—chemical substances—were entities with many faces. They were not unambiguously constituted as philosophical objects, as were the atom, the vacuum, the magnetic and electrical fluids, and other imperceptible objects studied in experimental philosophy. They did not fit exclusively into the category of perceptible natural objects, such as plants, animals, and minerals. And they were not exhaustively defined as useful materials applied in the arts and crafts and the wider society. Studying chemical substances as applicable materials, perceptible natural objects, and things that carry imperceptible features, eighteenth-century chemists' inquiries moved from the perceptible to the imperceptible dimension of substances, and vice versa. In so doing they constituted multidimensional objects of inquiry that resist our common distinction between perceptible and imperceptible things, and scientific and technological objects.[27] The substances in eighteenth-century chemistry were all of this at once. We will come back to this point after our discussion of chemists' three strands of inquiry into chemical substances, which will further illuminate the different ways in which chemists constituted their objects of inquiry.

27 For these distinctions see, in particular, Sellars [1963]; Bachelard [1996] (first publication 1957); and Rheinberger [1997]. Sellar's distinction between manifest and scientific images largely coincides with the distinction between perceptible and imperceptible objects (see also van Brakel [2000] pp. 41–46).

Practices of Studying Materials in Eighteenth-Century Chemistry

According to a widespread picture of eighteenth-century learned inquiry into nature, there existed two separate domains of such inquiry. On the one side were the naturalists painstakingly observing objects in nature. They would go out in the field to attentively observe plants and animals, rocks, mountains and glaciers, thunderstorms and shapes of clouds, and collect plant, animal, and mineral specimens in order to accurately describe, depict, and order them in their natural cabinets. Sometimes they would also use instruments, such as the microscope, to observe the minute inhabitants of nature or the fine structure of animal organs and mineral crystals. But even so, naturalists would still stick with the observation of objects created and overtly displayed by nature. In distinction to this first domain—natural history—the second one, experimental philosophy, heavily relied on instruments and intervention into nature in order to render manifest its imperceptible hidden entities. By means of the air-pump, the electrical machine, and other philosophical instruments, the experimental philosopher would procure hitherto unknown phenomena standing in a causal nexus to their imperceptible objects of inquiry, such as the vacuum, electrical fluids, and so on. Whereas the naturalist scrutinized visible objects presented to our senses by nature, the experimental philosopher turned to a subtler world of things and processes. Instead of studying the great multiplicity of visible things encountered on our globe, experimental philosophers explored the underlying invisible structures, processes, and causes, few in number, that brought about that multiplicity. This distinction corresponded with profound differences in the types of objects of inquiry. Whereas the naturalist's objects of inquiry possessed the self-evidence familiar from quotidian objects, the experimental philosopher dealt with elusive entities that were difficult to grasp and circumscribe and totally unknown in everyday life.

Our description is, of course, a caricature that simplifies eighteenth-century historiography and certainly does not do justice to the many much more sophisticated historical studies on that period that have appeared in recent years. We wish to use this caricature as a historiographical model to carve out as clearly as possible a third domain of learned experiential inquiry in the eighteenth century that has been largely obliterated in our grand picture of Enlightenment sciences: experimental history. We argue in the following that eighteenth-century chemists constituted multidimensional objects of inquiry by interconnecting three styles of experimentation and observation: experimental history, technological improvement, and experimental philosophy. Eighteenth-century chemistry was neither exclusively an experimental philosophy nor a natural history. It was further not merely an academic culture and teaching discipline but also an ensemble of forms of knowledge, techniques, instruments, and materials at artisanal sites. Our still-dominant distinction between eighteenth-century chemical science and technology may be convenient for historical research and writing, yet it is an obstacle for the proper historical understanding of the specificities of the chemistry of that time.

In the following we argue for this new historical understanding of eighteenth-century chemistry and the revised larger picture of the eighteenth-century sciences that comes with it by first discussing chemists' experimental history. Experimental history bridged natural history, technology, and experimental philosophy. We then present an example of chemists' attempts to improve chemical technology, along with a brief discussion of the material culture of eighteenth-century chemistry. Finally we deal with seventeenth- and eighteenth-century chemists' experimental inquiries into the imperceptible dimension of substances, or experimental philosophy. We wish to emphasize at the beginning of our discussion that our division into experimental history, technological improvement, and experimental philosophy is an analytical one and that in practice these strands were often intertwined.[1] Furthermore, the significance of each of the three strands of inquiry and the ways they actually interacted were not always and everywhere the same. As our analysis refers to the communal level of chemistry, covering different countries and an entire century, we pay attention to patterns of collective practices but leave aside the more fine-grained local and individual differences.

2.1 Experimental history (*historia experimentalis*)

The "experimental history" institutionalized during the seventeenth century was a tradition of experimentation and observation that evolved around the multiplicity of natural and artificial things. Like natural history, experimental history collected, described, and ordered facts relating to the perceptible dimension of particular objects and processes. And like natural history, experimental history was also regarded as "the basis for the agricultural, commercial and colonial improvement of the human estate."[2] But whereas natural history was concerned with the observation and collection of things "given by nature," experimental history reported phenomena procured by intervention into nature, both in arts and crafts and academic sites. For example, late seventeenth- and eighteenth-century chemists' experimental histories of substances reported phenomena observed at many different places, ranging from households and everyday life to the fields, the workshop, and the academic laboratory. Their experimental histories of substances ended with the collection and classification of phenomena, leaving inquiries into their causes to "experimental philosophy."

An explicit program of an "experimental history" first arose in the early seventeenth century when Francis Bacon (1561–1626) became its most prominent spokesman.[3] Bacon outlined his ideas of an experimental history (*historia experimentalis*) in a text entitled *Preparative towards a natural and experimental history*, which was published in 1620 in the same volume with the *Novum Organon*. Experimental his-

---

1 An example for the interconnection of all three strands of inquiry in the early nineteenth century is Louis Jacques Thenards's series of experiments on ethers; see Klein [2005a].
2 See Jardine and Spary [1996] p. 3.
3 We use here the term "spokesman" because Bacon gave voice to a cultural movement that had already begun to develop in the fifteenth and sixteenth centuies—see below.

tory in Bacon's original sense was, first of all, a collection and description of existing factual knowledge developed in the arts and crafts. It was an inventory of artisanal operations and experiments in the broadest sense, which complemented natural history.[4] "Experimental history," Bacon stated, is "the history of Arts, and of Nature as changed and altered by Man."[5] This definition embraced what were known as the "mechanical arts," "the operative part of the liberal arts," and "a number of crafts and experiments which have not yet grown into an art properly so called, and which sometimes indeed turn up in the course of most ordinary experience."[6] The core of experimental history, however, was formed by "the collection of experiments of arts" that "exhibit, alter, and prepare natural bodies and materials of things; such as agriculture, cookery, chemistry, dyeing; the manufacture of glass, enamel, sugar, gunpowder, artificial fires, paper and the like."[7] Similarly, Robert Boyle (1627–1691), a keen follower of Bacon, argued that learned men must collect as many facts as possible from craftsmen and merchants. "Learned and Ingenious Men have been kept such strangers to the Shops and Practices of Tradesmen," he complained, and further recommended that an "Inspection into these may not a little conduce, both to the Increase of the Naturalist's knowledge, and to the Melioration of those Mechanical Arts." The "Phaenomena afforded by Trades," he proclaimed, "are (most of them) a Part of the History of Nature."[8]

In Bacon's view, experimental history complemented natural history, since it explored the possibilities and limits of movements and other alterations of things. The history of arts, he emphasized, "exhibits things in motion, and leads more directly to practice."[9] Boyle echoed Bacon's words, asserting that "there are very many things made by Tradesmen, wherein Nature appears manifestly to do the main parts of the Work;"[10] and further he stated that

> many of the Phænomena of Trades are not only parts of the History of Nature, but some of them may be reckon'd among its more noble and useful Parts. For they shew us Nature in *motion*, and that too when she is (as it were) put out of her Course, by the strength or skill of Man, which I have formerly noted to be the most instructive condition, wherein we can behold her.[11]

Robert Boyle in particular made efforts to demarcate experimental history from its philosophical counterpart; that is, experimental philosophy. For example, in his *Experimental History of Colours* (1664), he asserted that his present work would

---

4 On Bacon's broad notion of "experiment" that extended to operations in the arts and crafts, see Klein [1996b].
5 Bacon [1986–1994] vol. IV p. 257.
6 Ibid.
7 Ibid. p. 257f.
8 Boyle [1999] vol. VI p. 467. In another essay on the *Usefulness of Natural Philosophy* (1671), Boyle stated that it is "my main business, to take all just Occasions, to contribute as much, as without indiscretion I can, to the history of Nature and Arts." Of the truth of experiments, he continued, "one may be easily satisfyed, by inquiring of Artificers about it, and the particular or more circumstantial accounts I give of some of their experiments, I was induc'd to set down by my desire to contribute toward an experimental History" (ibid. p. 396).
9 Bacon [1986–1994] vol. IV p. 257.
10 Boyle [1999] vol. VI p. 467.
11 Ibid. p. 468.

excite its readers "by the delivery of matters of facts." "Finding my self as *unfit to speculate*, as unwilling to be altogether idle," he continued,

> I have the less scrupled to set down the following *Experiments, as some of them came to my mind*, and as the Notes wherein I had set down the rest, *occurr'd to my hands*; that by *declining a Methodical way of delivering them*, I might leave you and my self the greater *liberty and convenience to add to them*, and transpose them as shall appear expedient.[12]

In this statement, Boyle first epitomized the epistemological credo of experimental history; that is, the absence of speculation and explanation. He further added remarks about the method and the literary style of experimental history, which served its further demarcation from experimental philosophy. Experimental history did not require a structured presentation of facts. If the experimenter was not, or not yet, able to create order among the experimental facts and to discover regularities, he may present them as they came to mind and hand, by "declining a methodical way." Furthermore, as experimental history in its most rudimentary stage was a mere collection of phenomena engendered by experiments, the extension of experiments must be possible with the greatest "liberty" of action; the experimenter may add new experiments and thereby collect new facts without knowing where the journey will go. Unlike experimental philosophy, experimental history abstained from reduction, conceptual unity, and inquiries into hidden movements and other causes.

Boyle's emphasis on the absence of any speculation and preconceived methods in experimental history, his insistence on the collection of facts without any intellectual and methodical constraints, resonated with another broad cultural movement we have not yet mentioned: the *historia* tradition. The *historia* tradition had gained momentum in the Renaissance, when physicians and other learned men revalued the empirical description of objects of nature and of human action vis-à-vis speculation about causes. *Historia* offered thorough descriptions of "how things are" without explaining why they are so. It sought to base knowledge on sense perception and aimed at knowledge of particulars. As Gianna Pomata and Nancy Siraisi pointed out, "the Baconian reform of natural history may be seen as the culmination of the reappraisal of *historia*'s epistemic value set in motion by the early humanists. What Boyle called 'natural experimental history'—the detailed, fully circumstantial report of experiments—seems to have grown out of the rich humus provided by *historia*'s soil."[13]

Furthermore, Bacon's and Boyle's emphasis on the importance of technical artifacts and artisanal operations for the writing of an experimental history was embedded in another ongoing cultural movement that revalued the role played by the methods and accomplishments of artisans for the acquisition of natural knowledge. For example, in a treatise published in 1531 Juan Luis Vives (1492–1540), friend of

---

12   Boyle [1999] vol. IV p. 25; our emphasis.
13   See Pomata and Siraisi [2005] p. 27. It should be noted that Pomata and Siraisi do not distinguish here between "experimental history" and "experimental philosophy," as we do. Experimental philosophy also highlighted detailed description, but its objects of inquiry were not particulars. Furthermore, the meaning of the objects of inquiry in experimental philosophy was constituted in philosophical discourse and hence thoroughly underdetermined by observation. See also below in this chapter the section on "experimental inquiries into the imperceptible dimension of chemical substances."

Erasmus and of Thomas More, invited European scholars to study the technical problems of machines, weaving, agriculture, and navigation. As Paolo Rossi pointed out, he urged scholars to overcome their traditional disdain for manual labor and "enter workshops and factories, ask craftsmen questions and try to understand the details of their work."[14] The technological treatises of the fifteenth and sixteenth centuries on architecture, machines, shipbuilding and navigation, military instruments and ballistics, the art of fortification, mining and metallurgy, alchemy, the art of distillation, and so on gave voice to this new attitude, which rejected the Scholastic divide between practice and theory, nature and art, and certain knowledge (*episteme*) and technology (*techne*).[15] Both the *historia* tradition and the revalution of artisanal knowledge stabilized experimental history as a collective style of experimentation and contributed to its institutionalization as an acknowledged academic practice that persisted well into the nineteenth century.

Historians of science have discussed Bacon's program of an experimental history in connection with the Royal Society's endeavor of a *History of trades* in the seventeenth century.[16] But this program also had an impact on the encyclopedic ventures of the Paris *Académie Royale des Sciences*, such as the large seventeenth-century project on the history of plants,[17] which also included chemical experiments, and the *Descriptions des arts et métiers*.[18] The more successful *Encyclopédie ou dictionnaire raisonné des sciences, des arts et métiers* by Denis Diderot and Jean D'Alembert (1751–1780) hinged no less on the Baconian program, as D'Alembert's preface to the *Encyclopédie* manifests clearly. In addition, the Baconian program of an experimental history also lent intellectual authority to a distinct style of experimentation in the seventeenth and eighteenth centuries that differed from experimental philosophy. This latter significance of "experimental history" for an adequate understanding of the institutionalization and development of the experimental sciences from the early modern period until the early nineteenth century has been ignored almost completely in the existing historical literature.[19]

The distinct style of experimental history can be discerned especially well in the history of chemistry from the seventeenth century until the early nineteenth century. In the chemistry of this period, "experimental history" meant a collection of facts about a great number of substances from all possible practical areas, ranging from artisanal sites and everyday life to the academic chemical laboratory. Of course, only a few eighteenth-century chemists ever read Bacon's writings or subscribed explicitly

---

14 Rossi [2001] p. 30.
15 On this movement and the Renaissance technological literature, see Darmstaedter [1926]; Eamon [1994]; Lefèvre [2004]; Long [2001]; Newman [2004]; Olschki [1965]; Rossi [1968], [1970], [2001]; and Smith [2004].
16 See Houghton [1941]; Hunter [1992] pp. 87–112, and [1995] pp. 74–80, 106–107; Merton [1970] pp. 137–159; Ochs [1985]; Rossi [2001]; Stewart [1992]; and Smith [2004].
17 See Stroup [1990].
18 See Cole and Watts [1952].
19 "Experimental history" is omitted in our most recent compendia on the history of science, such as the *The Cambridge History of Science*, the *Companion to the History of Modern Science*, the *Encyclopaedia of the Scientific Revolution*, and *The Oxford Companion to the History of Modern Science*. See Applebaum [2000]; Heilbron [2003]; Olby et al. [1990]; and Porter [2003].

to the Baconian program of natural and experimental history. In this respect the Leiden professor of chemistry, Herman Boerhaave, was an exception. As Boerhaave was an ardent supporter of Bacon's natural philosophy and one of the most influential chemists of the eighteenth century, he contributed to the reception of the Baconian program in European chemical communities.[20] Boerhaave explicitly used the term *historia experimentalis* in the practical part of his famous chemical textbook *Elementa Chemiae*.[21] But the majority of chemists was influenced by this program only indirectly via its different cultural and institutional forms of transmission, such as the Royal Society's project of a history of trades, the *Descriptions des arts et métiers* and Diderot's *Encyclopédie*. Chemists' experimental history was concerned with the preparation, the practical uses, and the properties of substances; that is, their color, smell, taste, consistency, measurable physical properties, and their plethora of chemical properties. It meant an extension of objectives of natural history to a laboratory science, which, like the classical domains of natural history—botany, zoology, and mineralogy—was concerned with a great multiplicity of things. Its target was neither hidden causes nor imperceptible entities (such as the vacuum, forces, corpuscles, electrical fluids, and typical philosophical objects of experimental philosophy), but the perceptible dimension of particular materials and operations. And its objective was not philosophical knowledge, but connoisseurship of materials, their varieties, properties, chemical transformations, and practical uses.

Well into the nineteenth century, chemists often performed experiments on a broad variety of substances, knowing that they would not, or not yet, be able to unravel regularities and general chemical laws, or to improve chemical theories. One day they would study a mineral water from a nearby spring, the next day an iron ore from a new ore deposit, then test the quality of a dyestuff produced in a local manufactory, distill rosemary to reproduce the essential oil of rosemary sold in apothecary's shops, and afterwards study the chemical properties of apothecaries' ordinary ether and compare it with ethers prepared in their own laboratories. Their experiments turned from the study of a material belonging to one class to that of another class, and from the kingdom of minerals to vegetable and to animal substances, and vice versa. Compared to experimental philosophy in the seventeenth- and eighteenth centuries, and compared also to the "experimental systems" in the twentieth-century laboratory sciences that evolve around one coherent scientific object and cluster of questions,[22] this style of experimentation may at first glance appear as aimless artisanal tinkering or mere cookery.[23] As it contributes little to heroic historiography it has been largely obliterated from systematic historical research.[24] Instead, most histori-

---

20 See Klein [2003b].
21 See Boerhaave [1732] vol. II p. 124; see also Boerhaave [1753] vol. II p. 78.
22 For the concept of "experimental systems," see Rheinberger [1997].
23 The view that the style of experimentation described here as "experimental history" was "cookery" can be found in many older histories of chemistry. Needless to say, in our view the designation "cookery" is mistaken, given the collectively accepted agenda of an experimental history, and given also its conceptual underpinning, which will be described in detail throughout this book.
24 This style of experimentation has not entirely escaped the attention of historians. W. Brock pointed out about J. Liebig, that prior to the 1830s "there was little coherence in his programs; research was an eclectic mixture"—see Brock [1997] p. 62.

ans of chemistry have highlighted episodes of eighteenth-century chemical experimentation in which experiments were more systematically focused on one scientific object and interconnected to a coherent "investigative pathway." However, F. L. Holmes' conception of an "investigative pathway," developed exemplarily in his path-breaking work on Lavoisier, describes a historical exception rather than a typical style of chemical experimentation in the late eighteenth century.[25] The scientific careers of the vast majority of chemists from the seventeenth century until the first decades of the nineteenth century show that, as a rule, chemists' experiments studied a great number of different substances, and often changed from one substance to the other without organizing their experiments into a systematic investigative pathway. This experimental practice may be confusing to historians, or conceived as too trivial to be studied in any historical detail. But if we historicize our notion of experiments, we must acknowledge that it was an established collective style of experimentation in chemistry well into the nineteenth century.

Another characteristic feature of experimental history in eighteenth-century chemistry was the frequent repetition and extension of experiments performed with one particular substance, and the continuous accumulation of factual knowledge about the ways of its preparation, its perceptible properties, and its practical uses. In the course of the eighteenth century it was especially the testing of chemical properties—such as combustibility, acidity, solubility in various solvents, and interaction with reagents—that contributed to the extension and refinement of the experimental histories of substances. "Chemical property" referred to the observable phenomena that were created when a substance was heated or mixed with a reagent. Experimental histories reported such phenomena without seeking to explain them by referring to invisible movements of substance components and chemical affinities. In the second half of the eighteenth century chemists' testing of the chemical properties of substances with a growing number of solvents and reagents led to an enormous increase in the size of experimental histories, sometimes covering dozens of pages for one single substance. At the same time chemists addressed a broader and more diverse audience than in the early eighteenth century, when physicians, students of medicine, and pharmaceutical apprentices constituted the majority of practitioners interested in chemistry. New groups of practical men interested in learning chemistry, such as dyers, manufacturers, and officials of the state bureaucracy, demanded detailed descriptions and analyses of a broad range of materials. This development contributed to changes in the presentation of experimental histories, especially in chemical textbooks, in the last decades of the eighteenth century. Chemists then often presented long experimental histories of substances as a collection of facts that ignored the more specific practical uses and techniques in the chemical arts. This move was reinforced by the separation of textbooks into parts on "pure chemistry" and "applied chemistry," with the inclusion of the histories of substances in the part on "pure chemistry" and the descriptions of artisanal techniques and practical uses of materials in the part on "applied chemistry."[26] Nevertheless, the observation, repetition, and

---

25 See Holmes [1985], and [2004].

modification of artisanal operations in the academic chemical laboratory remained an important source for chemists' experimental histories of substances well into the nineteenth century, as can be seen more clearly in experimental reports than in chemical textbooks.[27] Likewise, observations on visits to mines, foundries, assaying shops, mints, arsenals, distilleries, dyeing manufactories, workshops of glass makers, chemical factories, and so on were a persistent source not only for texts on "applied chemistry" but also for chemists' experimental histories of substances. Furthermore, both early and late eighteenth-century chemists gathered facts for the writing and teaching of experimental history in their own artisanal occupations as apothecaries, mining officials, inspectors of dyeing, porcelain makers, manufacturers of beet sugar, and other kinds of chemical entrepreneurship. Eighteenth-century chemists' dual careers as savants and technologists contributed considerably to their experiential knowledge and the enrichment of their experimental histories.[28]

Experimental history in teaching

In the eighteenth century, the teaching of chemistry was to a large extent the teaching of experimental histories of substances. All eighteenth-century textbooks of chemistry contained long sections on the history of substances, although the style of their presentation changed in the course of the eighteenth century. Similar to its sibling natural history, experimental history made students acquainted with a multifarious world of particulars (substances, in this case), their preparations, perceptible properties, and various applications in the arts and crafts. As most of the eighteenth-century students of chemistry would later become apothecaries, physicians, assayers, chemical manufacturers, and other practitioners, connoisseurship of a broad range of materials and knowledge about applicable, useful chemical operations were dominant pedagogical goals.[29] Hence, the histories of substances contained in the "theoretical parts" of early eighteenth-century chemistry textbooks, and the recipes for operations contained in their "practical parts" also added considerations about practical uses. In the second half of the century, descriptions of artisanal techniques were often transferred to the parts of textbooks dealing with "applied chemistry." Chapters on applied chemistry presented detailed accounts of the manufacture of chemical remedies; metallurgical smelting and assaying; the fabrication of glass, ceramics, and porcelain; the making of soaps; dyeing, bleaching, and calico-printing; large-scale distillations of

26 For the distinction between "pure" and "applied chemistry" in the second half of the eighteenth century, see also Meinel [1983], and [1985].
27 For an example, see Klein [2005a].
28 We use the term "technologist" as an analytical term to highlight the specific social and epistemic status of eighteenth-century chemical practitioners, who often were not merely apprenticed artisans but also academicians. On eighteenth-century chemists' dual career patterns as savant-technologists, see Klein [2005b] and the secondary literature quoted there.
29 It should be noted that our statement about students of chemistry in the eighteenth century refers to many European countries. Especially in Sweden and Germany, assayers, mining officials, and practitioners other than apothecaries and physicians already belonged to the audience of chemistry teaching back in the early eighteenth century. On Swedish chemistry in the early eighteenth century and its context of mining and metallurgy, see Fors [2003]; and Porter [1981]; on the role of mining and metallurgy in eighteenth-century German chemistry, see Hufbauer [1982]; and Klein [2005b].

acids and spirit of wine; the making of beer and wine; the extraction and purification of sugar; the fabrication of salts and gunpowder; the examination of soils and mineral springs; and so on. At the same time the experimental or chemical histories of substances were presented in separate chapters, often designated "pure chemistry." This convention was less the result of an actual bifurcation of experimental history and chemical technology than a way of handling and ordering the increasing amount of subjects of teaching.

To give a brief example of chemists' teaching of experimental history in the early part of the century, we take Herman Boerhaave's *Elementa Chemiae* (1732) as a particularly influential model for this kind of teaching.[30] The student of chemistry, Boerhaave wrote, "should at least understand the principal ways of procuring the useful things. And it would be wrong in this art to leave any one unacquainted with the useful methods of working."[31] Boerhaave was firmly convinced of the usefulness of chemistry in medicine and the "mechanical arts," among which he mentioned in particular painting, enameling, staining glass, manufacturing glass, dyeing, metallurgy, the art of war, natural magic, cookery, the art of winemaking, brewing, and alchemy.[32] In the "practical part" of his textbook he presented a collection of 227 "processes," which were, apart from some dozens of experiments devoted unambiguously to chemical analysis, recipe-like descriptions of "the actual operations of chemistry"[33]—that is, familiar operations performed in pharmaceutical and chemical laboratories all over Europe, both for the acquisition of knowledge and for the manufacture of useful goods. Much of the fame of Boerhaave's *Elementa Chemiae* relied on this second, practical part of the book. Indeed, as late as 1782, when Johann Christian Wiegleb (1732–1800) again translated the book into German, he omitted its theoretical chapters and published only its practical part. In his preface Wiegleb highlighted the particular value of the book's collection of recipes, "which teaches how to prepare from all realms of nature excellent remedies against all kinds of diseases." He pointed out that this part was of "great use" for apothecaries and all other practitioners concerned with the art of separating (*Scheidekunst*) and distilling.[34] Wiegleb's emphasis on common knowledge and the applicability of the experiments presented in the practical part of the *Elementa Chemiae* coincided with Boerhaave's own claim that he wanted to present well-known and useful experiments.

Almost all 88 experiments with plants and plant materials described in the practical part of Boerhaave's textbook aimed to acquaint students with the extant techniques, instruments, and materials necessary to produce chemical remedies. Boerhaave's descriptions were short and operational, addressing the student in the second person, as a potential applier of the recipe. For example, his recipe for the composite "*elixir proprietas*" reads as follows:[35]

30 See Boerhaave [1732]. In the following we will quote from the English translation, Boerhaave [1753]. On Boerhaave's plant chemistry, see also Klein [2003b].
31 Boerhaave [1753] vol. II p. 2.
32 Ibid. vol. I pp. 178–205.
33 Ibid. vol. I p. 4.
34 Boerhaave [1782] *Vorrede*.
35 Boerhaave [1753] vol. II p. 173.

Take choice aloes, saffron and myrrh, of each half an ounce, cut and bruise them, put them into a tall holt-head, pour twenty times their own weight of the strongest distilled vinegar thereon; let them simmer together in our little wooden furnace for twelve hours: now suffer the whole to rest, that the faeces may subside, and gently strain off the pure liquor thro' a thin linen. Put half the quantity of distilled vinegar to the remainder, boil and proceed as before, and throw away the faeces. Mix the two tinctures together, and distil with a gentle fire till the whole is thickened to a third; keep the vinegar that comes over for the same use; and what remains behind is the elixir proprietatis, made with distilled vinegar.

To this description, Boerhaave added information about the practical application of the elixir, praising it as an "aromatic medicine, of great use in the practice of physic."[36]

Boerhaave's collection of chemical operations also contained examples for applications other than pharmaceutical. For example, syrup, jelly, and other kinds of extracts procured by the decoction of plants were not merely used as chemical remedies but also as nutritious goods. Boerhaave recommended them especially for long travels, remarking that "perhaps nothing would more conduce to the health of the British and Danish sailors, than a due provision of this kind."[37] Essential oils obtained by the distillation of roses, lavender, lilies, jasmine, hyacinths, and other flowers were used in the eighteenth century both as chemical remedies and precious ingredients for fabricating perfumes. These oils, Boerhaave informed the prospective apothecaries and chemical entrepreneurs, "for their excellent fragrance, are valued by great personages, and sold at a high price; whence it is worth while to study them."[38] Similarly, he acquainted his students with the fact that the "butter of wax," fabricated through the distillation of wax collected from the surface of plants, was not only used as a remedy against nervous diseases but also as a cosmetic that "successfully preserves the skin from roughness, dryness, and cracking in the cold."[39] In his collection of facts, Boerhaave even mentioned applications of plant materials for dyeing and metallurgical purposes. Tartar from wine that had rested for a while, he remarked, was not only a purging medicine but also applied by dyers and silversmiths on many occasions, and the assay masters used the ashes remaining from the combustion of plants to make their test crucibles.[40]

In eighteenth-century chemical experimental history, teaching, writing, and experimenting interacted and mutually stabilized each other. The expectations of students of chemistry played an important role in this practice. It was, in particular, practitioners—in the case of Boerhaave, mainly students of medicine, physicians, and apothecaries—who were keen to acquire connoisseurship of materials. For this audience of practitioners, which in the second half of the eighteenth century broadened considerably to include dyers and colorists, assayers and mining officials, chemical manufacturers, and so on, it was natural to proceed from descriptive experimental

36 Ibid.
37 Ibid. p. 18.
38 Ibid. p. 70.
39 Ibid. p. 99.
40 Ibid. pp. 140, 20.

histories to more analytical modes of technological inquiry and to active attempts of technological improvement. This second strand of eighteenth-century chemical experimentation, which in practice was often intertwined with experimental history, is the subject of the next section. The imagination of the Enlightenment public, however, was less captured by details of experimental history and artisanal techniques. For this distinctive audience of chemistry the more natural way to proceed from experimental history was experimental philosophy, which is discussed below in section 2.3.

2.2 Technological improvement

Like cultures of natural history, experimental history focused on the perceptible dimension of particular materials in order to acquire knowledge about their properties, preparation, and various practical uses. These epistemic goals were far from being separated from the realm of artisanal production. Rather, chemical experimental history collected facts about materials and operations from all practical areas, including pharmacy, metallurgy, and other arts and crafts. It further played an important role in the education of practitioners such as apothecaries, mining officials, state commissioners, and chemical manufacturers. In so doing, it contributed to the social reproduction of eighteenth-century technology. Being primarily a collection and description of existing materials and operations, it did not actively contribute to the improvement of the arts and crafts. And yet it was only a small step from experimental history to active technological contributions and improvements by chemists. In the following we wish to show by way of example, taken from apothecaries' preparation of ether, how chemists' descriptive experimental histories sometimes switched to experimental attempts at improving materials and techniques in pharmacy and other arts and crafts.

Beginning in the middle of the eighteenth century, descriptions of the properties of impure ordinary ether, the technique of its preparation, and its medical uses were included in almost all experimental parts of chemical textbooks or chapters concerned with the history of chemical substances. The chemical experiments on ether had a longer history, going back at least to the sixteenth century.[41] If spirit of wine was mixed with sulfuric acid and the mixture was distilled, a liquid product was obtained that from the middle of the eighteenth century was called "ether." In the seventeenth century, when chemically prepared remedies became more widely accepted in Europe, this operation was already applied in pharmacy. Around 1720, ether diluted with spirit of wine was mentioned as a remedy in the official pharmacopoeias under names such as "Naphtha" and "*Hoffmanns Tropfen*."[42] The properties of the pure ether were first described in an article by the German chemist Siegmund August Frobenius (?–1741), published in the *Philosophical Transactions* in 1730. Frobenius also introduced the name "ether."[43] The recipe for its preparation was initially kept

---

41 For the following account, see also Kopp [1966] vol. IV p. 309ff.; Priesner [1986].
42 See Schneider [1968–1975] vol. III p. 82, vol. VI p. 32.

secret because of Frobenius' commercial interests. Together with his English colleague Godfrey Hanckwitz (1660–1740), he wanted to sell pure ether as a novel, highly effective remedy. But immediately after Frobenius' publication, several British, French, and German chemists and apothecaries tried to reproduce pure ether and to study the process of its formation; among them were Macquer, G. F. Rouelle, Baumé, Cromwell Mortimer (c. 1698–1752), Jean Hellot (1685–1765), and Friedrich Hoffmann (1660–1742).[44] In 1741, Cromwell Mortimer made the production process public. From that time on, descriptions of varieties of ethers and variations of their pharmaceutical preparation proliferated in chemists' experimental histories. The ingredients originally used to make pure ordinary ether were spirit of wine and sulfuric acid. Twenty years later, chemists and apothecaries had already tested alternative possibilities to produce ethers from spirit of wine and acids other than sulfuric acid, such as nitric acid, muriatic acid (today hydrochloric acid), and acetic acid. In Macquer's famous *Dictionnaire de chimie* (1766) all of these ethers, which then became demarcated by more specific names such as "ordinary" or "vitriolic ether" and "nitric ether," were described in detail along with their preparation methods.[45] Macquer also described the medical use and virtues of ordinary ether as a sedative and anti-spasmodic remedy.[46] The liquid substance was dropped on a piece of sugar that was swallowed in the hope of relieving stomach aches, hysteric convulsions, and other nervous diseases.[47]

But apart from their experimental-historical goal of acquiring knowledge about the perceptible properties of ethers and the techniques of their preparation, eighteenth-century chemists were also keenly interested in improving the manufacture of ethers. As Macquer declared in 1766, the most important obstacle for the broader application of ordinary ether in arts other than pharmacy was its high price:

> Ether is not yet employed in the Arts, although it appears capable of being usefully employed in many cases, and particularly in the dissolution of certain concrete oily matters [contained] in varnishes; however its high price is a considerable obstacle to its introduction into the Arts.[48]

Therefore chemists were eager to find ways to minimize useless by-products of ether production. Varying the kind and proportions of ingredients, altering temperature during distillation, and observing the time when by-products first occurred in the distillation process, they hoped to improve the commercial technique of ether production.

By the end of the eighteenth century Antoine F. Fourcroy (1755–1809) and Nicolas Louis Vauquelin (1763–1829) undertook collaborative efforts to study the manufacture of ordinary ether more closely.[49] "The preparation of ether," they observed,

---

43 See Macquer [1766] vol. I p. 456.
44 For life and work of Hellot and Hoffmann, see Gillispie [1970–1980] vol. VI pp. 236f. and 458ff., resp.
45 Macquer [1766] vol. I pp. 455–470.
46 On the medical virtues of ethers, see also Fourcroy [1801–1802] vol. VIII p. 179.
47 Macquer [1766] vol. I p. 462f.
48 Ibid. vol. I p. 461.
49 For life and work of Vauquelin, see Gillispie [1970–1980] vol. XIII p. 596ff.

"is a complicated pharmaceutical operation, the results of which are as well known as its theory remains obscure."[50] As becomes clear from this remark, in 1797 the dominant goal of Fourcroy and Vauquelin's experiments was epistemic; they even spoke of a "theory" of ether formation. But the theory that they eventually proposed was regarded by the next generation of Parisian chemists—among them the famous chemist Louis Jacques Thenard (1777–1857) and the chemist-pharmacist Pierre François Guillaume Boullay (1777–1869)—as a promising route to improving the commercial manufacture of ordinary ether.[51] In 1807 Boullay made the following suggestion:

> The use of sulfuric ether is today quite extensive, and its consumption is considerable; it has become a true product of art produced on a grand scale. The operation, although it is quite simplified, still deserves attention, and *it seems it could be improved, especially with regard to economics and in terms of the purity of the product* [...]. According to the wise research and *theory of Messrs. Fourcroy and Vauquelin,* the attraction of sulfuric acid for water, with the help of heat, determines the transformation of alcohol into ether. This reaction of the principles of alcohol, which takes place under the influence of sulfuric acid, precedes the carbonization of the mixture, the formation of sweet wine oil, the release of sulfurous acid and other phenomena towards the end of the operation. [...] Therefore, it would be advantageous *to prevent or at least delay the appearance of these products* that announce the complete decomposition of alcohol.[52]

Based on Fourcroy and Vauquelin's theory of ether formation,[53] which also explained the formation of the many by-products, Boullay looked for ways to avoid unwanted by-products in order to increase the efficiency of the manufacture of ordinary ether. For that purpose he suggested that the alcohol be periodically replenished during the distillation in order "to preserve the proportions of ether formation" and to extend the first phase of pure ether formation in which only few by-products were built.[54] His suggestion was in turn taken up by Thenard, who suggested replacing the consumed alcohol from time to time in order to block the formation of by-products and increase the amount of ether produced. Thenard added that "this has been confirmed by experience, and precisely this is what is practiced by many pharmacists in the laboratory."[55]

Laboratory and workplace

How was it possible that eighteenth-century academic chemists repeated artisanal operations in their laboratories, and even performed experiments aimed directly at improving chemical technology? Eighteenth-century chemists' repetition and modification of artisanal chemical operations relied on an institutional and material precondition, which should be mentioned briefly. It was conditioned to a large extent by a

50 Fourcroy and Vauquelin [1797] p. 203.
51 For life and work of Thenard, see Gillispie [1970–1980] vol. XIII p. 309ff. For biographical information on P. F. G. Boullay, see Partington [1961–1970] vol. IV p. 345.
52 Boullay [1807] p. 242f.; our emphasis.
53 For this theory, see Klein [2003a] pp. 91–95.
54 Boullay [1807] p. 243.
55 Thenard [1817–1818] vol. III p. 278.

material culture of academic laboratories that overlapped strongly with the realm of instruments, reagents, techniques, and materials applied and produced in apothecary's laboratories, assaying shops, and distilleries. Eighteenth-century chemists not only shared a few single instruments with apothecaries and other artisans, but implemented an arsenal of instruments used in the arts and crafts. Their smelting and testing furnaces, bellows, crucibles, calcination dishes, and balances were similar to the instruments used by assayers. The same types of mortars, pestles, filters, vessels, boxes, glass tubes, vials, retorts, alembics, pelicans, receivers, and transmission vessels that academic chemists used in their laboratories were also used by distillers for producing nitric acid, alcoholic spirits, and fragrant oils, and by apothecaries for making medicines (see figures 2.1 and 2.2).[56] There was even agreement in the size of vessels and instruments used by academic chemists, apothecaries, assayers, and distillers. The small-scale trial was intrinsic to assaying that studied the composition of ores and other minerals for calculating the productiveness of mining and metallurgy. As pharmacy was still a handicraft in the eighteenth century, it also produced remedies on a small scale and for a comparatively small local market. The distilling of essential oils for the making of perfumes and alcoholic spirits was performed on a small scale too, even though there were enormous modifications made to the distilling apparatuses used by commercial distillers.

As a consequence, almost all of eighteenth-century chemists' instruments could be bought from ordinary merchants and artisans. The chemist-apothecary Antoine Baumé enumerated the following suppliers of chemical instruments: *faïenciers* (earthenware makers and glass makers), *marchands potiers de terre* (shopkeepers for potters), *fournalistes* (furnace makers), *chaudronniers* (boilermakers), *balanchiers* (balance makers), *marbiers* (marble masons), *marchands de fer* (iron merchants), *boisseliers* (bellows makers), and *fondeurs en cuivre* (copper foundry owners).[57] More exceptional "philosophical instruments," such as apparatus for the creation of and experimentation with gases, were developed and collectively accepted by chemists only from the middle of the eighteenth century. But as F. L. Holmes observed: "Until late in the eighteenth century no major technological changes altered the character of the chemical laboratory as a material or social setting."[58] Even in the second half of the eighteenth century the balance and thermometer remained the main physical instruments widely distributed in chemical laboratories.[59]

The resemblance of the equipment of eighteenth-century chemical laboratories with that used at certain artisanal workplaces accorded with the fact that most of chemists' types of experimental techniques corresponded with artisanal operations, in particular those of apothecaries, assayers, and distillers. Dissolutions, distillations,

---

56 See also Eklund [1975]; Holmes [1989]; Holmes and Levere [2000].
57 Baumé [1773] vol. I pp. cxxx–cxliv.
58 Holmes [1989] p. 18.
59 Baumé's list quoted above also includes the makers of thermometers and barometers, who were concentrated in Paris. It should be emphasized that in the early eighteenth century, and even before that time, chemists already used balances, but balances were also used by apothecaries and assayers. On the use of balances in the earlier alchemical practice, see Newman [2000]. On the thermometer in eighteenth-century chemistry, see Golinski [2000].

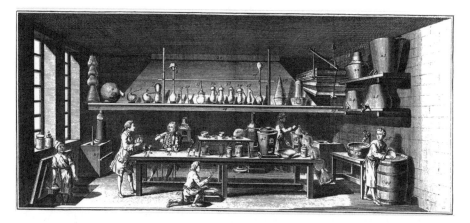

Figure 2.1: An eighteenth-century chemical laboratory. From Diderot and d'Alembert [1966] XXIV Chimie Planche 1.

Figure 2.2: Chemical-pharmaceutical instruments from the seventeenth until the nineteenth centuries (courtesy of the *Deutsches Apothekenmuseum*, Heidelberg).

evaporations, precipitations, combustions, and smelting were types of operations performed by both chemists and artisans. To be brief, we quote the historian of chemistry Ernst Homburg: "If we consider the manipulative skills which could be learned in an academic context and those which could be acquired 'in practice,' we must conclude that there were no radical differences on this point."[60] As many of the minerals and plant and animal materials eighteenth-century chemists explored in their laboratories were also bought from merchants and applied in practice by apothecaries and other artisans, the overlap of the material culture of academic chemistry and that of the mundane world of the artisanal crafts was quite strong.

In the eighteenth century the term "laboratory" referred both to academic sites where chemical and pharmaceutical operations were performed and to the workplaces of apothecaries, assayers, gunpowder makers, and distillers. The Latin word *laborare*, from which laboratory is derived, points to the similarity between academic laboratories and workshops. For a chemist, a laboratory was a necessity. "Anyone who would become a chemist," Macquer pointed out, "must indispensably have a laboratory."[61] This was different in the core areas of experimental philosophy, where "physical cabinets" and "physical theaters" (*theatrum physicum*) were established.[62] As is almost apparent from these latter terms, these institutions frequently served as locations for the collection and exhibition of instruments and for the demonstration of curious experimental effects rather than as actual places of work. By contrast, an eighteenth-century chemical laboratory was a site of daily experimental work.[63]

In many cases these sites more resembled a kitchen than a carefully designed place for learned experimenters. In the middle of the laboratory stood a large table for performing chemical operations that did not require fire. Shelves were fastened to its walls, containing hundreds of glass vessels filled with reagents and products of chemical operations. The room also often contained a cistern for the water supply. A chimney was a necessity, with various furnaces placed under it, and "ovens like those in kitchens."[64] Of constant concern were moisture and fresh air. As laboratories needed a water supply and tons of charcoal and wood, many chemists preferred the ground floor as the most convenient place. But as a consequence of the humidity on the ground floor, Macquer warned, "saline matters become moist in time; the inscriptions fall off, or are effaced; the bellows rot; the metals rust; the furnaces moulder, and every thing almost spoils."[65] Hence, Macquer recommended choosing a dry place for the laboratory above the ground floor; he further pointed out that two opposite windows were necessary to carry off poisonous vapors by a current of air.

Eighteenth-century chemists' experimental histories and technological improvements were conditioned to a considerable extent by the material culture of academic laboratories and their intersection with laboratories in the arts and crafts. This would

60   Homburg [1999b] p. 6.
61   Macquer [1766] vol. II p. 3.
62   Wiesenfeldt [2002].
63   See also Crosland [2005].
64   Macquer [1766] vol. II p. 5.
65   Ibid. p. 4.

change only in the course of the nineteenth century when the chemical laboratory acquired a new appearance with the introduction of small test tubes and other instruments that spurred the differentiation between chemical technology and academic laboratory practices.[66] Additionally, the establishment of the large-scale pharmaceutical industry further increased the gap between type and scale of instruments and apparatuses used by academic and industrial chemists.

2.3 Experimental inquiries into the imperceptible dimension of substances

So far we have studied eighteenth-century chemical practices concerned with the perceptible dimension of chemical substances. We now turn to a third strand of inquiry that was deeply embedded in experimentation and at the same time pursued philosophical or theoretical objectives. Eighteenth-century chemists studied substances not only as perceptible objects and in the context of experimental history and technological improvement. They also regarded them as natural objects that carry imperceptible features. The invisible composition and components of substances, the hidden processes taking place in their chemical transformations, and the driving forces of chemical transformations were persistent issues on the chemists' agenda. Depending on the kind of conceptual network in which these issues were defined, eighteenth-century chemists linked their study of the imperceptible dimension of substances with a broader philosophical discourse that transcended chemistry, or, alternatively, restricted it to chemistry. We address here a very significant differentiation, which we will explain in more detail in the following discussion: the differentiation between an atomistic approach—which relied on philosophical schools and traditions that took up experiential knowledge but was far from being determined exclusively by observation and experimentation—and a chemical approach highlighting substance components, affinities between substances, and the hidden movements of substance components in chemical reactions.[67] The latter approach developed in the context of artisanal and experimental practices, even though it was by no means entirely disconnected from philosophical discourse. It became firmly established around 1700 in connection with experiments on salts—performed mainly by Parisian chemists in the institutional context of the Paris Academy of Sciences and the Royal Botanical Garden—and attempts by Parisian chemists to draw together and systematize in tables existing experiential knowledge, including pharmaceutical operations with salts and chemical operations in metallurgy.[68]

Studying the decomposition and recomposition of natural salts, the chemical preparation of new artificial salts and their subsequent decomposition, the separation

---

66 See Homburg [1999b].
67 Most of the secondary literature on eighteenth-century chemical theories of composition does not differentiate between these two approaches; for recent examples, see Kim [2003] and Siegfried [2002]. With respect to the seventeenth century, arguments concerning the underdetermination of atomism by experiments and other forms of experience are presented in Chalmers [1993]; Meinel [1988].
68 For the latter, see, in particular, our tables in chapter 8, and Klein [1994a], [1994b], and [1996a]. For chemical tables, see also Cohen [2004]; and Roberts [1991].

of metals from ores, the making of alloys, and the recovering of single metals from alloys, as well as a few additional chemical operations, chemists developed an understanding of chemical transformations that implemented a substance building block model. They regarded all of these reversible chemical transformations as re-combinations of chemical building blocks—certain kinds of chemical substances—that were preserved during the chemical transformation and directed by elective chemical forces, designated "chemical affinities." The substances made from such stable substance building blocks were "chemical compounds"—such as salts and alloys—which contained the preserved substance building blocks as their "chemical components," interconnected by chemical affinities between pairs of substances.

This understanding of chemical transformations and of the meanings of "chemical compound," "composition," and "chemical affinities" was most clearly represented in the eighteenth-century tables of chemical affinities, the first of which was published in 1718 by E. F. Geoffroy, and in the table of chemical nomenclature published in 1787. This understanding also informed the mode of identification and classification of chemical compounds according to chemical composition, which is analyzed in part II of our book. Throughout the eighteenth century, the identification and classification of chemical substances according to composition always referred to substance components rather than atoms. It was only with the emergence of carbon chemistry in the late 1820s and the application of chemical formulae that this began to change and a chemical-atomistic image of composition became implemented in the practice of identifying and classifying substances.

A second distinctive feature of the eighteenth-century chemical approach to composition must be highlighted. The components this chemical approach referred to—mostly acids, alkalis, earths, and metals—were chemical substances that could be separated from chemical compounds by "chemical analysis" and, when set free, were ordinary substances that could be perceived and handled themselves. These kinds of substance components could also be used to resynthesize the original compound. That is, they differed not only from atoms defined in the philosophical tradition of atomism, but also from the simple elements and principles defined in sixteenth-century and seventeenth-century Aristotelian and Paracelsian philosophies.[69] The following discussion of the concepts of "element" and "principle" in seventeenth-century chemical philosophies will further illuminate differences in meaning between these older philosophical concepts and the eighteenth-century concept of chemical substance components as represented in chemical tables. But the strongest argument

---

69 This distinction is missing both in the older and the more recent secondary literature on eighteenth-century chemical theories of composition (for an example of the most recent literature, see Kim [2003] pp. 72–110). An argument for treating the approach to chemical composition as represented in Geoffroy's table on a par with the philosophies of elements and principles may be that many of the experiments and publications by French chemists around 1700—especially those by Wilhelm Homberg—interwove the philosophies of elements and principles with the chemical approach focusing on ordinary substances. Our distinction is in indeed an analytical one. Yet it is indispensable for an adequate historical understanding of the many tensions and frictions that occurred when eighteenth-century chemists attempted to extend their new understanding of chemical composition, analysis, and re-synthesis to areas of chemistry where it was not originally developed, such as plant and animal chemistry, and the study of principles. See also our discussion in chapters 6.3, 12, and 13.

for our distinction is the following: all of the substances included in the chemical table by Geoffroy were not merely perceptible substances, applicable for the resynthesis of the original compound, but were also commodities already well known in seventeenth-century pharmacy and metallurgy.[70]

In the following, we seek to highlight the characteristics of the new approach of eighteenth-century chemists to chemical composition by comparing it with alternative approaches that had been broadly accepted in seventeenth-century alchemy and chemistry, or "chymistry."[71] This comparison is all the more necessary as some very important terms—such as "compound," "composition," and "analysis"—were in use both in seventeenth-century chymistry and in eighteenth-century chemistry, yet, as we will see, with utterly different meanings. A historization of the meaning of terms such as "compound" and "composition," which have long been treated as universals, is still a historiographical challenge.

Spiritual and corporeal substances in seventeenth-century chymistry

The epistemic transformation of perceptible substances and mundane materials into objects that carry imperceptible features was not a novelty of the eighteenth century. In the centuries before, alchemists also attributed philosophical meaning to materials. Yet their approaches differed markedly from the eighteenth-century chemical approach. Their names for chemical remedies—such as magisteria, arcana, essences, quintessences, spirits, and elixirs—indicate to some extent their philosophical understanding. In the eyes of the early modern chymists chemical remedies were not unambiguously perceptible materials, but also participated in a world of exalted substances that were spiritual rather than rough corporeal objects. Magisteries, spirits, elixirs, specific extracts, essences, and so on, most of which were obtained by distillation, had nobler virtues and subtler qualities than ordinary materials. As Andreas Libavius proclaimed in his *Alchemia* (1597), essences contained all virtues (*virtus*) of substances and made them "live" (*vivere*) and "blossom" (*in vigore esse*).[72] By contrast, the bodily residue that remained after the extraction of the essence was a dead and rough material designated *caput mortuum*.[73]

Like Libavius, the influential seventeenth-century French chymist and follower of Paracelsus Nicaise Le Febvre (c. 1615–1669)[74] asserted that the "essences do not only differ in body, from that of the Compound whence they were extracted, but are advanced also to nobler and more efficacious qualities and virtues."[75] Referring to

---

70 See Klein [1994a], and [1994b] and the tables included in chapter 8. There is one exception to our general statement, the sulfur principle. On this see our arguments in chapter 8.
71 In the seventeenth century the terms alchemy, chymistry, and chimia as well as alchimist, chymist, chymicus, and so on were used more or less synonymously; see Homburg [1993] pp. 51–82; Newman and Principe [1998]; and Principe [1998] pp. 8–10. In the following we use the terms chymistry and chymists for this time period.
72 Libavius (1597) [1964] p. 72. Libavius used "essence" as a generic term that he further subdivided into the kinds of substances mentioned above.
73 Ibid.
74 For life and work of Le Febvre, see Gillispie [1970–1980] vol. VIII p. 130f.
75 Le Febvre [1664] vol. I p. 55.

Paracelsus, he emphasized that in order to obtain a pure essence it "must be freed of that gross body wherein it is imprisoned."[76] In the Paracelsian chemical philosophy, which was widely accepted by seventeenth-century chymists, "purity" meant freedom from coarse bodily qualities.[77] In their purest form, essences and other exalted, distilled substances were spiritualized to such a degree that they acquired the nature of fire. As Le Febvre made clear, this purest stage of spiritualization was not what chymists actually intended, for if that stage was actually achieved, experimenters were no longer able to perceive and handle the substance:

> For as Nature cannot communicate its Treasures unto us, but under the shade of Bodies, so can we do no more then to devest them by the help of Art from the grossest and most material part of that Body, to apply to our uses: for if we urge them, and spiritualize too much, so that they should flye from our sight and contact, then do they lose their bodily Idea and character, and return again to the Universal Spirit.[78]

These two examples may provide a first idea of early modern chymists' philosophical understanding of chemical remedies. Essences, elixirs, magisteries, and similar chemical remedies occupied a place between the ordinary, perceptible materials and the imperceptible, sublunar spirits that were deprived of any corporeality. As they were freed from rough corporeal qualities, they penetrated the human body more easily and thus were much more effective remedies than the ordinary raw materials.

"Compound," "composition," and "analysis" in the Paracelsian understanding

Early modern chymists' distinction between corporeal and spiritual substances of natural bodies raises questions concerning their understanding of the constitution of natural bodies. To discuss this understanding, we quote again the Paracelsian Nicaise Le Febvre:

> Her [chymistry's] operation is based upon these so different *Compounds*: for she may choose any of these bodies, either to divide and resolve it into its Principles, by *making a separation of the Substances which do compound it*; or she uses them, to *extract the mystery of Nature* out of them, which contains the Arcanum, Magistery, Quintessence, Extract and Specifick, in a much more eminent degree, than the body from when it is extracted.[79]

Le Febvre defined two goals of chymistry. The first goal—the "separation" of parts ("principles") of the "compound"—sounds quite modern. Yet its combination with the second goal—the extraction of the more eminent, spiritualized arcana and other essences—rings bells in the ears of historians. What did Le Febvre and other

---

76 Ibid.
77 It should be noted that the Paracelsian philosophical meaning of purity was also linked with considerations about the medical effects of the chemically prepared substance. A chemical remedy was more effective than a raw material, since it was purer. Hence Le Febvre could also claim that "by *purity* we will understand, all what in the Mixt or Compound can be found to serve our end and purpose; as to the contrary, by *impurity* all what opposeth it self and contradicts to our intention in the work." (Le Febvre [1664] vol. I p. 50).
78 Ibid. p. 34f.
79 Le Febvre [1664] vol. I pp. 70–71; our emphasis.

seventeenth-century Paracelsian chymists mean when they claimed to study the "composition" of natural "compounds" or "mixts" and to "separate principles" from them? That words change their meaning and reference in history is a commonplace for historians and philosophers, and historians of chemistry have long known that the meaning and reference of "compound" and "composition" changed several times from the late eighteenth century onward, in developments following the Lavoisierian Chemical Revolution. But they have often presupposed that, prior to that period, there were no significant changes in the meaning of these terms. Even in the most recent literature on seventeenth-century and eighteenth-century chemistry, we find little awareness of the historical specificity of the meaning and polysemy of terms like "compound," "composition," and "separation" prior to the Chemical Revolution.[80] In the following we argue for a historization of these significant chemical categories and outline meanings of "compound" and related terms in the seventeenth century that differed from their meaning in the eighteenth century and afterwards. Our outline is neither a dense description of all facets of meaning nor one of minor individual variations. Rather, it depicts recurrent patterns of understanding shared by many seventeenth-century chymists.

In the Paracelsian theory of matter, which informed not only Le Febvre's chemical philosophy but also that of many other seventeenth-century chymists, a compound or mixt was a unified whole, the same in all its bodily parts.[81] The different principles, which constituted a mixt, were not preserved in the process of a mixt's generation and hence not distinct bodily parts, but sets of qualities that invested the whole mixt with perceptible properties and virtues. This understanding, which shared features with the Aristotelian and Scholastic conceptions of a homogeneous natural

---

80   One of us (Klein) has argued in detail for a historization of the chemical concepts of "compound" and "composition," and related chemical categories (see Klein [1994a], [1994b], [1994c], [1995], [1996a], [1998b]). Similar arguments have been presented by William Newman with respect to van Helmont's philosophy of matter and its difference from the modern chemical conceptions of compound, composition, analysis, and synthesis (Newman [1994] pp. 141–143). In collaboration with Betty Jo Dobbs, Robert Siegfried also pointed out the importance of historical studies of "chemical composition" prior to the Chemical Revolution as early as 1968 (Siegfried and Dobbs [1968]), and in a more recent monograph he further developed his line of argumentation. But Siegfried's main analytical tool for tracing changes in the meaning of "composition" prior to Lavoisier has been the distinction between a "metaphysical" meaning of composition and a "materialist" or "operational" one. He has attributed these labels to the different Aristotelian, Paracelsian, corpuscular, atomic, and other chemical philosophies and theories without presenting more detailed historical analyses of differences of meaning (Siegfried [2002]).
      Another recent monograph by Mi Gyung Kim on the history of the concept of "chemical composition" has also covered both the seventeenth and eighteenth centuries. In her discussion of the "analytic practice" (or "chemical analysis," as she has also designated it) of seventeenth-century French chemists, Kim has addressed problems of reference, such as whether the products of distillation were actually identical with the postulated five chemical principles. She has also discussed discrepancies between the seventeenth-century French chemists' "analytic/philosophical ideal" and their actual practice. But she has not paid attention to the embeddedness of the meaning (not reference) of the terms "separation," "analysis," "composition," and "principles" in seventeenth-century chemical philosophies, and the difference of their meaning from that of eighteenth-century chemical categories. As a result, Kim has proposed that the seventeenth-century French chemists "stabilized composition as a major theory domain of chemistry and shaped the analytic ideal of chemical elements and principles, which was later endorsed by Lavoisier" (Kim [2003] p. 19). In the following we will present arguments that question the latter view.
81   For Paracelsus' theory of matter, see also Debus [1966], [1967], [1977], and [1991]; Pagel [1982]; and Newman [1994] pp. 106–110.

mixt defined by its substantial form, constitutes a stark contrast to the eighteenth-century concept of chemical compound established around 1700. The eighteenth-century concept of chemical compound meant a substance consisting of different substance components held together by chemical affinities. Not only could these substance components be separated from the compound, they also could be used as ingredients to resynthesize the original compound. They were thus regarded as building blocks, which are preserved in certain kinds of analyses and syntheses. By contrast, Le Febvre left no doubt that his understanding of a chemical mixt or compound was that of a homogeneous union which did not preserve the original, constituting substances: "natural mixture properly so said, is a strict union of the substances, whence some things substantial doth result, and *yet different from the other Substances which constitute it, by the help of Alteration.*"[82] He highlighted his view by demarcating proper and natural mixts from those "improperly called." The latter were "mixed together, but without change or alteration of the whole substance."[83] When wheat and barley, for example, were mixed together, they were "divided in parcels scarce perceptible to the eye." Yet in this case, the original wheat and barley did not change chemically, that is, they underwent no "alteration," so that the result was a mere mixture by "apposition." Le Febvre extended the range of improper or mechanical mixtures to "confusions" like alloys and water and wine, in which the "mixt parcels are not only imperceptibly divided, but also confounded together."[84] The reason for regarding solutions and other "confusions" as mechanical mixtures was that the original substances were preserved rather than altered qualitatively in their substance.

But why did Le Febvre and other Paracelsian philosophers designate homogeneous natural bodies as "mixts" and "compounds?" A clue for the meaning of these terms can be found in their theory of the "generation" of natural mixts. In the process of generation different constituents contributed to the coming into being of a new unified natural mixt. Paracelsus had taught his followers that all natural bodies—plants, animals, and minerals—were created from an "element"—either earth, water, air or fire—and the three different principles sulfur, mercury, and salt. Drawing an analogy to the generation of living beings, Paracelsian philosophers conceived of the element as the generating "matrix" or "womb" and the principles as "semina" carrying different sets of qualities.[85] Whereas many older alchemists assumed that in the act of generation the male semina entered the female elemental womb from outside, Paracelsus's proposed that the semina were always potentially present in the womb, "like a picture in wood."[86] Le Febvre, who presented himself explicitly as a follower of Paracelsus, differed slightly from his master in this latter respect.

---

82 Le Febvre [1664] vol. I p. 59; our emphasis.
83 Ibid. p. 58.
84 Ibid. p. 59. For Le Febvre's understanding of alloys see also below, in connection with our discussion of van Helmont's understanding of chemical transformations.
85 See Klein [1994a] pp. 38–46, and the primary and secondary literature quoted there. Newman has emphasized the "intensely biological character" of this doctrine, which he has designated "hylozoism" and "vitalism." See Newman [1994] pp. 106–108.
86 See Paracelsus [1996] vol. III pp. 778–779. In this respect Paracelsianism was informed by the Scholastic dualism of matter and form and of actual and potential components, which cannot be translated adequately into modern scientific language.

According to Le Febvre, the principles originated in a universal seminal spirit, which was "divested of all Corporeity."[87] As this universal spirit was homogeneous, it contained the different principles only in the potential form of a "denomination."[88] In the generation of natural bodies, the universal spirit descended from heaven to earth, where it was corporified by the specific matrixes or wombs; that is, elemental water or earth. At the same time the different principles were actualized.[89] In the formation of the natural mixt, the female elemental matrix contributed its corporeal qualities, like consistency, solidity, and hardness. All other properties and virtues of natural bodies stemmed from the male seminal principles. This mode of generation was considered to be the same in principle for all natural bodies, plants, animals, and minerals alike. Its result was a unified whole that was fully analogous to a living organism or even conceived to be a living organism. Hence, natural bodies were "mixts" or "compounds" not in the sense of being composed of different parts, but in the sense of being descendants of different constituents. Like a living organism, such offspring possessed one coherent body, the same in all its bodily parts, which was invigorated by a spiritual essence. And just as the souls of animals were not localized parts of the body, neither did it make sense to think of the spiritual essences of minerals or of materials stemming from the vegetable or animal kingdoms as discrete parts juxtaposed with its bodily parts. Rather, the spirits expanded over and penetrated the entire body.

Furthermore, when Le Febvre and other chymists claimed that the generating principles could be "separated" from natural mixts, the meaning of "separation" also differed strongly from the modern concept of chemical analysis. As natural mixts were homogeneous, the same in all their bodily parts, separation did not mean a resolution into *preexisting* physical parts or substance components. Rather, it meant first a purifying process by which the spiritual essence of a mixt was freed from its corporeal prison. Second, similar to the actualization of the principles contained in the universal spirit, separation meant a kind of creative act which reconstituted the original seminal principles. In this reconstitution and subsequent separation by chemical art, the seminal principles were not totally spiritualized, but retained some corporeal qualities of the natural body. Hence the separated principles, whose number were five—phlegm or water, spirit or mercury, sulfur or oil, salt, and earth—were not fully identical with the original, generating spiritual principles.[90] But Le Febvre emphasized that they were nonetheless natural, meaning not transmuted by chemical art.

Taking Le Febvre's philosophy of compounds and principles as an example, we have discussed the Paracelsian view of the invisible constitution of compounds at

---

87 Le Febvre [1664] vol. I p. 13.

88 Le Febvre also spoke of "distinct but not differing substances": "This spiritual substance, which is the primary and sole substance of all things, contains in it self three distinct, but not differing substances. [...] otherwise, as Nature is one, simple and homogeneous, if the seminal principles were *heterogeneous*, nothing would be found in nature one, simple, and homogeneous; [...] Let us then conclude, that this radical and fundamental substance of all things [the universal spirit], is truely and really one in essence, but hath a threefold denomination" (Ibid. p. 15). It is probably impossible not to be reminded of the Holy Trinity by this philosophy of substance.

89 Ibid. pp. 16ff., 29f., 39, 41f.

90 Ibid. pp. 17–20.

some length in order to highlight the historical embeddedness and differences in the meaning of terms like "compound," "composition," and "separation" or "analysis." In the context of the Paracelsian natural philosophy, compounds were offsprings of different constituents, but they did not consist of heterogeneous physical parts or substance components. This meaning of compound differed profoundly from the conceptual network of chemical compound, composition, affinity, analysis, and synthesis represented in the eighteenth-century tables of affinity and the 1787 table of nomenclature by Lavoisier and his collaborators. Our following outline of three competing seventeenth-century approaches to compounds and composition can be much briefer, as our main goal here is to carve out the distinctive features of the eighteenth-century conceptual network, which, as a shorthand, we will designate in the following as the "conceptual network of chemical compound and affinity."

"Compound," "composition," and "analysis" in seventeenth-century atomistic and corpuscular theories

In the first half of the seventeenth century, scholars like Daniel Sennert (1572–1637), Angelo Sala (1576–1637), Joachim Jungius (1587–1657), Sebastian Basso (fl. 1620), Johann Chrysostomus Magnenus (fl. 1645), Claude Guillermet de Bérigard (c. 1578–1664), and Etienne de Clave (fl. 1640) propagated a qualitative atomism that linked ancient atomism with the Aristotelian and Paracelsian philosophy of elements or ultimate principles.[91] These chemical philosophers proposed that natural mixts consisted of indivisible atoms, and that there existed different qualitative kinds of elemental atoms. The French chymist Etienne de Clave, for example, asserted that all natural mixts were composed of the atoms of the five principles phlegm (or water), spirit (or mercury), sulfur (or oil), salt, and earth.[92] The five kinds of imperceptible and indivisible atoms were "actual" bodily parts of all mixts (*actuellement inclus*), the difference between mixts being caused only by quantitative differences in the five kinds of atoms. Clave's atomism agreed with the Paracelsian philosophy in its assumption of a small number of ultimate principles that constituted all natural mixts. Like Paracelsian philosophy, it also ascribed a peculiar ontological status to the ultimate principles. Principles were causal agents that engendered the properties of the natural mixts. As causes their number had to be small. Furthermore, these hidden entities were never encountered in nature in a free, isolated state, but only as constituents of plants, animals, and minerals. However, de Clave's chemical philosophy differed profoundly from its Paracelsian counterpart in the assumption of atoms and the understanding of compounds as objects consisting of heterogeneous physical parts, the atoms of the principles.

---

91   An excellent study of these atomists is presented in Lasswitz [1984], part I. See also Meinel [1988]; Newman and Principe [2002]; and the essays collected in Lüthy et al. [2001]. For a recent study, especially on the relation between seventeenth-century Paracelsianism and chemical atomism, see Clericuzio [2000]. For the life and work of Sennert, see Gillispie [1970–1980] vol. XII p. 310ff.; of Sala, ibid. p. 78ff.; of Jungius, ibid. vol. VII p. 193ff.; of Basso, ibid. vol. I p. 495; of Magnenus, ibid. vol. IX p. 14f.; of Bérigard, ibid. vol. II p. 12ff.

92   See de Clave [1641], and [1646].

When we compare this kind of seventeenth-century qualitative or chemical atomism to the eighteenth-century conceptual network of chemical compound and affinity as represented in chemical affinity tables and the 1787 table of nomenclature, we find similarity, but also significant differences in reference and meaning.[93] In both cases "composition" meant heterogeneous composition from different parts. The main difference consisted in the way reference to experiments was intertwined with metaphysics. The seventeenth-century atomists referred quite unambiguously to imperceptible entities when they spoke of "atoms." By philosophical definition, atoms were imperceptibly small bodies, and no seventeenth-century atomist ever claimed that atoms could actually be separated from natural bodies and made visible by chemical analysis. Instead, they thought it possible that the principles or elements, each of which was made up of the same kind of atoms, could be separated from natural mixts. Yet for very similar reasons as in the case of atoms, the question of whether the simple principles or elements could actually be separated by chemical analysis from the natural mixts was open to philosophical debate. By definition the three or five Paracelsian principles and the four Aristotelian elements were not ordinary natural mixts but constituents thereof. Paracelsian principles and Aristotelian elements were natural matter pertaining to an ontological level that caused the different species of natural mixts, and no chemist and natural philosopher of the time would have confused cause and effect. Paracelsian principles and Aristotelian elements belonged to an ontological reality beyond the multifarious, changeable, and contingent world of natural mixts. Therefore, as Robert Boyle argued in his *Sceptical Chymist* (1661), it was impossible to demonstrate with certainty that the undecomposable substances discovered by chemical experiments were identical with the postulated principles and elements.

By contrast, in the area of investigation mapped by chemical tables, "components" meant entities that were both imperceptible parts of chemical compounds and perceptible substances that could be separated from compounds and manipulated as ordinary materials in the laboratory. When separated from a salt, for example, the component vitriolic acid was no less an ordinary chemical substance than the compound, the salt—despite the fact that eighteenth-century chemists, too, conceived of components as agents that caused the properties of the compound. The ontological status of components and compounds was exactly the same. This understanding was embedded in chemists' analytical practice—the fact that chemical analysis actually took apart compounds such as salts, isolated their components in a pure form, and resynthesized the original compound from the components. Whatever the more specific definitions of substance components such as acids, alkalis, earths, and metals were in the eighteenth century, the existence of such kind of components was never

---

93 It should be noted that we refer here exclusively to the meaning of chemical compound and related concepts in the specific area of eighteenth-century chemistry that was mapped by the chemical affinity tables and the 1787 table of nomenclature. We will argue in part II that the reference and meaning of "chemical compound" and related concepts in the context of this specific area differed to some extent from their reference and meaning in other contexts. It differed, in particular, when eighteenth-century chemists were seeking the ultimate principles or elements of substances.

controversial among chemists—in contrast to that of atoms and of simple elements and principles. On the contrary, these kinds of substances were quotidian commodities familiar even outside of academic institutions.

The chemical atomists' concepts, methods, and style of argumentation came under attack in the second half of the seventeenth century. Robert Boyle's *Sceptical Chymist*, in particular, launched a devastating critique based on both theoretical and experiential grounds.[94] We again examine briefly Boyle's views as an example for a widespread kind of seventeenth-century corpuscular philosophy that considered mechanical qualities of particles such as size, shape, and orientation in space to be the most fundamental qualities of matter. According to Boyle's corpuscular philosophy, natural bodies and substances consisted of compound corpuscles made up of smaller particles of a universal matter, the "primitive fragments" or "prima naturalia." Corpuscles were "clusters" or "little primitive concretions" of the prima naturalia, which differed only in their mechanical qualities—size, shape, and movement.[95] The substance-specificity of the compound corpuscles was determined by the size, shape, and relative position in space of their primitive particles, which brought about as an emergent result the distinctive structure or "texture" of a corpuscle. It was the structure or texture of corpuscles that determined the species of a substance.

This conception of substance-specific textures of corpuscles enabled Boyle to take a "middle course," as he called it, between the idea of homogeneous mixts, the same in all their parts, and the chemical atomists' idea of juxtaposed atoms of mixts. On the one hand, it was possible theoretically—and Boyle also advanced experiential arguments for that possibility—that chemical changes took place in such a way that the corpuscles of the initial substances were fused into a new kind of corpuscles so that the resulting substance consisted of only one kind of corpuscles having one unified texture. The result of such a kind of chemical change—which was a true alchemical transmutation—was a compound that was the same in all its (corpuscular) parts. On the other hand, it was also possible on theoretical grounds that the corpuscles of reacting substances were preserved in chemical transformations and merely juxtaposed very closely, so that the resulting new compound consisted of different kinds of

94 See Klein [1994a] pp. 54–83, Klein [1994c]; and Principe [1998]. In his interpretation of Boyle's *Sceptical Chymist,* published in 1998, Lawrence Principe came to largely the same conclusions, though in a different way, as one of us, U. Klein, in 1994. Principe has argued, in particular, that the real targets of Boyle were not the alchemists (and alchemical beliefs in transmutation) but the Paracelsian chymists (Principe [1998] chapter 2 in particular). In 1994, Klein presented a detailed re-interpretation of the *Sceptical Chymist*, which demonstrated that Boyle presented both empirical and theoretical arguments for the very possibility of alchemical transmutation (Klein [1994c]). Klein's re-interpretation also showed that Boyle's discussion of the question as to whether simple elements (or any other species of substances always preserved in chemical transformations) existed or not concluded in a stalemate. Boyle saw himself unable to decide that latter question. Klein's interpretation differs from Principe's inasmuch as she has argued that the target of Boyle's criticism and skepticism concerning the notion of simple elements was not only Paracelsians, but also atomists who had adopted the Aristotelian or Paracelsian idea of elements. Many of Boyle's specific arguments in favor of the assumption of elements refer not to the Paracelsian notion of homogeneous mixts (or compounds in the Paracelsian sense) but to a notion of mixts consisting of different elemental atoms or corpuscles. This becomes most evident at the end of the second part of the *Sceptical Chymist* in the context of Boyle's discussion of "the genuine notion of mixts" (see Boyle [1999] vol. 2 pp. 267–276).

95 For more details concerning the following account, see Klein [1994c] and the primary sources and secondary literature quoted there.

corpuscles. That second theoretical possibility converged with the seventeenth-century atomists' image of ultimate elemental atoms that are preserved in chemical transformations, and Boyle was seeking empirical evidence for this possibility as well. In so doing he explicitly referred to well-known operations in metallurgy and pharmacy, all of which shared one outstanding feature: they were reversible in the sense that the original substances used in the formation of a new compound could be recovered in subsequent operations. That is, unlike many chemical atomists, Boyle did not refer to the dry distillation of substances when studying the question of whether ultimate elements existed, but to reversible decompositions and recompositions. Yet in the end Boyle did not decide clearly in favor of either of his two theoretical alternatives, leaving open the question as to whether ultimate chemical elements existed or not. His *Sceptical Chymist* ended in a stalemate.

As we will see in the following discussion, and in more detail in part II, around 1700 chemists began to pay increased attention to reversible chemical transformations. It was in this practical context that they accepted a new concept of chemical compound and composition that was intimately linked with another very important chemical concept: the concept of chemical affinity. The eighteenth-century concept of an elective chemical affinity allowed chemists to demarcate mechanical mixtures that merely appeared to be homogeneous from true chemical changes and novel chemical compounds. This concept, however, was absent from Boyle's corpuscular theory. As a consequence, in the context of this philosophy it was not possible to distinguish the mechanical apposition of corpuscles from processes of chemical transformation, or mechanical mixtures from novel compound chemical species. Considerations like these may have spurred another influential seventeenth-century chemical philosopher—Joan Baptista van Helmont (1579–1644)[96]—to draw completely different conclusions than the chemical atomists. In van Helmont's philosophy the possibility of recovering an ingredient from a newly procured substance was a clear indicator that the apparently new chemical species was in reality merely a mechanical mixture. A very similar view, though without corpuscular arguments, was held by seventeenth-century Paracelsians like Nicaise Le Febvre. As true mixts were homogeneous, it was not possible to recover the original substances from a newly made perfect mixt. Le Febvre explained this understanding using the example of metal alloys.[97] Referring to the fact that a metal alloy could be separated again into the original metals from which it was made, he raised the question of whether the alloy was a new chemical species. His answer was in the negative, since the original metals could be recovered from the alloy:

> The first Question raised, is: If when several Metals are melted together, any Metallic Species, differing from the Metals, which do make the whole compound, is thereby raised? To which, the Answer is negative, because *it is not a true mixture*, much less a strict and exact Union, but *rather a Confusion, since they may again be separated asunder.*[98]

---

96 For life and work of van Helmont, see Gillispie [1970–1980] vol. VI p. 253ff.
97 For further examples, see our discussion above.
98 Le Febvre [1664] vol. I p. 66; our emphasis.

As William Newman has observed, van Helmont's corpuscular philosophy shared many features with the Paracelsian philosophy of homogeneous mixts and the Paracelsian understanding of chemical change. Van Helmont distinguished sharply between mixtures resulting from the "mere apposition" of particles, on the one hand, and the products of transmutation or true "wedlock," on the other.[99] Apposition of preserved corpuscles did not procure a new kind of chemical species but rather a mixture that only appeared to be different from the original species. For example, glass was made by fusing potash with sand, and the result of that fusion seemed to be an entirely new substance that had none of the properties of the two original substances. Nonetheless, van Helmont argued that glass was *not* a genuine product of chemical change and thus a new species of substance. The argument he presented for this was the following: "by means of art glass returns into its original ingredients (*pristina initia*) once the bonding holding them together is broken: the sand can even be regained in the same number and weight."[100] Since it was possible to recover wholesale the sand used as an ingredient for the making of glass, van Helmont conceived of glass as the product not of true chemical transformation, but of a mechanical mixture. By contrast, in a true chemical transformation a "marriage" took place in which the original corpuscles were "subtilized" and transmuted into integral corpuscles of a novel species. Transmutation and true wedlock into novel, unified corpuscles implied that it was impossible to recover the original substances: "to the Helmontian, two substances that had undergone transmutation could not be reduced into their original constituents."[101]

For van Helmont there was no third way between mixture and mere juxtaposition of preserved particles and alchemical transmutation in which the original particles were not preserved. Putting it in the words of Newman: "The upshot of Van Helmont's theory is that *there is no intermediate state* between mechanical mixture and transmutation." And: "Van Helmont's theory *replaces the real combination* of two volatile substances *with transmutation* of the two substances into something else. [...] two substances that had undergone transmutation could not be reduced into their original constituents." Newman added that van Helmont's concept of transmuted compounds "violates the very notion of the chemical compound as we know it."[102]

Why does the Helmontian philosophy violate the notion of the chemical compound "as we know it"—or, as we argue, as the majority of chemists accepted in the course of the first half of the eighteenth century? We will see in the subsequent chapters of this book that the vast majority of eighteenth-century chemists did not accept van Helmont's view. Rather, they established a conceptual network linking the concepts of chemical compound, separation or analysis, recomposition or synthesis, and affinity in new ways that did not exist in the seventeenth century. We argue that the eighteenth-century concept of chemical compound (as represented in the chemical tables) and the concept of chemical analysis and synthesis coupled with it was indeed

99 See Newman [1994] pp. 110–114, 141–143.
100 We quote from Newman and Principe [2002] p. 77.
101 Newman [1994] p. 143.
102 Newman [1994] p. 143; our emphasis.

a third way of understanding chemical transformations and their material products. The eighteenth-century chemical concept of analysis and synthesis meant a true chemical transformation, but a transformation in which the original substances were preserved as components of the new compound and held together by chemical affinity. Chemical affinity coupled with electivity was the centerpiece demarcating imperceptible movements of building blocks of preserved substances and their aggregation to new chemical compounds from mere mechanical mixtures. As a consequence of their preservation, the original building blocks could be recovered from the new compound. The eighteenth-century concept of analysis and resynthesis meant exactly what van Helmont had excluded as a third way between transmutation and mechanical mixture.[103]

Seventeenth-century classification of substances

Our discussion above has shown that the meanings of the terms "compound," "composition," and "analysis" in seventeenth-century chymistry were different from the meanings of these terms in the chemistry of the eighteenth century and afterwards. In the following we further substantiate our historization of these chemical categories by examining seventeenth-century chymists' mode of classifying substances. We present two representative examples of philosophically informed classification in seventeenth-century chymistry: a systematic one presented in Andreas Libavius's *Alchemia* (1597), and a more implicit one contained in Nicaise Le Febvre's *Traicté de la chymie* (1660). Both Libavius and Le Febvre were chymists who had a great impact on the emerging communities of chymists in Germany and France. Their work thus elucidates the seventeenth-century mode of chemical classification in the philosophical context beyond the level of individual chemists.

In his *Alchemia*, Andreas Libavius (1560–1616) presented a comprehensive taxonomic system of "chymical species," which he also depicted in the form of a diagram (see figure 2.3).[104] The system comprised all kinds of "chymical species" that is, substances "exalted" (*perficere*) by chemical operations.[105] The fact that Libavius's chemical taxonomy ordered exclusively chemically processed substances is important. He excluded natural raw materials for a very specific reason, which is telling for his ontology of substances. Libavius conceived of ordinary materials or natural mixts as substances endowed with rough corporeal properties, whereas he regarded chemical preparations as exalted substances.[106] Chymistry was concerned in the first place

---

103 It should be noted that there is a strange discrepancy between Newman's interpretation of van Helmont quoted above and later interpretations presented in *Alchemy Tried in the Fire* (Newman and Principe [2002]). In *Alchemy Tried in the Fire*, Newman and Principe also presented van Helmont as a seventeenth-century chymist "who put great emphasis on the paired chymical analysis and synthesis of bodies"—an idea which they interpreted as an "outgrowth" of Paracelsian chymistry. Furthermore, they argued that "it was this definition of chymistry as the art of analysis and synthesis that would provide a disciplinary identity to the field lasting well into the nineteenth century" (Newman and Principe [2002] p. 90; see also p. 303).
104 For life and work of Libavius, see Gillispie [1970–1980] vol. VIII p. 309ff.
105 See Libavius [1964] p. 119.
106 See our discussion above.

TABVLA LIBRI SECVNDI ALCHEMIÆ.

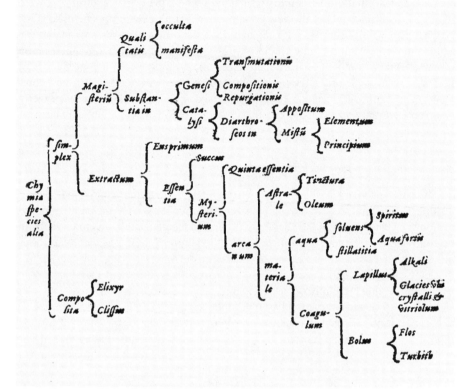

Figure 2.3: Libavius's classification of chemical species. From Libavius [1597].

with the exalted chemical preparations.[107] The exaltation and ennoblement of substances was the distinctive goal of chymistry, which rendered it superior to all other arts and crafts.[108] This overall goal of chymistry, shared with the Paracelsians, also determined Libavius's mode of identifying and classifying substances.

Unlike hermetic Paracelsians, Libavius, the defender of humanist Lutheran culture, strove for a public and methodized discipline of chymistry.[109] His methodological ideals were made apparent in his taxonomic system. Following the Aristotelian classificatory method, he arranged his classes of chemical substances in the binary, encaptic fashion of a taxonomic tree.[110] The first, most basic bifurcation distinguished between simple (*simplex*) and composed chemical species (*composita*) (see

---

107 It should be noted that, in addition to chemically processed and exalted substances, Libavius also included a small number of other products of the chemical arts and crafts, which he considered to be mixtures in the sense of mechanical apposition (*per appositionem*); see Libavius [1964] p. 267.
108 Libavius used the term "alchemy" for all kinds of chemical arts, as was common at the time (see Newman and Principe [1998]).
109 See Hannaway [1975].

figure 2.3). This distinction may at first glance appear to presuppose knowledge of chemical composition. Yet, as we have shown in our discussion of Paracelsianism and other seventeenth-century chemical philosophies, use of the terms "simple" and "composed" does not imply that their meaning was the same as in the eighteenth century. Indeed, we will show in the following that Libavius's term "composed" did not mean "consisting of" different substance building blocks, but rather "generated" from different substances.

This latter meaning of "composed" becomes clear when we study Libavius's more specific classes of composed chemical species. Libavius further subdivided the composed chemical species into elixirs and clissus and the simple chemical species into magisteries and extracts (see figure 2.3). Elixirs and clissus were not further subdivided, and comprised only a comparatively small number of chemical preparations, all of which were applied as chemical remedies.[111] Elixirs were made from several processed chemical substances. To prepare an elixir, vegetable and animal extracts, or several kinds of mineral and vegetable magisteries, or both magisteries and extracts of different natural origins were mixed together and further processed by distillation. The number of chemical ingredients for the preparation of elixirs was often quite impressive, and Libavius presented precise recipes for their manufacture, including information about the masses of the ingredients. The clissus, which formed the second class of composed chemical species, were also made from several different chemical preparations, but in this case the chemical preparations stemmed from one and the same natural raw material. For example, when the oil and salt extracted from cinnamon were mixed together and further processed, the chemical product was called a "clissus."

How did Libavius understand the chemical processes that yielded elixirs and clissus? An anachronistic reading would interpret the making of these *species composita* as a chemical synthesis and their results as chemical compounds consisting of different substance components. But this was not Libavius's understanding. Like the Paracelsians, Libavius did not consider *species composita* to be compounds in which the original substances were preserved as integral building blocks, held together by a chemical force.[112] *Composita* were descendants of different chemical species, but they did not contain their constituents in the form of physico-chemical components. Rather, in the chemical operations that engendered an elixir or clissus, the chemical ingredients became further exalted and fused together into one unified *compositum* that contained the purified essence of the original ingredients (*quidquid in ea est essentiale ad unum redigitur compositum*).[113] Hence Libavius could declare that a

---

110 Libavius's encaptic system did not achieve the ordering of all chemical species in an unambiguous way; that is, without overlapping taxa. But this aspect of his classification is not at stake here. Our goal is not to provide a comprehensive analysis of Libavius's taxonomic system, but merely to analyze his criteria for classification.

111 See Libavius [1964] pp. 543–563.

112 As Owen Hannaway has observed, Libavius criticized Paracelsus and his followers in many aspects (Hannaway [1975]). Yet he also shared with the Paracelsians some basic understanding of the constitution of substances and of chemical change. His understanding of chemical change as exaltation is only one example for the latter. See also our discussion of the historical meaning of chemical concepts above.

clissus, prepared from the oil, salt, or other extracts of one natural mixt, "stood for the entire substance of the [original, natural] matter and its complete essence" (*pro tota rei substantia est, & essentia completa*).[114]

What about Libavius's second large class, the "simple species," which were further subdivided into magisteries and extracts? Magisteries and extracts, which formed the two most comprehensive classes in Libavius's system, were not made from chemical preparations but from natural mixts; plants, animals, or minerals (metals, stones, half metals, and so on).[115] And whereas the composed elixirs and clissus were chemical products of at least two subsequent chemical operations, the simple magisteries and extracts were results of only one chemical operation. A "simple" (*simplex*) chemical species, as defined by Libavius, was made "in one single continuous process, in which it was brought to perfection out of an imperfect and raw state" (*simplex est quae uno continuo processu ad suam in se perfectionem perducitur proximé ex imperfecto & rudi*).[116]

What distinguished magisteries and extracts? Both classes had in common that they were products of "exaltations." But when a magistery was prepared, more or less the whole quantity of a given natural mixt was transformed into another one.[117] A typical magistery was made by transmutation, which transformed the entire mass of a sample of iron, for example, into gold.[118] Transmutation was an alteration of the qualities of a substance in such a way that one species of substance became transformed into another one without adding or separating further substances. By contrast, an "extract" was separated from the natural mixt, leaving behind a considerable residue of the latter.[119] A typical extract was the distilled spirit or the tincture obtained by the distillation or decoction, respectively, of a plant.

As can be seen in figure 2.3, Libavius also used the term "compositiones" for a subclass of magisteries. The significant feature of simple chemical species like magisteries was that they were made from natural mixts. But the number of natural mixts applied to prepare a magistery differed in chemical and pharmaceutical practice, and this difference provided another taxonomic criterion for the further distinction of subclasses of magisteries. The class of *transmuted* magisteries contained preparations that were made from one natural mixt only,[120] whereas the class of *composed* magisteries contained preparations made from several natural mixts.[121] Yet, as in the case of the elixirs and clissus, Libavius did not understand proper composite magisteries to be compounds consisting of different substance components, but "homogeneous and simple substances."

---

113 Libavius [1964] p. 559; Libavius [1597] p. 421.
114 Ibid.
115 See Libavius [1964] pp. 119–324 (on magisteries), and 327–540 (on extracts).
116 Libavius [1964] p. 119; [1597] p. 85.
117 See Libavius [1964] pp. 120, 179.
118 Ibid. pp. 179–187.
119 Ibid. p. 327.
120 Libavius [1964] p. 179.
121 Libavius also allowed that composed magisteries were made from other magisteries; that is, from chemical species (ibid. p. 238). As a consequence, his class of composed magisteries overlapped with his class of elixirs.

To sum up, Libavius considered all kinds of chemical preparations to be exalted substances. This understanding informed his major classificatory divide: the separation of chemical species from the raw natural mixts, and the exclusion of the latter from his taxonomic system. Within the large class of chemical species, he demarcated subclasses according to differences in their preparation and exaltation. Simple species resulted from one coherent chemical process, composite species from at least two subsequent chemical processes. This taxonomic criterion was not operational in a modern philosophical sense, but also implied a distinctive philosophical understanding of chemical transformations: repeated operations yielded more exalted chemical species. Libavius's taxonomic criteria systematically interwove perceptible features of chemical operations with the philosophical interpretation of the imperceptible transformation processes. The centerpiece of that interpretation was exaltation. In stark contrast to the modern understanding of chemical transformations as analysis and synthesis, an understanding which was part and parcel of the eighteenth-century conceptual network of chemical compound and affinity, Libavius understood chemical transformations as different forms of exaltations of substances. And in stark contrast to the eighteenth-century classification of chemical compounds according to composition, Libavius classified substances according to different chemical methods of exaltation. Our next example, Nicaise Le Febvre's mode of classification, displays a very similar pattern.

Nicaise Le Febvre's *Traicté de la chymie* (1660) belonged to a tradition of French chymistry textbooks concerned not only with the taxonomy of chemically processed substances, like Libavius, but also with the order of natural raw materials or "natural mixts."[122] As to natural mixts, Le Febvre took over the naturalistic division according to their origin from the three natural kingdoms: "all Mixt perfectly framed, is either Animal, Vegetable, or Mineral."[123] But what did Le Febvre mean when he spoke of natural mixts? Why did he add the predicate "perfectly framed?" Le Febvre's answer to these questions went as follows: a perfect mixt like an animal, vegetable, or mineral, was "endowed with a substantial principal form, distinguished from the Principles whereof it is compounded."[124] We have argued above that Paracelsian chemists conceived of natural mixts as descendants of heterogeneous constituents (an element and three or five principles), which, nevertheless, were unified wholes. In the quotation above, the Paracelsian Le Febvre used the Scholastic term "substantial form" to highlight precisely this Paracelsian understanding. Natural mixts were integral wholes resulting from the perfect union of their different constituents.

It is obvious that this kind of holistic understanding of natural mixts logically excluded their identification and classification via distinct physico-chemical components. An almost natural consequence of this understanding and of the entire Paracelsian philosophy was the further division of natural mixts into animated and unanimated beings. To the class of animated mixts, which Le Febvre also designated

---

122 In the following we will quote from the contemporary English translation of Le Febvre's textbook, Le Febvre [1664].
123 Le Febvre [1664] p. 51.
124 See Le Febvre [1664] p. 60.

"organical,"[125] belonged the vegetables and animals (and also men) as well as the minerals inasmuch as they remained connected with their elemental matrix; that is, in their natural state in the earth.[126] To the unanimated or dead mixts belonged the minerals extracted from the earth—which Le Febvre further divided into metals, stones, and middle minerals or marcassites—as well as materials "ordinarily" separated from plants and animals, such as fruits, seeds, roots, gums, resins, wool, cotton, oil, and wine. As becomes clear from the examples, "ordinary" here meant commodities not prepared in the chemical laboratory.[127] The further identification and classification of these mixts was achieved in the same way as in natural history and the arts and crafts, according to perceptible properties and virtues. For example, Le Febvre demarcated metals, marcassites (or half metals), and stone as follows:

> Metal is a Mixt, extending it self under the Hammer, and melting in the fire. Marcassites are fusible, but extend not under the Hammer; and Stones are neither fusible nor extensible.[128]

Like Libavius, in the theoretical part of his chymistry textbook Le Febvre demarcated the natural mixts, which included natural raw materials as well as materials not made in the chemical laboratory, such as metals, oils, and wine, from the class of chemically processed or "pure substances." And like Libavius's *Alchemia*, Le Febvre's chymistry was primarily a science and art of pure substances. "The chief ends of Chymistry," Le Febvre emphasized, were to "separate Purity from Impurity."[129] It is important to remember that in the context of the Paracelsian philosophy the meaning of "purity" differed from the modern chemical meaning of the term. In Paracelsian philosophy, "purification" meant the freeing of substances from corporeal qualities, their exaltation and spiritualization, which brought them nearer to the nature of fire and the universal spirit. Thus, Le Febvre's concern about purity and pure substances must not be confused with a quest for chemical analysis, the separation of single or pure components, along with the study of chemical composition.

What kind of substances did Le Febvre group together under the label "pure substances?" The relevant chapter in his *Traicté* starts as follows:

> Besides the five Substances or *Principles*, which we have formerly said may be extracted out of natural Compounds, by the ministery of fire, there may be yet some *Essences* drawn, by diversifying the Operations of Art, which exalts, and do ennoble the Principles of these Mixts, and raise it to their *purity*.[130]

To the class of "pure substances" belonged first the five principles, which were the natural constituents of mixts separated from them by fire—dry distillation. As we have seen above, Le Febvre conceded that chemical art did not completely reduce the principles into their original spiritual form, but it nevertheless separated these pure

125 See Le Febvre [1664] p. 62.
126 Le Febvre emphasized that "Minerals do live, as long as they are joined to their Matrix and Root." Ibid. p. 62.
127 Most of these materials belonged to the pharmaceutical class of *simplicia*. For this, see chapter 11.
128 Le Febvre [1664] p. 62.
129 Ibid. p. 73.
130 Ibid. p. 55; our emphasis.

substances from natural mixts without altering their natural qualities. In case of the second subclass of pure substances—the "essences"—alteration by chemical art was an explicit goal. Whereas the five principles were "not extracted from the Mixt by transmutation, but by mere natural separation," the essences were transmuted, exalted, and ennobled substances.[131] They were pure in the sense of having been freed from rough corporeal qualities. Hence, like Libavius, Le Febvre demarcated the class of principles from that of essences by means of a theory of the imperceptible chemical transformations underlying the perceptible chemical preparation of the substances. By way of the same ambiguous taxonomic criterion that interwove observation with interpretation, he further subdivided essences into "magisteries" and "elixirs."[132] Magisteries were products of exaltation and extraction from one mixt only, whereas elixirs were ennobled purified substances extracted from a mixture of several mixts. All further subdivision of magisteries and elixirs was based on properties and virtues observed or attributed to them. Tinctures, for example, were a specific kind of elixir that had color.

In summary, in the theoretical part of his *Traicté* Le Febvre divided the realm of materials first by grouping them into the two large classes of natural mixts and pure substances.[133] He further classified the natural mixts according to their origin from the three natural kingdoms. His classification of pure substances—certain kinds of chemical preparations—relied on a taxonomic criterion that combined perceptible features of chemical operations with their interpretation as either "purification" or "exaltation"—a mode of classification also chosen by Libavius. The further subdivision of essences and other kinds of pure substances relied on their perceptible properties and medical virtues.

Chemical composition in the modern sense played a role in neither Libavius's nor in Le Febvre's classification. Like Libavius and Le Febvre, many seventeenth-century chymists conceived of natural mixts and chemical species as unified wholes, engendered by different constituents but not consisting of localized components. Classification according to composition from distinct substance building blocks would have been incompatible with that kind of philosophy. Moreover, as seventeenth-century chymists regarded chemically processed substances as exalted materials, their classifications divided and grouped together substances according to chemical processes of exaltation.

---

131 Ibid. p. 22.
132 Ibid. p. 56.
133 It should be noted that in the practical part of his textbook, Le Febvre classified all materials—both natural raw materials and chemical preparations—according to their origin from the three natural kingdoms. The headings on the chapters dealing with materials stemming from the vegetable and mineral kingdoms actually read: "Of vegetables and their chymical preparation" (ibid. vol. I p. 161, and vol. II p. 1) and "Of minerals and their chymical preparation" (ibid. vol. II p. 79). For example, the long chapter on "Vegetables and their chymical preparations" collected descriptions of roots, leaves, flowers, fruits, seeds, rinds, and woods of various kinds of medical plants as well as a huge number of descriptions of chemical preparations of distilled waters, extracts, tinctures, quintessences, elixirs, and other kinds of chemical remedies made from the former raw materials.

The eighteenth-century conceptual network of compound and affinity

After our short detour through the chemical philosophies of the seventeenth century, we now present a rough historical outline of the conceptual network of chemical compound and affinity, which informed eighteenth-century chemists' mode of identifying and classifying substances according to composition. Reference to chemical transformations that can be reversed was a distinctive feature of that "modern" conceptual network, introduced in the decades around 1700 and proliferating in the European chemical community by the middle of the century.[134] In the late seventeenth century, French chymists in particular grouped together kinds of chemical alterations that they no longer explained as transmutations, extractions of essences, or other kinds of exaltations, but as interactions or reactions of two substances and regrouping of their stable substance components, directed by chemical affinities between pairs of substances. The result of such elective reactions was chemical compounds consisting of heterogeneous substance components, interconnected by chemical affinities. Important steps in this direction were made by the two Parisian chymists Nicolas Lemery and Wilhelm Homberg (1652–1715), with their classification and theory of *sels salés* and middle salts.[135] The next step was the table of chemical affinities or *rapports*, published by Etienne François Geoffroy in 1718.[136] Geoffroy grouped together almost all operations known in his days that could be interpreted as simultaneous decomposition and recompositions, or replacement reactions. With only one exception—the decomposition of sulfur into sulfuric acid and the oily principle, and the resynthesis of sulfur from these principles—all of the operations presented in Geoffroy's table were well-known operations performed in metallurgy and pharmacy.[137] Furthermore, the bulk of the substances and reactions mapped by Geoffroy's affinity table belonged to the mineral kingdom.[138]

We will study the eighteenth-century tables of chemical affinity and their historical context in more detail in part II. At this point we wish to further illuminate the distinctive type of experiments they referred to, and the systematization to which these experiments lent themselves. The core of eighteenth-century affinity tables was built by the salts, acids, alkalis, earths, metals, and alloys. Under largely the same physical conditions, especially at ordinary temperatures, salts, acids, alkalis, earths, metals, and alloys displayed a stable, reproducible pattern of chemical transformation. For example, when a salt, such as copper vitriol, common salt, or saltpeter, was mixed with certain ingredients and heated, it yielded a mineral acid. When the mineral acid obtained in this way was again mixed with calcareous earth, or another salifiable base, the original salt could be restored. Such kinds of reversible chemical transfor-

---

134 We use the term "modern" here only as a convenient shorthand for a rather stable conceptual network that still had a working life in the nineteenth century, the period usually seen as the beginning of modernity.

135 See Holmes [1989]; Kim [2003] pp. 86–90; Klein [1994a] pp. 228–232.

136 See Geoffroy [1718].

137 For this exception see chapter 8.

138 Exceptions are the watery solutions of alcohol and spirit of wine and the compounds of vinegar (acetic acid).

mations—decomposition and recomposition, as it were—were also performed with metal alloys, metal sulfides, and watery solutions.

When chemists began around 1700 to pay increased attention to such reversible chemical changes and compared cycles of decomposition and recomposition, they observed regularities.[139] For example, metal salts—compounds consisting (in the system of Geoffroy) of an acid and a metal—were decomposed by alkalis and earths, but salts containing alkalis or earths were not decomposed by metals. Chemists explained such elective behavior of substances in chemical reactions as replacement reactions of pre-existent substance components caused by differences in chemical affinities. Since the affinity, for example, between an alkali and an acid was greater than the affinity between a metal and an acid, alkalis replaced metals from their union with acids and instead combined with the acid into a new salt. Tables of chemical affinities ordered pairs of substances according to degrees or "laws" of affinity. As can also be seen in these tables, eighteenth-century chemists established laws of affinity between pairs of substances; that is, they referred to chemical operations that employed only two ingredients to form one reaction product, and which could subsequently be decomposed again into the two original ingredients.[140] Chemical operations involving more than two ingredients and yielding more than one reaction product either could be reduced to reactions taking place between pairs of substances or were too complex to be explained by laws of affinity. Thus, laws of affinity became coupled with the view of a binary constitution of chemical compounds, which became the bedrock of the new chemical nomenclature introduced in the *Méthode de nomenclature chimique* in 1787.

The conceptual network of chemical compound and affinity that emerged around 1700 was embedded in a specific cluster of reversible chemical operations, which allowed, so to speak, the movements of chemical substances to be traced in chemical reactions. There were substances that disappeared when a new substance was created but always reappeared in certain kinds of subsequent chemical operations. As a rule, these subsequent operations required the admixture of a third substance from the same class as one of the two original substances. And there was regularity and direction to be observed in these kinds of reversible operations. Not all admixed substances spurred the decomposition of a compound and the recovery of an original substance. There seemed to exist a kind of electivity of chemical affinity between pairs of substances, which directed the chemical reactions. When we compare all the materials chemists experimented with around 1700, there was only a small group of substances that displayed this kind of regular, traceable, and controllable chemical behavior. The bulk of materials, especially those stemming from plants and animals,

139 For an argument that the classification of substances implied in Geoffroy's 1718 affinity table manifests that increased attention, see chapter 3.
140 Chemical affinity tables were made up of columns containing symbols denoting pure chemical substances that behaved like stable building blocks in replacement reactions. The pragmatics of the table included the presupposition that the combination of two symbols, the one found at the head of a column and the other below, stood for a chemical compound. A chemical compound was thereby represented as a binary entity. The one-to-one correspondence between a symbol and the denoted component represented visually the binarity of constitution.

behaved differently. They would, for example, only decompose but not recompose in the chemical laboratory, yielding thereby a cascade of transformation products whose creation was difficult to reconstruct. The window of opportunity was small for the emergence of the modern concept of chemical compound and the understanding of chemical transformations as reactions, analysis, and synthesis.

Almost all of the traceable substances mapped by chemical tables belonged to the mineral kingdom. In hindsight, the reason why these substances were traceable is easy to explain: all of them were pure chemical substances rather than mixtures of different kinds of chemical species, and all of them formed simple, binary compounds. However, the modern concept of chemical purity in the sense of single, stoichiometric compounds or single chemical elements (in the modern Lavoisierian sense) did not yet exist around 1700. The increased attention chemists paid to reversible chemical reactions from that time onward contributed to the formation of this chemical concept, but we must not presuppose it in the early eighteenth century. Around 1700 chemical purity was a material, not a conceptual resource for the emergence of the modern conceptual network of chemical compound and affinity. Moreover, it was a social and technological resource. As we will argue in more detail in part II, the pure substances that underwent reversible chemical transformations—transformations that lent themselves to their interpretation as analysis and synthesis—were not unambiguous products of academic chemical laboratories, but rather had a long tradition as commodities, applied in the apothecary trade and metallurgy. In the following we will designate these distinctive materials "pure substances," since they allowed chemists to trace and control chemical reactions in their laboratories. "Purity" here refers to a feature of materiality, to which we will come back in the conclusion of the book.

2.4 The many dimensions of material substances

We have argued that eighteenth-century chemists shifted their inquiries from the perceptible to the imperceptible dimension of chemical substances, and vice versa, and that their substances were multidimensional objects of inquiry bridging chemical technology and chemical science, and natural history and philosophy.[141] This argument was substantiated by studying three strands of inquiry into chemical substances: experimental history, technological improvement, and experimental philosophy. Experimental history studied the perceptible properties of materials, modes of their production, and their practical uses by collecting facts from the arts and crafts, repeating and varying artisanal operations in the chemical laboratory, and further test-

---

141 The concept of "boundary objects" by Star and Griesemer (see Star and Griesemer [1989]) differs from our concept of "multidimensional objects of inquiry" inasmuch as we are not primarily concerned here with the question of how different social groups were interconnected by dealing with the same kind of objects, but rather how chemists studied materials from different perspectives and invested them with different meanings. In addition, materials were also boundary objects linking different social groups such as merchants, apothecaries, metallurgists, academic chemists, and so on. See also our discussion in the Conclusion.

ing of properties with chemical reagents. It shared with natural history the emphasis on observable facts and on the multiplicity of things and processes—an emphasis reinforced by chemists' classification of materials according to their origin from the three natural kingdoms. Furthermore, we have argued that the experimental-historical inquiry into substances could be extended to more analytical modes of technological inquiry, linked with attempts at technological improvement. Or, alternatively, it spurred experimental inquiries into the imperceptible dimension of substances, such as their composition, affinities, and invisible movements in chemical reactions.

But how did the interplay between the perceptible and imperceptible dimension of chemical substances actually work? How and why did eighteenth-century chemists shift their inquiries from the study of the color, smell, taste, consistency, solubility, combustibility, and practical uses of a substance to the analysis of its components or the regrouping taking place in replacement reactions, and vice versa? First, studies of chemical properties, such as solubility and combustibility, implemented the same kinds of instruments and techniques as chemical analysis and other experiments studying chemical reactions. Hence, with respect to material resources, there was only a small step to be made from studies of chemical properties and the mere registering of observable effects procured by reagents to the actual separation and identification of reaction products for analytical purposes or for reconstructing other kinds of chemical reaction. When chemical analysis was a mere extension of familiar analytical techniques from one kind of substance to a similar one it seemed obvious to take this next step.[142]

Second, when eighteenth-century chemists performed chemical analyses for the acquisition of knowledge about the invisible components of chemical compounds, the immediate result of analysis was not data, counts, diagrams, or other forms of inscriptions. Rather, the acquisition of knowledge about the composition of chemical compounds and the representation of that imperceptible object went hand in hand with material production. Chemical analysis was performed by letting the substance under analysis interact with another substance and transform into new substances. This kind of analysis, which was utterly different from modes of analysis known in other eras of experimental philosophy, actually took apart substances, thereby producing perceptible substances that chemists interpreted as the components of the compound. Likewise, the immediate experimental traces of the imperceptible movements taking place in chemical reactions were not inscriptions but material things: reaction products. They were, in the language of the philosopher Charles Sander Peirce, "indices," that is, signs physically connected with their object.[143] Based on the examination and identification of the reaction products and the comparison of their composition with the composition of the original substances, chemists drew conclusions about the invisible regrouping of substance components during the chemical reaction. That is, their study of invisble composition and regroupings in

---

142 But often this was not done and experimental history had a life of its own, or switched to technological inquiry.
143 Peirce [1931–1958] vol. II pp. 156–173.

chemical reactions required the study of visible indices, the material reaction product. In this way investigation of the imperceptible features of chemical substances always returned to that of their perceptible dimension.

Throughout the entire period of classical chemistry from the middle of the eighteenth century until the middle of the twentieth century, chemists' studies of the composition of chemical substances and of regrouping in chemical reactions switched from the imperceptible to the perceptible dimension of substances.[144] The interplay between perceptible and imperceptible objects of inquiry was not merely a matter of chemists' interests and intentions, but was built into the very technique and collective understanding of chemical analysis. The imperceptible objects—chemical components and their movements in chemical reactions—could not be separated technically and epistemically from the perceptible dimension of substances. Rather, the study and representation of imperceptible features of a substance always entailed the material production of new perceptible substances that had first to be examined and identified before conclusions concerning the imperceptible objects could be drawn.

When we compare this to twentieth-century scientists' studies of the microstructure of bodies and materials by spectroscopic methods, for example, the difference consists first in the kind of experimental tracing, that is, the difference between inscriptions and proper material traces. Second, by means of the new physical analytical methods of the twentieth century it was also possible to study the imperceptible structure of things separate from their perceptible dimension.[145] Spectroscopic analysis of a species of material requires a macroscopic bit of that material, a target. Yet in this case the macroscopic target is no more than a technical surface, so to speak, of the imperceptible scientific object: the molecular structure or nanostructure. Preparations of targets are technical endeavors, even though they may entail problems of their own. And the technical surface of the target is left behind as soon as the imperceptible object of inquiry becomes represented. A twentieth-century chemist studying the spots produced by x-ray diffraction in order to unravel the molecular structure of a protein does so without returning to the perceptible properties of the target. Likewise, a molecular biologist studying the form of ribosomes using the electron microscope is no longer interested in the macroscopic target when she has photographs at hand that represent her scientific object. Even in light microscopy, one left behind the visible surface of the specimen as soon as one looked through the microscope to explore the specimen's minute structures. By contrast, in classical chemistry this kind of separation between the perceptible technical surface of a target and the imperceptible structure of a substance did not exist, since the experimental traces of the imperceptible objects were perceptible substances; and information about the imperceptible structure could be obtained only by further examination of the perceptible reaction products.

---

144 For the notion of classical chemistry and its periodization see also the essays in Morris [2002].
145 On the deep changes in chemical analysis occasioned by the introduction of spectroscopic methods by the middle of the twentieth century, see Baird [1993], [2004]; Morris [2002]; Reinhardt [2006]; Taylor [1986]; and White [1961].

It is of great importance for our concept of multidimensional objects of inquiry to highlight subtle distinctions in eighteenth-century chemists' approaches to the imperceptible dimension of substances and chemical processes. Depending on the kind of discourse and investigation, chemists were concerned with substance components known also as useful materials, or with simple principles and elements defined in the traditions of Aristotelian and Paracelsian philosophies, or with atoms and corpuscles. Due to the peculiarity of chemical analysis—the fact that chemical analysis actually yielded material substances understood as components, rather than data or other inscriptions—they persistently shifted from studies of invisible components to perceptible laboratory substances. Yet the same cannot be stated with respect to Aristotelian elements, Paracelsian principles, atoms and corpuscles. When chemists studied problems concerning atoms and simple principles rather than ordinary substances—which they did, in particular, in the theoretical parts of textbooks—they entered a new world, so to speak. There was, in particular, no smooth continuous way, or shift, from experimental and classificatory practices, or from experimental history and technology, to atomic and corpuscular theories. Atoms and corpuscles were central entities of chemists' theories of matter throughout the eighteenth century. Although these theories were underpinned to some extent by observation and experiments, they were not "embedded" in experiments, or "derived" from experiments, or whatever kind of empiricist model and terminology may be used to explain their existence. Atoms and corpuscles were entities of a philosophical discourse that extended to issues far from being treatable in an experimental way, such as the creation of the universe or the generation and corruption of natural bodies. Eighteenth-century atomism and corpuscular theories were utterly underdetermined by observation and experiments.[146]

Eighteenth-century chemists' experimental reports referred primarily to substances, not only when they described ingredients and techniques, but also when they studied the imperceptible movements taking place in chemical reactions and the building blocks of chemical compounds. Talk about analysis, synthesis, replacement reactions, and affinities was concerned, first of all, with what substances do. Hence, the symbols applied in the eighteenth-century chemical affinity tables referred to substances and affinities between pairs of substances, rather than atoms and Newtonian attractive forces between atoms. There is no question that most eighteenth-century chemists had adopted an atomistic image to account for the homogeneous appearance and specificity of properties of chemical compounds, that is, the fact that the properties of the substance components separated from chemical compounds and used for their resynthesis could no longer be observed in the chemical compound. That atomistic image was sometimes also expressed in experimental reports. Even so, it was no more than an image of invisibly small particles that, as a rule, accompanied debate about the movements of substances. This atomistic image had no additional heuristic function that would have contributed to a distinctive understanding or interpretation of the experiments different from that referring to substances and their properties.

---

146 For a similar argument with respect to the seventeenth century, see Chalmers [1993]; Meinel [1988].

As to eighteenth-century chemists' ways of identifying and classifying substances and their writing of histories of substances, atomism and corpuscular theories were totally irrelevant. The many ways of identification and classification of substances we have studied in the different fields of eighteenth-century chemistry never included a classificatory order according to composition from atoms or atomic structure. In the eighteenth century, identification and classification of substances according to chemical composition always meant classification according to substance components rather than number, size, figure, or any other parameter of atoms and atomic structure. To be sure, some specific chemical theories did play a role in classification according to chemical composition. To these theories belonged the Stahlian theory of a graduated order of chemical composition, the theory of phlogiston, Lavoisier's theory of oxidation and acidity, physiological theories, and so on. All of them are treated in parts II and III in the context of our discussion of chemists' classificatory practices. Yet atoms and atomic structure were absent in these practices, in mineral, plant, and animal chemistry. This historical fact also sheds light on the epistemological status and significance of atomism and corpuscular philosophy in eighteenth-century chemistry. This latter type of theory was part of chemists' philosophical activities, which contributed to their reputation in the Enlightenment elite. But it played no significant role in classificatory and experimental practices. This began to change only in the 1830s, and these changes took place not in mineral or inorganic chemistry, which had the theoretical lead in the eighteenth century, but in organic chemistry. In the new experimental culture of organic or carbon chemistry, chemical formulae, which embodied a theory of invariable quantitative portions or bits of substances, became indispensable paper tools for the classification of chemical substances and for the interpretation and modeling of chemical reactions.[147]

---

147 See chapter 16; Klein [2003a]; and Rocke [1984].

3

Why Study Classification?

Tables of chemical affinity (the first of which was published in 1718 by the French chemist and apothecary Etienne François Geoffroy), the 1787 table of chemical nomenclature by Lavoisier and his collaborators, lists of compound components or "proximate principles of plants" published from 1750 onward, and chemists' sorting out of "organic substances" from the latter lists in the 1790s are landmarks of chemical classification in the eighteenth century. What can be learned from a historical and philosophical study of classification? In this chapter we will discuss this question by addressing the following aspects, both generally and more specifically, of classificatory practices: selectivity, ontological shifts, productivity, diversity, and modes of representation. We do so by way of examples from eighteenth-century chemistry that are discussed at length in parts II and III.

3.1 Selectivity

When we think of classification in the eighteenth century, what comes to mind almost immediately are the comprehensive taxonomic systems in botany like those of Linnaeus. But unlike the eighteenth-century botanists, the chemists of that period did not attempt to group together and order in one single taxonomic system all known chemical substances and raw materials. Apart from their large divide between natural bodies and substances according to the three natural kingdoms, the classificatory practices of eighteenth-century chemists were highly selective. The outstanding document of chemical classification in the Chemical Revolution—the 1787 table of chemical nomenclature—omitted the bulk of plant and animal substances. The class of proximate priniciples of plants and the later one of organic compounds was even more restricted to a small group of substances. The relevant substances grouped together in the class of proximate priniciples of plants, for example, had to be recurrent and natural; that is, they were chemically untransformed and often found components of plants procured by the life processes of the plants. Likewise, E. F. Geoffroy's table of chemical *rapports*, or affinities, provided a classification of only a comparatively small number of substances known at the time. We argue that chemists' singling out of substances among a larger range of materials manifests the particular importance that these substances had for them. Classification selects relevant objects and relevant features of objects, and classificatory practices order things according to distinctive aspects. Linnaeus, for example, chose a very distinctive set of features of plants for his taxonomic system—namely, the number, figure, position, and proportion of the parts of the fructification organs of plants. As we will see in the following, eighteenth-century chemists' classificatory practices were far less rigorous than Linnaeus's. But eighteenth-century chemists also focused on distinctive features of substances when they created chemical order, whereas they ignored others. The following discussion of E. F. Geoffroy's table provides an example for the selectivity of chemists' classificatory practices.

E. F. Geoffroy's table of chemical *rapports* grouped together simple replacement reactions—that is, chemical decompositions of binary compounds by means of a third substance that combines with one of the two original components into a new binary compound—along with kinds of chemical substances that undergo such types of chemical transformations.[1] Compared with the whole range of substances and chemical transformations familiar to early eighteenth-century chemists, this kind of classification was highly selective. In a time when transmutations and extractions of essences, magisteries, quintessences, elixirs, and similar kinds of exalted substances did not yet belong to the past, it was far from a matter of course to pick out the chemical substances which could be made from two original substances and decomposed into these two original substances. Why did Geoffroy pay attention to this particular kind of substances and chemical transformations? We argue that Geoffroy's table manifests an increase in the importance of these chemical substances and their reversible chemical reactions for the French academic chemists. The fact that Geoffroy, and before him Wilhelm Homberg and Nicolas Lemery, actually grouped together binary compounds and replacement reactions should be taken as a manifestation of an ongoing transformation of chemists' understanding of chemical transformations and ontology of substances.

Knowledge about reversible chemical transformations was not an absolute novelty around 1700. Nor did the building-block image of substances that undergo such transformations emerge suddenly at that time. The metallurgical literature back in the sixteenth century already mentioned such kinds of chemical transformations and substances, as did the seventeenth-century chymistry textbooks and treatises describing the making of chemical remedies, and R. Boyle's *Sceptical Chymist*.[2] Yet the authors of the chemical textbooks did not put much emphasis on reversible chemical transformations. Neither explicit statements nor classification would have directed the readers' attention to this distinctive kind of chemical transformation. Instead of grouping together the reversible decompositions and recompositions of chemical compounds, they separated them from each other by interspersing numerous descriptions of different kinds of chemical transformations, understood for the most part as exaltations and transmutations. Hence, if historians want to track down the reversible chemical transformations of chemical compounds that were described in the chemical textbooks of the sixteenth and seventeenth centuries, they have to read through hundreds of pages of text. Exaltations and transmutations were at the center of sixteenth- and seventeenth-century chymists' interests, whereas reversible decompositions and recompositions either were regarded as mere mechanical mixtures or as a type of chemical change that was open to philosophical debates.[3]

The epistemic status of chemical compounds and reversible chemical transformations changed significantly in the late seventeenth century, when Nicolas Lemery and Wilhelm Homberg, both members of the Paris Academy of Sciences, studied the syn-

1   Geoffroy's table will be discussed in more detail in chapters 8 and 9.
2   See Klein [1994a], [1994b], and [1996a].
3   See above, chapter 2.3.

thesis of salts from an acid and a salifiable base as well as cycles of decomposition and recomposition of natural salts, summarized in their theories of salts.[4] Their approach was extended by E. F. Geoffroy's grouping together of an even larger number of substances considered to be binary compounds that undergo reversible transformations, namely salts, metal alloys, compounds of sulfur, and solutions of water.[5] By the middle of the eighteenth century, Geoffroy's table and similar tables of affinity proliferated all over Europe.[6] We take the dissemination of these tables along with their chemical order as an indication of the collective acceptance of the new chemical ontology of compounds, affinity, analysis, and resynthesis in the European chemical community.

Historians of science who tend to focus on single historical events and individual actors may argue that chymists knew reversible chemical transformations long before 1718, and thus the table did not introduce a significant novelty. By contrast, we argue that reversible chemical transformations and the substances that undergo such transformation existed only at the margins of the collective attention of chymists before they were grouped together in tables and thereby highlighted as an important domain of inquiry. Classification is a selective and constructive epistemic activity. It is determined neither by compelling objects nor by our disinterested perception of essential features of objects, but involves the selection of relevant objects and features of these objects. We have to actively pay attention to such objects and features of objects to create order among them. Therefore changes in modes of classifying objects provide precious information to historians and philosophers about shifts in historical actors' points of view and their ontology.

From a methodological point of view the use of Geoffroy's table can hardly be overestimated. Had that table not been published, historians might have easily missed the deep change in chemists' ontology taking place around 1700; for neither Geoffroy nor any other chemist explicitly reflected upon these transformations or represented a corresponding theory of chemical compounds in the form of an elaborate text. Experimental reports, which are a more direct route to chemical practice and objects of inquiry, also manifest these transformations. Yet historians needed to study a large number of such reports in a comparative way, covering chemical experiments performed at the Paris Academy of Sciences around 1700 as well as experiments performed earlier in the seventeenth century and in different institutional contexts, in order to grasp a pattern of historical change in chemists' experimental goals and kinds of scientific objects. Geoffroy's table enormously facilitates the recognition of such a pattern. It presents a concise summary of the new understanding of chemical transformations and chemical compounds, thus emphasizing the ongoing changes in chemistry. In so doing, it raises questions concerning shifts of practice, understand-

---

4   See above, chapter 2.3.
5   Although Geoffroy did not introduce a specific name for these operations, such as "replacement reaction," he defined them as "hidden movements" (*mouvemens cachés*) in which the substance A that had a greater "disposition to unite" with the substance B than the substance C that is united with A lets go (*lâcer prise*) the substance C. See Geoffroy [1718] p. 203.
6   See Duncan [1996] and chapter 9.

ing, and ontology. Why did chemists around 1700 pay increased attention to reversible decomposition and recomposition? Why did Geoffroy select salts, metal alloys, compounds of sulfur, watery solutions, and the components of these compounds from the bulk of materials handled in chemical laboratories, representing in his table the series of replacement reactions that these substances undergo? Similar questions can be asked with respect to the classifications in eighteenth-century plant chemistry studied in part III. Why did chemists group together the more compound, proximate principles of plants from the middle of the eighteenth century onward? Why did they select "organic" substances among this former class around 1790? Why did they highlight features of organic substances in the 1830s that differed from the features they paid attention to earlier? Attempts to fully answer these questions require, of course, additional historical studies that go beyond the study of modes of classification. But studies of classification and of changing modes of classification pave the way to obtain a large picture of ontologies of the past.

3.2 Ontological shifts

Selection of relevant objects and features of objects in classification is tied to interests, goals, and meaning entrenched in collective practices.[7] Studies of modes of classification, in particular when they are comparative and cover a long time period, allow historians to grasp patterns in historical actors' understanding of the types of objects they deal with. The way historical actors actually selected and ordered objects illuminates their most basic points of view. Classification does not require explanation and an elaborate theory of the classified objects. It may also precede the closer experimental investigation of things and processes and the acquisition of refined experiential knowledge. On the simplest taxonomic level of species, classification assembles singular things and events into kinds of objects of attention and recognition. In higher taxonomic ranks it further establishes relationships between kinds of objects by ordering them into higher classes. For this activity it is not necessary to present systematic definitions of the kinds of objects under investigation, or to fully understand the established relationships and explain them in theoretical form. Geoffroy, for example, neither argued for his selection of substances and reactions nor explained the relationships ordered in his table; as to the latter he merely claimed to represent laws. But the fact that he, as well as Lemery and Homberg, selected and grouped together distinct kinds of substances along with a specific type of chemical transformation, which differed from earlier modes of classification, provides insight into changes in their world of materials. "Worlds differ in the relevant kinds they comprise" (Goodman).[8]

This insight spurred not only Nelson Goodman, but other philosophers and historians as well, to study modes of classification and structures of kinds and to compare taxonomic changes over time in order to unravel the historical actors' points of view,

---

7   For an excellent discussion of this social context of classification, see Barnes et al. [1996].
8   See Goodman [1978] p. 10.

paradigms, episteme, or ontologies.[9] Particularly in his last years, Thomas Kuhn studied changes of structures of kinds as a way of coming to grips with scientific revolutions and incommensurability. In *The Structure of Scientific Revolutions,* Kuhn had argued that "after discovering oxygen Lavoisier worked in a new world," and that the chemical systems before and after the Chemical Revolution were incommensurable.[10] In the 1980s, Kuhn further developed his view on incommensurability by studying changes in taxonomic structure. He recast his concept of incommensurability by arguing that scientific systems can be commensurate with one another, or mutually translatable, only if they share the same taxonomic structure.[11] In our study we bypass the problem of incommensurability, but we share with Kuhn the view that taxonomic structures provide insight into historical actors' approach to objects, or what we call their ontologies, and ontological shifts in history, and we further use his criterion for scientific revolutions to examine in part II the Chemical Revolution.

Our scrutiny of the 1787 table of chemical nomenclature by Lavoisier and his collaborators and its comparison with the earlier tables of chemical affinity in part II will show that the structure of kinds established in the Chemical Revolution differed from the earlier structure of kinds, mainly in its clear distinction between a class of simple and a class of compound substances. However, it also agreed in important structural features. The 1787 table of chemical nomenclature preserved the mode of classification according to chemical composition that was previously represented by tables of affinity. It also preserved the type of selected substances; plant and animal substances and raw natural minerals continued to be excluded. And it even took over most classes of chemical compounds previously distinguished in tables of affinity. Although the "new chemists" added a number of new substances to their chemical table, especially gases, and clearly demarcated simple from compound substances, these changes did not engender a complete restructuring of the existing mode of classifying substances. Despite Lavoisier's revolutionary rhetoric, the Lavoisierian system of classification and nomenclature was not the result of a *change of* the mode of classification, but rather a *change in* the existing mode of classification. It did not overthrow the existing taxonomic structure. Our scrutiny of the classificatory structure of affinity tables and the 1787 table demonstrates that, contrary to Kuhn's view, the ontology of the old and the new chemists did not differ profoundly.

Taxonomic changes and ontological shifts may be driven by diverse agents and intricate interaction among agents. New goals and points of view for classifying objects may be entirely or partly self-referential, going hand in hand with changes in social institution and power.[12] For example, eighteenth-century chemists' classification of materials according to their origin from the three natural kingdoms was linked with the increase in the cognitive authority of natural history and chemists' self-rep-

---

9   See, in particular, Buchwald [1992]; Buchwald and Smith [1997]; Kuhn [1989], and [1993]; Foucault [1970]; Hacking [1993], and [2002].

10 Kuhn [1996] p. 118.

11 On Kuhn's later views about incommensurability and the translatability of taxonomic structures, see Kuhn [1989], [1993]; and Buchwald and Smith [1997].

12 On self-reference and related social terms (S-terms), see Barnes [1983].

resentation as participants in cultures of natural history and Enlightenment discourse. Inversely, the exclusion of Galenic composite remedies from eighteenth-century chemists' pharmaceutical inquiry and from the class of plant materials was conditioned by their restriction of disciplinary boundaries. Galenic composite remedies were left entirely to apothecaries, since they were not prepared by chemical operations and hence did not fit into the disciplinary matrix of chemists. Or, to give another example, when after 1830 plants and animals, fruits, leaves, seeds, and other organs of plants, as well as non-stoichiometric compounds and mixtures, became excluded from the range of scientific objects studied in carbon chemistry, this exclusion was also spurred by chemists' collective demarcation of general or pure chemistry from special areas of applied chemistry, such as medical and physiological chemistry, as well as from other academic disciplines, like botany and the emerging academic discipline of pharmacy. Self-positioning and self-representation in the academic system and the broader culture affected chemists' ways of classifying, although in chemistry and the laboratory sciences more general changes in classification and ontology have rarely been driven by self-referential, cultural, and social processes alone.

In addition to new epistemic, cultural and institutional constellations, the invention and discovery of new objects, too, may spur new ways of classifying, especially in the experimental sciences. Eighteenth-century chemists' experiments on air and the isolation, identification, and demarcation of different kinds of air, for example, were a precondition for the Lavoisierian Chemical Revolution. Like water, until the 1770s chemists considered air to be a simple chemical element, and the subdivision of air, material and epistemic, into different kinds of air challenged the existing classificatory system. A much more far-reaching impact on chemists' way of classifying was brought about by the creation of new carbon compounds through substitution reactions in the 1830s. Production of these new types of organic substances spurred chemists' dismissal of the traditional natural historical and artisanal mode of individuation, identification, and classification in organic chemistry and its replacement by a new type of individuation and chemical order embedded in a large network of experiments and work on paper. The new mode of classifying organic substances in the 1830s manifested not only a new point of view on already existing objects, but also a new material world of objects.

3.3 Productivity

Plutonium and other transuranic elements, PVC, Teflon, lasers, and genetically altered organisms are familiar techno-scientific objects to twentieth-century people. But the laboratory sciences were materially productive long before the twentieth century, and this material productivity had consequences for classificatory practices. In the laboratory sciences, changes in modes of classification may be conditioned by both alterations of epistemic regimes and the material culture of a science. Chemistry has certainly been the most productive laboratory science in history. From the second half of the eighteenth century onward, the production and individuation of new chemical substances that were unknown in the world of workshops, trade, and consumption, accelerated in chemical laboratories. Hence in parts II and III we tackle questions such as: did eighteenth-century chemists merely change names and meaning when they introduced new classifications but preserved the referents? Or did the referents change, too? If the latter was the case, was it the result of a change of knowledge, goals, and social interests, in such a way that chemists were no longer concerned with the traditional materials and instead turned to alternatives already existing in other fields? Or did new objects come into being in a literal sense—that is, were new substances produced?

In the recent historical and philosophical discussion about discursive formations, structures of kinds, and historical ontology, taxonomic change has mostly been regarded as the result of changes of knowledge, episteme, lexical structure, social interests, power, and ethics.[13] Though material productivity of the laboratory sciences has not been ignored, it was far from a prominent theme in that discourse.[14] In part III, in particular, we will show that chemists' introduction of a new mode of classification in organic chemistry was intertwined with the production and individuation of new kinds of organic substances. In the following we discuss the interconnectedness of ontology, material productivity, and mode of classifying in the laboratory sciences by way of an example that has had some prominence in the philosophical discourse on natural kinds: water. We argue that beginning in the 1780s, chemists produced new kinds of water, and, as a consequence, changed their practice of classifying and naming samples of substances as "water."

In classical chemistry, that is, the chemistry from c. 1750 to 1950, chemical analysis meant the physical alteration of the scientific object through the actual separation of its components, and, as a proof of analytical results, the resynthesis of the original compound. Classical chemical analysis materially changed its object of inquiry and thereby prompted the individuation, identification, and demarcation of new kinds of objects. When Lavoisier and Jean Baptiste Meusnier (1754–1793) analyzed water in their famous gun-barrel experiment in the winter of 1783 and 1784,[15] they actually took apart water, thereby creating hydrogen and oxygen (in form of iron oxide) as analytical products.[16] A year later, Lavoisier repeated the experiment, this time supplementing the analysis by the resynthesis of water from oxygen and hydrogen. Only then Lavoisier asserted that "water is not an element, that it is on the contrary composed of two very distinct principles," namely oxygen and hydrogen.[17] Yet knowl-

---

13 See Daston [2000]; Foucault [1970]; Hacking [1991], [1993], and [2002]; Kuhn [1989], and [1993].

14 See, in particular, Hacking's discussion of the experimental production of objects, or what he calls "creation of phenomena," in Hacking [1983], [1993], and [2002] pp. 14–16. Although Hacking highlights the material productivity of the laboratory sciences, he does not regard the experimental production of objects to be part of his "historical ontology." The creation of phenomena, he states, does not mesh with Foucault's "three axes of knowledge, power, and ethics." Hacking concedes that the Hall effect, for example, was "produced" by Hall, that "Hall produced his effect only because he was networked into a microsociology of power," and had some "ethical concerns." But he argues that "there was no constituting of anything" (Hacking [2002] p. 16; see also ibid. pp. 2, 5, and 11). We consider experimental production and individuation of objects to be part of their "constitution," and our concept of "historical ontology" differs in this respect from Hacking's.

15 For life and work of Meusnier, see Gillispie [1970–1980] vol. IX p. 342ff.

16 On this experiment, see Holmes [1985] pp. 211–213; Poirier [1993] pp. 150–152.

edge, texts, or inscription were not the only result of Lavoisier's experiment. Its immediate result and the first experimental traces were not inscriptions but materials: first, hydrogen, collected in a bell jar, and oxygen, which was combined with iron; and second, resynthesized, chemically pure water. The latter substance was no longer the same referent as natural water, but rather a chemical preparation or artifact made in the laboratory, which differed from natural water by the absence of minerals and other admixtures.

Some twenty years later, chemists defined chemically pure water as a stoichiometric compound, denoted by the formula H_2O. They then clearly demarcated stoichiometric compounds from non-stoichiometric compounds and mixtures such as watery solutions and the kinds of water stemming from springs, lakes, rivers, and other natural sources. The difference they then made between chemically pure water—H_2O—and samples of ordinary, natural water was not only a difference in chemical meaning but also a difference in the material referents. In the twentieth century, chemists again changed their view that chemically pure water was always H_2O. They then distinguished between different molecules of water, consisting of varying isotopes of hydrogen and oxygen, among which H_2O, D_2O, DOH and $H_2{}^{18}O$ prevail quantitatively. Again, this new understanding of water was not merely a matter of knowledge, meaning, and importance of characteristics of a stable referent, but came with the production of corresponding new referents in the laboratory. Today pure D_2O, for example, is used in chemical experimentation as well as in nuclear reactors.

The case of water shows that experimental scientists have often created new material objects not only in purposeful technoscientific practices, but also in their study and analysis of natural objects. Material productivity is inherent in the laboratory sciences. This has many consequences, no less for the broader technology and society in which laboratory sciences are embedded than for the specific practices of the laboratory sciences themselves. What is the consequence for scientists' ways of classification?[18] In the philosophical discussion about natural kinds, water has been presented as an outstanding example of a natural kind. Philosophers who defend the concept of natural kinds argue that people have always recognized samples of water—taken from springs, rivers, lakes, oceans, rain, dew, the tap—as "water."[19] In so doing, they have preserved a sameness-of-kind relationship, which is cross-cultural and immune to historical change. In the words of Willard V. Quine: "For surely there is nothing more basic to thought and language than our sense of similarity."[20] In his causal theory of reference, Hilary Putnam has presented a slightly different argument, proposing that the centerpiece of the sameness-of-kind relationship is not perception of sameness but the abstract act of reference. Putnam takes into account that the meaning of natural kinds may be different in different cultures and time periods. For exam-

---

17   Quoted after Poirier, see Poirier [1993] p. 151.
18   For a discussion of technoscientific productivity and related issues concerning the relationship between science and society, see Klein [2005b]; and Pickering [2005].
19   For the concept of "natural kinds" see the classic papers by Kripke [1971], and [1980]; Putnam [1975b]; Quine [1969], and the more recent papers by Harré [2005b]; Putnam [1990]; and van Brakel [2000], and [2005].
20   See Quine [1969] p. 116.

ple, water has been defined as a transparent, colorless, and odorless liquid that quenches thirst, rains from the sky, flows from springs, and fills rivers, lakes, and oceans. In addition to this pervasive common understanding of water, sixteenth-century Paracelsians considered water to also be an elemental womb that generates minerals and other natural bodies. In the early eighteenth century, most chemists understood water as an ultimate, unchangeable element of more compound substances. In the last third of the eighteenth century, Lavoisier and his group again changed the meaning of water, defining it not as an element but as a chemical compound consisting of hydrogen and oxygen.

Putnam argues that in cases like water and other natural kinds we can abstract from these different historically and culturally situated meanings of water when we study reference. Putting aside differences in meaning, there remains something that does not change over time, culture, or different social groups: the natural object or referent water and the abstract fact that we refer to it. According to Putnam, reference is constituted by an original, abstract act of labeling an object. Once a sample of water has been labeled by the name "water" or another sign, members of a community learn the connection between the label and the referent and transmit it to subsequent generations. Hence, people talking about "water" always refer to the same samples of a kind of object. They preserve a sameness-of-kind relationship between the individual samples of water, and what changes historically or culturally is only the meaning of "water" and the importance they attribute to specific features of the stable referent water.

Philosophers of natural kinds have further highlighted the role played by science in the taxonomy of natural kinds. Science, so goes their argument, eventually discovers the objective essence of natural kinds, so that all samples or specimens of a given natural kind can be identified clearly by that essential character. Quine and Putnam have argued, for example, that the science of chemistry made it possible to define objectively the sameness-of-kind relationship between individual samples of water. Based on chemical analysis, they argue, we now know that all samples of water at any place and time have always been H_2O, independent of our beliefs about water, and that it is "impossible that it isn't H_2O."[21] It is, in particular, scientists' analysis of the imperceptible microstructure of objects that has been regarded as the proper route to the discovery of natural essences and subsequently to objective, unique classification. In a recent essay on chemical taxonomy, Rom Harré argued that chemists' analysis of the microstructure of chemical substances led to the discovery of the real essences of substances, and that we can indeed observe in the history of chemistry a move toward real essences: "An examination of chemical taxonomies shows that there is a strong tendency to move from a nominal essence based classification system to one which is based upon hypotheses as to real essences."[22] Here "nominal essences" are defined as clusters of perceptible properties, whereas "real essences"

---

21 See Putnam [1990] p. 54. Putnam concedes, however, that "future investigation might reverse even the most 'certain' example" (Putnam [1975b] p. 142).
22 Harré [2005b] p. 8f.

are primary qualities, defined as corpuscular composition and structure.[23] Harré further argued that "real essences are independent of human experience," and that "real essence taxonomy would be final."[24]

It is not an objective of this book to study the question of whether objective classification is possible.[25] Rather, the question we are interested in is whether the philosophy of natural kinds matches the history of chemists' actual practice of classification. Does the history of chemistry confirm the view that chemists have aimed at a unique classification based on knowledge of the real essences of substances, that is, their molecular composition? And does the key idea of the causal theory of reference—that the referents of "water" and other natural kind terms do not change in history—apply to the laboratory sciences?

It is a central argument of this book that eighteenth- and early nineteenth-century chemists classified substances in many different ways, depending on the practical and conceptual contexts of classification and the different significance they attributed to specific substances or to features of substances in these different contexts. Even in conceptually driven inquiries, which are the focus of our study, chemists accepted different modes of classifying chemical substances. In the next paragraph and the subsequent parts of this book we will present more details supporting this argument. Here we wish to add some considerations concerning the time period after the 1830s, which demarcates the end of the time period we scrutinize in our studies. Do the historical developments in chemistry after the 1830s justify Harré's claims? Our historical analyses presented in part II and especially part III show that in contexts of conceptual inquiry (and only in such contexts) chemists indeed moved from classification according to natural origin and perceptible features of substances to classification according to composition. Beginning in the late 1850s, they further accepted classifications of substances according to molecular structure. Yet we must not conclude from that observation that knowledge of molecular structure would be finite, and hence settle the question of chemical taxonomy. Rather, what actually happened was that research on chemical structure diversified, engendering new questions, concepts, and modes of representing chemical structure. As a consequence, there were, and today still are, different ways of studying molecular structure, and it has always depended on scientists' objectives which aspects of structure they highlighted. Today, submicroscopic studies of chemical substances extend from quantum-chemical research to nanotechnology. The concepts and modes of representation of molecular

---

23   Ibid. pp. 11 and 13.
24   Ibid. p. 27. It should be noted that Harré's idea of a "final taxonomy" relates to chemistry; it does not necessarily mean a complete, exhaustive taxonomy of all "natural kinds" in the universe.
25   It is worth mentioning in passing the case of biological species; more precisely, species that reproduce sexually. The biologist Ernst Mayr defines biological species as follows: "A species is an interbreeding community of populations that is reproductively isolated from other such communities" (Mayr [1992] p. 222). Since the reproductive isolation of such species is due not to external circumstances but inner causes, mostly to behavior, one could state that it is living beings, and not botanists or zoologists that discriminate members from non-members of a kind. Systematists who base their demarcation of species on reproductive isolation could thus claim with good reason to identify natural kinds or, as Immanuel Kant put it, "physical genera." (In the context of this book, we can pass over the fact that adherents of a phenetic approach to systematics do not share Mayr's view that biological systematics must be evolutionary; that is, must rest on demarcations brought about in the real process of speciation.)

structure implemented in these research practices are as different as the scientific objects themselves. As a matter of fact, chemical practice and theory have not spurred chemists to select one "real essence" of chemical structure, and, as a consequence, there have been different possibilities of classifying substances based on knowledge of molecular structure. Likewise, the classical representation of chemical composition by referring to the constituting elements of a chemical compound did not determine all aspects of classification. To take a very simple example from Lavoisierian chemistry, knowledge about the composition of water from hydrogen and oxygen left open the question of whether water should be ordered into the class of hydrides or oxides.

Arguments about natural kinds derived from the causal theory of reference pose additional problems with respect to the laboratory science. Ever since the first synthesis of chemically pure water in the 1780s, for example, the referent water has changed in the chemical laboratory. Depending on their research questions, chemists may not distinguish samples of ordinary water stemming from a lake or the tap from chemically pure water synthesized from hydrogen and oxygen.[26] However, in at least as many contexts of inquiry, it has indeed been crucial for them to clearly distinguish samples of ordinary water from chemically pure water. The results of many chemical experiments have been utterly dependent on the material difference between these two kinds of water. Today, that difference in referents is important not only to chemists. We can buy (almost) chemically pure, distilled water in drugstores, and if we decided to substitute this kind of water for ordinary water from the tap this would have serious consequences for our health.

As a result of their chemical analyses of water, chemists have produced, individuated, and demarcated many different kinds of water. What is natural and what is artificial in sciences that intervene into nature and even produce objects, such as lasers and transuranic elements, which do not exist in nature? Are D_2O, DOH or $H_2^{18}O$ less natural than H_2O? If one considers D_2O to be less natural than H_2O, H_2O is also less natural than the samples of water we encounter in rivers, lakes, and elsewhere in nature. That is, H_2O is not a natural kind. The argument that substances like D_2O, DOH, and $H_2^{18}O$ are not natural kinds amounts to the concession that the theory of natural kinds does not apply to laboratory sciences.[27]

---

26 Referring to the example of water, Putnam states that the difference between chemically pure water and water from Lake Michigan is a difference of "meaning" and "importance" of characteristics. Asking the question of what is meant by "sameness" of samples of a given natural kind, he emphasizes that the samples of the object must agree in "important" physical properties, and he further states that "importance is an interest-relative notion." "Normally," he continues, "the important characteristic of a typical bit of water is consisting of H_2O. But it may or may not be important that there are impurities; thus, in one context 'water' may mean *chemically pure* water, while in another it may mean the stuff in Lake Michigan." (Putnam [1975b] p. 157). We argue that what is at stake in Putnam's example is not merely a difference of importance of characteristics and meaning but also of the material referents. The water in Lake Michigan and chemically pure water are not the same referents, and in experiments the material difference of the referents may cause different outcomes. In a similar case Putnam seems to concede the latter objection. A sample of mono-isotopic iron that has been produced by a cyclotron, he states, behaves "slightly differently from a 'natural' sample." But asking the question of whether the mono-isotopic sample and the "natural" sample are samples of different substances he gives the quite elusive answer: "Well, it may depend on our interests;" see Putnam [1990] p. 68.

The material productivity of the laboratory sciences challenges the demarcation between objects given by nature versus objects constituted by us. It also challenges the focus of historical ontology on the "axes of knowledge, power, and ethics," and the separation of "historical ontology" from studies of creation of new experimental phenomena and objects.[28] The latter point concerns our own approach in this book. A comprehensive study in historical ontology would also require us to study in detail modes of production and individuation of objects and their institutional conditions. As experimental production and individuation is both a material and an epistemic activity, actions of identification and ordering of objects may be interwoven with experimental production and individuation. These two types of activities are not, however, the same. We will come back to this point in part III with respect to chemists' new individuation and identification of organic materials in carbon chemistry. But we will not explore in this book modes of production and individuation of materials in any detail or in a comparative way. Our focus is chemists' identification and classification of materials, and we largely restrict conclusions concerning chemists' ontology of materials to that area of chemical practice.

3.4 Diversity

In the first section of this chapter we discussed selectivity, the active epistemic process of singling out distinct objects or features of objects. We argued that classification goes hand in hand with selection, and therefore modes of classification provide information to historians about the significance historical actors attributed to various kinds of objects. Studies of classification are an avenue to historical ontology. Based on the results of our historical analyses in parts II and III, we further argued that chemists' modes of classification and their ontology of materials changed in history. Diversity of modes of classification is the overall image that emerges from historical studies of classification in chemistry.

Diversity of classification has recently been highlighted by sociologists, anthropologists, historians, and a few philosophers of science.[29] Most of these studies have analyzed differences in classificatory practices over time or between different social groups. In connection with his concept of scientific revolutions, Thomas Kuhn, for example, was mainly concerned with diachronic diversity of taxonomies.[30] Jed Buchwald has extended Kuhn's approach to different synchronic social groups in

---

27 In his recent paper, Harré raised the following question: "Is the concept of 'natural kind' indefensible in chemistry? Are the myriads of artificial kinds that chemists have synthesized subject to the same 'grammar'? If so is the distinction between artificial and natural kinds redundant?" (Harré [2005] p. 8). Later in the paper, he states that in organic chemistry after the late 1820s "the natural/artificial distinction melted away," to conclude with "the question of whether there is a viable concept of chemical kind at all, once the artificial/natural distinction has been blurred" (ibid. p. 24f). Yet despite these insights, for most of his paper Harré is a strong advocate of natural kinds and real essences.

28 For that separation see Hacking [2002], and our earlier description of Hacking's view.

29 See Barnes et al. [1996]; Bowker and Star [1999]; Douglas and Hull [1992]; Dupré [1993], and [2001]; Hacking [1993], and [2002]; and Kuhn [1989], and [1993]. Although Dupré has argued for a plurality of taxonomies, he seems to assume that chemistry is an exception in this respect (see Dupré [2001] p. 215).

30 See Kuhn [1989], and [1993].

order to explain "the frequent failure of groups of scientists to understand the work of other groups: the problem, that is, of incommensurability."[31] Our historical analyses in parts II and III manifest diversity of modes of classification in an even stronger sense: one and the same scientific community may accept different modes of classification of largely the same domain of objects. The community of eighteenth-century European chemists is a case in point.[32] Eighteenth-century chemists accepted a broad variety of different modes of classifying substances. They lived with different coexisting ontologies and different ways of knowing and making tied to them. We do not find just one ontology, discursive formation, or paradigm in eighteenth-century chemistry—though we do find conjoining agencies in the culture of chemistry of that time.[33]

The image of taxonomic diversity and ontological disunity in eighteenth-century chemistry arises first from the comparison of classificatory practices in all areas of chemistry, ranging from sites of academic chemical experimentation and teaching to the practical, commercial subcultures of chemistry or "applied chemistry."[34] In the different practical contexts of artisanal chemistry—mineralogy and metallurgy, chemical pharmacy, the chemistry of dyeing, pottery and the production of porcelain, glass-making, fermentation and the making of wine and alcoholic spirits, and so on— chemists classified materials in different ways, depending on their practical interests. In contexts of chemical pharmacy, for example, they grouped together indigo with opium into the class of juices of plants.[35] But in contexts of dyeing, indigo was grouped together with cochineal into the class of *bon teints*, or durable colors, which were characterized by resistance to sunlight, atmospheric air, and ageing, and by their high price.[36]

It comes as no surprise that chemists classified materials differently in different practical and commercial contexts. Our analyses in parts II and III, however, will not focus on these types of differences. Rather, we are mainly concerned with taxonomic diversity in contexts of conceptual or philosophical inquiry, and we will show that even in contexts of predominantly conceptual concerns chemists classified substances in different ways. The most important classificatory difference in the context of conceptual inquiry was classification according to chemical composition versus classification according to perceptible properties and provenance. Classification of substances according to chemical composition was restricted to a comparatively small range of substances, which were then mapped by chemical tables.[37] The classi-

---

31 See Buchwald [1992] p. 40.
32 For arguments that there was a European community in the eighteenth century see Holmes [1995]; Kragh [1998]; and Nieto-Galan [2001] pp. 123–136.
33 Our notion of "culture" implies the assumption of some coherence (see Klein [2003a] pp. 81–82) but not that of a unifying umbrella of ideas as expressed in the concepts of episteme, discursive formation (Foucault), and paradigm (Kuhn). Among the conjoining agencies we wish to highlight, in particular, the material culture of eighteenth-century chemistry (see our discussion in chapter 2) and the conceptual network of compound and affinity.
34 For eighteenth-century chemists' distinction between "pure chemistry" and "applied chemistry" see Meinel [1983], and [1985].
35 See, for example, Geoffroy [1736] p. 343.
36 See Nieto-Galan [2001] pp. 1–9.

fications of all substances not mapped by the chemical tables, which included many raw minerals and the bulk of plant and animal substances, implemented criteria of classification other than chemical composition. In eighteenth-century plant and animal chemistry, analytical knowledge of composition played a role inasmuch as chemists grouped together components of plants and animals: first, the simple components or principles, and from the middle of the eighteenth century onwards the more compound, proximate principles. But chemists identified and ordered the substances regarded as components of plants and animals according to their natural origin, mode of extraction, and observable properties.

A second difference in chemists' ways of classifying in conceptual contexts concerns the selection of materials grouped together in one class. For example, the range of substances included in the class of proximate principles of plants changed during the second half of the eighteenth century, both on a collective level and individually. On the collective level, by 1790 chemists dismissed mineral substances extracted from plants from the class of proximate components, which then was composed of only organic substances. By that time the conceptual interests of many chemists were captured more by the distinction between the organic and the inorganic and by physiological issues than by the chemical analysis of the composition of plant and animal substances. On the individual level, chemists differed both in their ways of ordering varieties of very similar plant substances into kinds of proximate components of plants, and with respect to the number of the kinds.

3.5 Representing chemical classifications

Eighteenth-century chemistry is usually not regarded as a taxonomic science, as most chemists were mainly occupied with experimentation and less with systemic classification than, say, botanists of the time. Experimentation, the hallmark of chemistry, has even been regarded as a style of inquiry evolving around concepts, practices, and methods that differed markedly from the taxonomic style.[38] Yet eighteenth-century chemical laboratories were replete with chemical substances, similar to the collections of dried plants, stuffed animals, and minerals in natural cabinets and botanical gardens. Chemists had to order a huge amount of substances, kept both in the bottles and on the shelves of their laboratories and manipulated on paper, for reasons of research, exchange and communication in general, and for chemical teaching in particular. One therefore finds all kinds of sorted registrations of the ever increasing eighteenth-century wealth of chemical substances, from inventories in alphabetical order to classifications based on chemical analysis. As alphabetical inventories and other forms of arbitrary, conventional, or artificial classification of materials can

---

37 We discussed preconditions and experimental contexts of this mode of classification in chapter 2 and will do so more extensively in part II.

38 For the distinction between an "experimental style" and a "taxonomic style" of reasoning, see Hacking [1992]. The assumption that the early modern and modern experimental sciences were not taxonomic goes hand in hand with the ignorance of "experimental history;" on experimental history, see above, chapter 2.

hardly provide information about the historical actors' ontologies, they are not part of our studies. For our purpose, we concentrate on those classifications which chemists considered to rely on inherent or natural features of chemical substances, be they perceptible properties, natural origin, composition, medical virtues, reactivity, or something else.[39]

Classifications that reflect the historical actors' ontology may be carried by ordinary language or constructed in the form of overtly formulated systematic texts, lists, tables, or diagrams. As Thomas Kuhn pointed out, the structured vocabulary, or "lexicon" of a language classifies objects.[40] It does so, however, only in an implicit way; that is, only insofar as classification is embodied in the structure of the terms of a language.[41] Speech and text often pragmatically apply a taxonomic structure along with communication without explicitly reflecting upon that structure as a distinct subject matter of its own. From such implicit modes of representing classifications various explicit forms may be constructed; for example, when people consider the extant terminology and its classificatory structure problematic—a case discussed in part II. Scientists in particular have often reflected upon and restructured classifications, but non-scientific practitioners also sometimes reflect upon their way of ordering and subsequently change it.[42]

The methodological advantage of explicit modes of classification is obvious, in particular in connection with comparative studies like ours, which cover a long period and use the historical actors' classifications as a source of information about the historicity of their objects of inquiry and ontological shifts. With respect to eighteenth- and early nineteenth-century chemistry, two main forms of more or less explicit classification can be distinguished: first, the organization of chemical lectures and textbooks, and second, classification by lists and tables. These two forms of representing classifications of chemical substances are at the center of our studies. Whereas the organization of chemistry textbooks, as a rule, represented classification in a manifest form but without additionally thematizing it, classificatory lists and tables were mostly accompanied by verbal explanations, which dealt explicitly with issues of classification.

Teaching was a particularly important activity spurring chemical classification. With respect to plant and animal chemistry in the first half of the eighteenth century, the semi-explicit mode of representing classificatory structures through the organization of lectures and textbooks is the most important source for studying classification. Chemistry textbooks and manuscripts of chemical lectures are also important historical sources for studying chemical classification in later periods.[43] As textbooks were (and still are) a means of social reproduction of scientific communities, they allowed changes in classification to be traced on the communal level of chemistry. However,

---

39 It should be noted that the majority of eighteenth-century chemists did not use the terms "artificial" and "natural" classification.
40 See Kuhn [1989], and [1993].
41 On this problem, see also Buchwald [1992] p. 55.
42 A trivial example is the classificatory order embodied in the arrangement of goods in stores, which is often changed—for some of us unfortunately—in order to meet new expectations of customers or to advertise new products.

these advantages also come with some disadvantages. Classificatory structures often remain elusive in textbooks and lectures, becoming clearer only with the help of additional, more explicit modes of representing chemical order. Among these, lists and tables were the most important ones in eighteenth-century chemistry.[44]

Beginning in the middle of the eighteenth century, lists of names were the favorite mode of representing classes of significant substances in plant and animal chemistry, especially the proximate principles of plants. In addition to lists, chemists also explicitly tackled questions of classifying plant and animal substances in distinct texts published in chemical journals. But the most explicit and systematic representation of chemical classifications in the eighteenth century was the table. The tables of chemical affinity and the 1787 table of chemical nomenclature are prominent examples of systematic chemical ordering in the eighteenth century. Tables were also ubiquitous tools of classification in eighteenth-century mineralogy as well as in botany and zoology. They further proliferated in organic chemistry after the 1830s, when the number of clearly demarcated chemical species began to explode through chemists' production and new means of individuation and identification of substances—namely, quantitative chemical analysis, chemical formulae, and the study of chemical reactions.[45] Before that time period, the use of tables was largely restricted to a specific domain of mineral chemistry: the chemistry of traceable or pure substances. Therefore, chemical tables will be central to our analyses in part II.

Up to Mendeleev's famous periodic table of chemical elements of 1869, classificatory tables in chemistry often display a striking syntactic arrangement, which is of special interest with respect to the philosophical discussion about formal requirements of scientific classification, in particular those initiated by Thomas Kuhn. Are scientific modes of classifying always encaptic? That is, do they always unambiguously define the place of one species within a taxonomic system?[46] The form of a matrix found in the 1787 table of chemical nomenclature, as well as in Mendeleev's table, is the crucial point at stake. This arrangement is chosen for ordering classes of substances according to quite different criteria at the same time. For example, Mendeleev's periodic table ordered chemical elements according to their atomic weight and their chemical properties. The arrangement of elements according to their atomic weight had to be interrupted at some point in order to achieve classification according to shared chemical properties as well. The result was a classificatory scheme that was not encaptic; that is, it did not exclude overlapping of taxa, but rather intended it. Each element belonged both to a chemical period, constructed according to atomic

---

43 On the role of chemical textbooks for late eighteenth-century and nineteenth-century classification, see also Bertomeu-Sánchez et al. [2002]; and Bensaude-Vincent et al. [2003]. The need to classify substances in chemical textbooks may also spur more explicit reflection upon classification. For example, Mendeleev's periodic system was constructed in the context of writing a chemical textbook; see Bensaude-Vincent [2001a]; and Gordin [2004].
44 For some semiotic aspects of these tables, see Cohen [2004]. Diagrams—taxonomic trees and others—are to be found especially in eighteenth-century botany. In the late sixteenth-century, Andreas Libavius represented his alchemical taxonomy in form of an encaptic taxonomic tree (see chapter 2.3), but eighteenth-century chemists no longer used diagrammatic modes of representing taxonomies.
45 See chapter 16.
46 On this problem see Buchwald [1992]; and Kuhn [1989], and [1993].

weight, and to a class constructed according to chemical properties. As we will show in part II, the 1787 table of chemical nomenclature displays a similar formal structure. If the authors of classificatory matrixes wished to avoid overlapping, they had to decide in each single case into which of the equally applicable taxa they wished to place a substance or class of substances. As we can see in the table of chemical nomenclature, the consequence of this was that the alternative taxon had to remain empty. Classification according to chemical composition did not provide a formal clue to unambiguous classification.[47]

The representations of eighteenth-century chemical classificatory systems, both by the organization of chemistry lectures and textbooks and by lists and tables, differed from the comprehensive, systematic, and technically sophisticated taxonomies in botany. Unlike the Linnaean taxonomy of plants, none of the eighteenth-century chemical classifications provided a comprehensive system covering all known substances. None of them singled out an invariable set of characters for ordering all substances, such as number, figure, position, and proportion of the parts of fructification organs, the set used in Linnaeus's taxonomy. And none of them introduced a sophisticated technical terminology for the taxonomic ranks comparable to that of Linnaeus.[48] Linnaean classification and other highly technical and systematic modes of classification in the so-called taxonomic sciences of the eighteenth-century have attracted the attention of historians of science as a manifestation of the Enlightenment "*esprit géometrique*," or "quantifying spirit," manifested by measuring and calculating as well as "the passion to order and systematize."[49] Yet it is not an easy task to evince the claim that the systematic Linnaean approach was actually followed in classificatory practices outside botany. The classificatory practices of chemists in particular tell a different story. Eighteenth-century chemists' utterly diverse ways of classifying substances—in particular in plant and animal chemistry—neither followed the "model" of Enlightenment natural history, as John Lesch has proposed, nor match the historiographical ideal of a "quantifying spirit."[50]

---

47 On this see also our earlier argument in connection with "objective classification" of natural kinds.
48 For a short overview on the Linnaean system see Müller-Wille [2003].
49 See Heilbron [1990] p. 2.
50 See Lesch [1990] p. 73.

Part II
A World of Pure Chemical Substances

Introduction to Part II

In part I, we argued that the majority of the processed substances which eighteenth-century chemists studied were commodities used in the apothecary trade, metallurgy, pottery, distilleries, and other arts and crafts. We have further argued that academic chemists examined these commodities from both technological and experimental-historical points of view as perceptible quotidian objects as well as from a theoretical or philosophical perspective as objects bearing imperceptible features such as composition and affinity. Commodities became transformed epistemically when they entered sites of academic investigation.[1] New meaning was attributed to them. Chemists' ontology of useful materials thus switched to an ontology of chemical compounds that were natural objects and consisted of imperceptible substance components invested with qualities and forces. Our third assertion was that not all of the processed substances studied in eighteenth-century chemical laboratories lent themselves equally to this kind of ontological transformation. Whereas metals, alloys, earths, alkalis, acids, salts, and a few other substances became involved in eighteenth-century studies of composition and affinity, and were identified and classified in part according to composition, plant and animal materials resisted chemists' attempts to create a conceptual order based on the modern concepts of chemical compound and affinity.

In part II, we examine the substances that were actually subjects of the ontological transformation connected with the modern chemical conceptual system of compound and affinity. The topic of this part is thus the world of pure chemical substances along with eighteenth-century chemists' modes of classifying these substances based on their knowledge of chemical composition.

Although a construction of academic chemistry, this world of pure substances can be regarded as neither a genuine intellectual creation nor an academic institution, for two reasons. First, the materiality of the substances—that is, their actual behavior in certain chemical operations—played a pivotal role in chemists' processes of selection and grouping together of pure substances. Second, the operations that were crucial for a substance's inclusion or exclusion in the new chemical order did not pertain exclusively to academic laboratories, but were also performed in several fields of the chemical arts and crafts—actually, most of them had been operations of the workshops before they were implemented in academic laboratories. The seventeenth and eighteenth centuries' world of pure chemical substances thus was a world that comprised domains of artisanal chemistry no less than academic chemistry. In other words, it was a world constituted through both specific practices and a new conceptual system, and these practices and conceptualizations rested on a material prerequisite that was not determined by chemists—namely, the fact that certain kinds of substances lent themselves to these practices and understandings whereas others did not.

---

1   At this point we are not yet discussing issues concerning material transformation.

The concept of chemical compound which took shape around 1700 manifested a profound shift in chemists' ontology of substances compared to the earlier alchemists' philosophical understanding of chemically processed substances as exalted, ennobled, or transmuted matter. This ontological shift is most apparent in the ways in which chemists classified these materials. Classification according to composition grants privileged access to this particular world of pure chemical substances—its gradual emergence, its dimension, and, as we will see, its practical context as well. First attempts to order materials based on the modern conceptual system of compounds and affinity appeared in the form of affinity tables in the early eighteenth century. Although we will also go back to these early endeavors to order substances according to composition, our focus will be on a mature classification of this type from the late eighteenth century, namely the *Tableau de la nomenclature chimique* constructed and published in 1787 by Louis Bernard Guyton de Morveau, Antoine-Laurent Lavoisier, Claude Louis Berthollet, and Antoine François Fourcroy.

We will regard and make use of the *Tableau* as if it were a Leibnizian monad representing the world of pure chemical substances. By investigating the various facets of the *Tableau*, we will explore and discuss various aspects of this world. And, inversely, by referring to and bringing in this world, we will clarify what is represented by this monad, the content of this classification, its historical background, and its socio-technological context. Scrutinizing the *Tableau's* concrete clusters of classificatory units, we will discover old partial classifications which, for their part, indicate and grant access to those practical artisanal spheres in which the materials of this world obtained their distinguishing features. Following the traces of older modes of classification that can be found in the *Tableau* will make visible the historical depth of this world. Finally, the *Tableau's* role in the reform of chemical nomenclature proposed by Lavoisier and his collaborators in 1787 at the peak of the Chemical Revolution will bring in the struggle about phlogiston as a struggle about chemical systems appropriate for the substances that pertain to this world.

To treat the *Tableau* in this way and use it as a mirror of several significant facets of the world of pure chemical substances means to address and discuss in turn several different aspects and dimensions of this classificatory table. This implies that answers to questions that suggest themselves in the course of the investigation must often be postponed to later chapters that can build on a more complete picture. Therefore it may be particularly desirable at the outset to present a clear overview about how we will proceed in part II.

In chapter 4 we will discuss the concrete circumstances under which Lavoisier and his collaborators constructed the classification of substances represented in the tableau as a means of nomenclatural reform. Apart from information on the persons involved and the institutional conditions, we will pay particular attention to four issues: 1) the new nomenclature's role in propagating Lavoisier's new chemical system; 2) chemists' request for a new chemical nomenclature; 3) the question of whether the new nomenclature indicated a divorce of academic from practical artisanal chemistry; and 4) the function of the classificatory table for the nomenclatural enterprise.

In chapter 5, foundations will be laid for the subsequent inquiries by a minute description of the table. This description will be aided by diagrams to which one can come back in the course of the following investigations. These diagrams also provide a sketchy overview of the table for those readers who do not wish to follow each and every detail of the description. The description will end with an analysis of formal classificatory features such as the table's classificatory ranks, its remarkable matrix structure, and the significance of this structure for the construction of chemical names.

Chapters 6 and 7 are dedicated to the table's chemical content, especially the conceptual framework employed in its construction. Chapter 6 will deal with the classification's most general principle, the principle of classification according to composition. Classification according to composition is not a universal mode of chemical classification, but a historical one. Based on our principal discussion of the modern concept of chemical composition in part I, three of the historical preconditions of this classification will be studied in this chapter: 1) the kind of chemical substances presupposed by this classificatory principle; that is, pure chemical substances which became the subject of chemical practice in the early modern period; 2) the demarcation of the modern notion of composition from earlier concepts of the constitution of substances; and 3) the development of the analytical method of procuring certain knowledge of chemical composition.

The short chapter 7 discusses classificatory principles employed in the table's construction in addition to the principle of classifying according to composition. Additional principles had to be employed in the case of simple substances, which, by implication, cannot be subdivided in classes according to composition. However, the classification of compounds required additional chemical considerations as well. As will become apparent on closer inspection, the choice of the table's classificatory units was governed by the authors' interpretation of three paradigmatic chemical syntheses: that of neutral salts, that of metal oxides and alloys, and that of gases. It will be shown that the concrete classificatory structure of the table resulted from a coherent linkage of the three clusters of classes pertaining to these syntheses.

Chapters 8 and 9 are dedicated to the context and historical background of the *Tableau*. Chapter 8 explores the fields of chemical theory and practice in which the chemistry of pure chemical substances originated in the early modern period. We will particularly study operations with such substances in sixteenth- and seventeenth-century metallurgy and pharmaceutical salt production, as well as the interpretations of these operations by the historical actors, in which elements of the modern concept of chemical compound gradually developed. Around 1700, the bulk of these as yet independent and little-connected chemical operations using pure substances was integrated into a consistent system on the basis of the concept of replacement reaction. This integration, which established the chemistry of pure chemical substances as a special domain of chemical theory and practice, is affirmed by the affinity tables of the eighteenth century. Main features of this integration will be explored by discussing the first and paradigmatic affinity table, Etienne F. Geoffroy's table of 1718.

Chapter 9 will investigate aspects of the history of classifying pure chemical substances before the construction of the 1787 *Tableau*. First we will focus on classificatory patterns that shaped affinity tables, Geoffroy's and subsequent ones. Next, more explicit classifications of pure chemical substances will be our topic. We will study chemists' customs of ordering these substances by using a mature and comprehensive, though in part naturalistic, classification of the mineral kingdom as our starting point—namely, Torbern Bergman's classification of minerals of 1779. We will discuss each of its main classes—metals, "inflammable bodies," earths, and salts—thereby briefly tracing their histories and the conceptual changes accompanying their gradual formation in the seventeenth and eighteenth centuries.

In chapter 10 we will finally come back to the *Tableau's* classification and try to assess its achievements on the basis of the historical insights obtained in the previous chapters. In particular, we will try to determine precisely which of its features were shaped by Lavoisier's new chemical system and which were not. However, we will arrive at the conclusion that the new chemistry shaped essential features of this classification only on the basis of a classificatory deep structure underlying both the phlogistic and the anti-phlogistic conceptualizations of pure laboratory substances. The chapter will close by addressing some questions of nomenclatural reform that we had to leave unanswered until the completion of our analyses.

In structuring part II, one of our concerns was to allow for different levels of interest in its topic. Readers who are primarily interested in the analysis of the classification of 1787, and less in its background and contexts, can concentrate on chapters 5, 7, and 10. Those less interested in the details of this analysis can concentrate on chapters 4 and 10 for the classification's connection with the Chemical Revolution, on chapter 6 for its conceptual foundation, on chapter 8 for the domains of practical, artisanal chemistry that form its background, and, finally, on chapter 9 for eighteenth-century classifications of pure chemical substances upon which the four authors of the *Tableau* could build.

4

1787: A New Nomenclature

In 1787, the Académie Royale des Sciences in Paris published a book entitled *Méthode de nomenclature chimique*.[1] The front page names four authors: Louis-Bernard Guyton de Morveau, Antoine-Laurent Lavoisier, Claude Louis Berthollet, and Antoine François Fourcroy. In this book, one encounters the first comprehensive classification of chemical substances according to composition in the history of chemistry. Taking into account the stature of its authors, its proximity in time to Lavoisier's *Traité élémentaire de chimie* from 1789, and the fact that it first appeared in a book on chemical nomenclature, this classification immediately raises a variety of questions and issues.

How is the *Méthode de nomenclature chimique* related to the Chemical Revolution associated with Lavoisier? Why did these four men join forces for this enterprise? For what reasons were chemical nomenclature and classification urgent issues just then? Were the nomenclature and the classification proposed in the *Méthode* instrumental in or a fruit of Lavoisier's Chemical Revolution? Or were they merely a means of its propagation? How did naming and classifying interrelate in this case? Which realm of materials was then covered by classification according to composition, and how was it structured by this classificatory principle? Where did the classificatory principles come from, and what do they tell us about the handling of these substances?

To answer these questions, a thorough and detailed investigation of the chemical classification from 1787 is required. Before beginning our scrutiny of the table, it may be fitting to first inform ourselves more about the nomenclatural enterprise in the framework of which this classification came about.

We can start with the fact that the *Méthode* was published "*sous le Privilége de l'Académie des Sciences*," as the title page announces. Thus it was a kind of an official document of the Parisian Academy, containing the nomenclatural proposal submitted by the four authors to the Academy in spring 1787 along with a report by the four members of the Academy—Antoine Baumé, Antoine-Alexis Cadet de Vaux, Jean D'Arcet, and Balthazar Georges Sage—in charge of the proposal's evaluation. Although the tone of this evaluation was reserved rather than enthusiastic, the four authors had achieved their immediate goal: their nomenclatural proposal had been issued as a document of the Parisian Academy.[2] Indeed, this proposal required such a protective shelter, for not only did it have to reckon with the resistance with which every radical nomenclatural reform meets, but it further contained a chemical nomenclature conforming to the anti-phlogistic chemical system. As this system was hardly generally accepted at the time, the proposal was bound to provoke the majority of

---

1 For bibliographic information on French editions as well as contemporary translations of this book, see Abbri and Beretta [1995]; for assessments and discussions of the *Méthode*, see Anderson [1984] chapter 7; Bensaude-Vincent [1983], [1994], and [1995]; Beretta [1993] chapter 4, and [2000]; Crosland [1962] part III; Dagognet [1969] part I; Golinski [1992]; Riskin [1998]; Roberts [1992]; Siegfried [2002] chapter 11; and Smeaton [1954].
2 For the strategic significance of this fact, see Simon [2002].

chemists. It therefore seems quite natural that some historians believed that the chief aim of the *Méthode* was indeed the propagation of this new chemistry.[3] However, there is clear evidence that a reform of chemical nomenclature was a serious concern of the authors quite independent of the new chemistry. It thus appears that two intentions drove the creation of the *Méthode*: the intent to promote the new chemistry (this may have been the concern of Lavoisier in particular); and the intent to replace the chemical nomenclature that had developed over centuries without any rules with one that was methodically constructed, which was doubtlessly Guyton's main concern.

4.1 The anti-phlogistic task force

Turning to the intent to promote the new chemistry through a new nomenclature, what strikes us first is that the four authors of the *Méthode* can be and actually were regarded as "the antiphlogistic task force" in the late 1780s, as Henry Guerlac put it.[4] Adding to this quartet Pierre Auguste Adet and Jean-Henri Hassenfratz, who provided the appendix to the *Méthode*, and further, the physicist Pierre-Simon Laplace and the mathematician Gaspard Monge, yields the army that was at Lavoisier's disposal during this period.[5] No less striking is the fact that, except for Guyton, none of the authors had distinguished himself previously as particularly concerned with the state of chemical nomenclature or issued any proposals for its reform.

As regards Claude Louis Berthollet (1748–1822), no evidence that he had any special interest in nomenclatural issues is known.[6] His contribution to the *Méthode*, too, remains unrecognizable.[7] Perhaps his name appeared on the title primarily because of the high reputation he enjoyed as a chemist,[8] and because he was the first chemist of note to accept Lavoisier's anti-phlogistic chemical system.[9] Trained as a physician and instructed in chemistry by Pierre Joseph Macquer and Jean Baptiste Michel Bucquet, he had been not only an adherent of the phlogistic chemical system well into the 1780s, but also an acute critic of Lavoisier's new ideas, one who pointed out their weaknesses. Following Lavoisier's experiments closely and recognizing their correctness step by step, he eventually abandoned the assumption of phlogiston in 1785. However, it is important to see that he did not accept Lavoisier's theory *in toto*. Particularly, he opposed Lavoisier's understanding of acids—a point we will come back to when investigating the classification of the *Méthode* in detail.

---

3   See, for instance, Perrin [1981] p. 52f.; and Donovan [1996] p. 157f.
4   Guerlac [1981] p. 80.
5   For a discussion of this "Arsenal group" from a different angle, see Kim [2003] p. 335ff. These men were called the "Arsenal group" because of their informal but regular meetings in Lavoisier's laboratory in the Parisian Arsenal; see Crosland [1967] p. 235f.; and Goupil [1992] p. 43.
6   For Berthollet's biography and works, see Partington [1961–1970] vol. III p. 496ff.; Gillispie [1970–1980] vol. II p. 73ff.; and Sadoun-Goupil [1977].
7   Being appointed by the Academy, together with Lavoisier and Fourcroy, to examine Adet's and Hassenfratz's proposal of new chemical symbols, he signed the report that was published along with this proposal as an appendix to the *Méthode*.
8   See Crosland [1962] p. 173; and Sadoun-Goupil [1977] p. 130f.
9   See Kapoor [1981] p. 74f.; and Le Grand [1975] p. 59ff.

Antoine François Fourcroy (1755–1809),[10] the youngest of the four authors, had shown some casual concern for the suitability of particular chemical names without engaging in nomenclature in a systematic manner.[11] Similar to Berthollet, Fourcroy was introduced to chemistry by Macquer and Bucquet as a medical student, and started lecturing on chemistry in the latter's private laboratory. In 1774, he succeeded Macquer to the important chair of chemistry at the Jardin du Roi. By the time he joined the other three authors in the nomenclatural enterprise, he enjoyed high repute as a chemist, particularly through his well-attended lectures. He had become convinced of Lavoisier's new chemistry in a very gradual process. In 1786 he announced his conversion in a special introduction to his *Élémens d'histoire naturelle et de chimie* which, being a retitled edition of his *Leçons élémentaires d'histoire naturelle et de chimie* (1782), had been written before 1786. He rewrote the work in terms of the new chemistry, thus making this very popular book a main vehicle of Lavoisier's new chemistry. Since he contributed to the *Méthode* a detailed explanation of its classificatory table, one may assume that he assisted Guyton particularly in constructing this table.[12]

Antoine-Laurent Lavoisier (1743–1794)[13] himself had been well aware of the importance of a suitable chemical nomenclature since early in his career as a chemist.[14] On several occasions in his work on gases he was forced to invent names for newly identified gases and was confronted with neologisms proposed by other chemists. There is also no doubt that his interest in the logic of constructing names predates Guyton's suggestions for a new chemical nomenclature. As proven by notes from 1780, which eventually were used for the famous *Discourse préliminaire* of his *Traité* from 1789,[15] Lavoisier had realized the potential of Condillac's *Logique* for chemical nomenclature immediately, in the very year of its appearance.[16] However, there are no indications that he himself thought of a systematic and exhaustive reform of chemical nomenclature before Guyton traveled to Paris early in 1787 to consult with Lavoisier, Berthollet, and Fourcroy on the problems of the nomenclatural project he had been pursuing since the early 1780s.

It is not clear who must be regarded the *spiritus rector* of the nomenclatural enterprise of 1787—Lavoisier or Guyton de Morveau.[17] In the *Mémoire* he read to the Academy in April 1787 and which opened the *Méthode*, Lavoisier gave full credit to Guyton's previous efforts concerning nomenclature, which thus appear likely as the

---

10 For Fourcroy's biography and works, see Kersaint [1966]; Smeaton [1962]; Partington [1961–1970] vol. III p. 535ff.; Gillispie [1970–1980] vol. V p. 89ff.
11 See Crosland [1962] p. 171f.
12 Guyton de Morveau et al. [1787] pp. 75–100.
13 For Lavoisier's biography and works, see, out of a immense amount of secondary literature, Guerlac [1981]; Partington [1961–1970] vol. III p. 363ff.; Poirier [1993]; and Donovan [1996].
14 See Crosland [1962] p. 168ff.
15 See Guerlac [1981] p. 81.
16 For the impact of Condillac's *Logique* (1780) on the French reform of chemical nomenclature, see Albury [1972]; and Beretta [1993].
17 For a recent discussion of this question, see Kim [2003] p. 375ff.

starting point of the joint enterprise.[18] A letter by Guyton, written just at this point of time, corroborates this assumption:[19]

> During my stay here [in Paris] I was almost exclusively occupied with the design of a chemical nomenclature. The gentlemen Lavoisier, Berthollet, Fourcroy and some others proved to be willing to accept most of the chemical names proposed by me and to come into an agreement as regards the remaining ones. We had together several consultations, and the result of our joint efforts was presented to the [Parisian] Royal Academy by Mr. Lavoisier in a special lecture. This lecture will shortly appear in print along with a table that allows to overlook the system of this nomenclature [...].[20]

In later years Guyton stated that the initiative for the collaboration had come from Lavoisier and his Parisian colleagues.[21] In any case, this collaboration was so intense and fruitful that the four men were able to present the principal result after just two months of work.[22]

Louis Bernard Guyton de Morveau (1737–1816) was a lawyer by training and served as one of the public prosecutors and members of the parliament in Dijon from 1762 to 1782.[23] As regards chemistry, he started as a self-instructed amateur chemist and soon attained significance through his efforts to promote chemistry in his native city and in Burgundy. He became known to the community of French chemists in 1780 when he was commissioned to write the chemical volumes of the *Encyclopédie methodique*. Particularly in pursuit of this work, he more and more felt the need for a radical reform of chemical nomenclature. He published a by and large well-received *Mémoire* on this issue as early as 1782.[24] Although well acquainted with Lavoisier's chemistry since meeting him for the first time in 1775, he remained an adherent of the phlogistic chemical system until his visit to Paris in 1787. There, finally, he adopted the new doctrine that made it possible to commence the collective work on the new nomenclature.

---

18 See Guyton de Morveau et al. [1787] p. 3ff.
19 Since Guyton mentions Lavoisier's *Mémoire*, which was read to the Academy on April 18, but not his own one read on May 2, the letter can be dated to the second half of April 1787.
20 "Bey meinem Aufenthalte hierselbst habe ich mich beynahe blos mit der Entwerfung einer chemischen Kunstsprache beschäftigt. Die Hrn. Lavoisier, Bertholet [sic.], Fourcroy u.a.m. bezeigten mir ihre Bereitwilligkeit, die mehresten meiner angegebenen chem. Benennungen anzunehmen, und [we]gen der übrigen sich mit mir zu verabreden. Wir stellten zusammen mehrere Berathschlagungen an, und das Resultat unserer gemeinschaftlichen Arbeiten legte kürzlich Hr. Lavoisier in einer eignen Vorlesung, der königl. Akademie vor. Jene wird nächstens gedruckt werden, und enthält zugleich eine Tabelle, auf der das ganze System dieser Kunstsprache zu übersehen ist, [...]." Letter of Guyton de Morveau to Lorenz Crell published by the latter in his *Chemische Annalen* for 1787 (volume II p. 54f.) in the rubric "Vermischte chemische Bemerkungen, aus Briefen an den Herausgeber."
21 See Crosland [1962] p. 174.
22 That is, if Guyton indeed arrived in Paris in late January or early February 1787. On this question, see Crosland [1962] p. 174f.
23 For Guyton's biography and works, see Smeaton [1957]; Partington [1961–1970] vol. III p. 516ff.; Gillispie [1970–1980] vol. V p. 600ff.; and Viel [1992].
24 Guyton de Morveau [1782]. We will come back to this *Mémoire* below in chapter 10.

4.2 Chemists' request for a new chemical nomenclature

The four chemists' linking of the proposed new nomenclature to Lavoisier's new chemical system was, of course, a bold challenge to the chemical community, for a proposal of a radical new nomenclature was more likely to win over a scientific community if it carefully avoided all further controversial issues. Indeed, the four authors of the *Méthode* did not address the issue of the new chemical system openly in the *Mémoire* submitted to the Academy. Instead they stressed the principle of constructing names for chemical substances according to their composition and explained how they had employed this principle.

It is important to see that Lavoisier and his collaborators had only two choices in the given situation: either to dare such a partisan proposal, or to postpone their nomenclatural project altogether to an indefinite future. For if the new names were to indicate the composition of substances, adherents and opponents of the phlogistic chemical systems could not agree upon those names, since they did not agree upon the composition of the substances at hand. This concerned not only the names for the newly identified gases—should, for instance, the "vital air" be given the name "dephlogisticated air" (Priestley) or "oxygen" (Lavoisier)?—but also the names of substances of ancient dignity such as "metal calces," as metal oxides had been called before Lavoisier. Precisely this difficult and delicate situation is what made Guyton, still in favor of the phlogistic chemical system at the time, travel to Paris for consultation with Lavoisier and his group on how to pursue a reform of chemical nomenclature under these circumstances.

The need for a reform of chemical nomenclature had been felt much earlier and had become an urgent issue since the middle of the eighteenth century, long before Lavoisier challenged the then-prevailing phlogistic chemical system.[25] In general terms, the problem consisted in the fact that chemists used names of substances that originated in a variety of different historical situations and contexts. The ancient names highlighted a variety of features of substances: their provenance (*Vitriol de Chypre*—sulfate of copper), their perceptible properties (*Sucre de Saturne*—acetate of lead) and chemical behavior (*Alkali minerale effervescent*—sodium carbonate), their uses (*Sel febrifuge de Sylvius*—potassium chlorate), their methods of preparation (*Précipité rouge*—mercuric oxide[26]), and their discoverers or creators (*Alkaest de Van Helmont*). Moreover, since distinctions and identifications of chemical substances developed over time, many substances had been assigned several names, while one single name sometimes designated different substances. This state of affairs was not only an obstacle in teaching, learning, and memorizing chemical names, but also a serious hindrance for the understanding of chemistry. Under such conditions, it was a matter of chance whether names indicated relations among chemical substances in a way that was meaningful for knowledgeable chemists.

---

25 On the following, see Smeaton [1954]; and Crosland [1962] p. 133ff.
26 For the preparation of *red mercury precipitate*, which did not actually involve precipitation, see Crosland [1962] p. 89.

A series of major steps toward a reform of chemical nomenclature commenced in 1766 with the first edition of Macquer's *Dictionnaire de Chymie*. Macquer's proposals, which focused on the names for acids and salts, were a first attempt at constructing names according to composition.[27] Baumé, who was to become a member of the commission of the Paris Academy of Sciences that evaluated the proposed *Méthode* in 1787, followed a similar path in his considerations about chemical nomenclature.[28] Much more systematic in character, and also much more comprehensive than those of these two French chemists, were the nomenclatural efforts that the Swedish chemist Torbern Bergman pursued from the early 1770s.[29] In the 1780s these efforts assumed features of a self-reinforcing movement, which gained strength through mutual awareness and response among the actors. In his *Mémoire sur les dénominations chimiques* of 1782, Guyton was clearly influenced by the nomenclatural proposals that Bergman had presented up to that time.[30] This *Mémoire*, inversely, influenced Bergman in the development of one of his mature contributions to both nomenclature and classification in chemistry, namely his *Meditationes de systemate fossilium naturali* of 1784.[31]

All of these efforts focused mainly on well-established classes of substances, in particular salts, acids, and the so-called bases of salts; that is, alkalis, earths, and metals. Though also often concerned with the question of how to name newly identified substances, these men tried above all to bring order to the mess of names they had inherited from their ancestors. With the chemistry of gases, which emerged in the second half of the eighteenth century, the situation became even more complicated. The finding of appropriate names for these new substances proved to be an affair of unknown delicacy, since hardly any of these gases fitted into the established classes of chemical substances. To make things worse, at just this time the community of chemists became divided over the phlogistic chemical system, thus losing the common ground on which problems of chemical nomenclature could be settled unanimously.

4.3 The new nomenclature: A divide?

Against this background, it seems rather surprising that the nomenclature proposed by Lavoisier and his collaborators did not meet with fierce resistance, but became accepted in almost all European countries and even in the United States of America within fifteen years. The new chemical nomenclatures that emerged in the different vernaculars around 1800 were to a large extent highly uniform, since they followed the same model, that of the nomenclature of the *Méthode* of 1787.[32]

---

27 For Macquer's nomenclatural efforts, see Smeaton [1954] p. 88f.; and Crosland [1962] p. 136ff.
28 See Crosland [1962] p. 133f.
29 For Bergman's nomenclatural efforts, see Smeaton [1954] p. 97ff.; Crosland [1962] p. 144ff.; and Beretta [1993] p. 138ff.
30 Guyton translated into French the first two volumes of Bergman's *Opuscula physica et chemica* (1779 and 1780).
31 In chapter 9 we will come back to Bergman's classification of the mineral kingdom.

Although the Lavoisierian reform of nomenclature did not extend to the entire culture of chemistry — in particular not to plant and animal chemistry — its impact was deep. In the words of Bernadette Bensaude-Vincent:

> Il y eut avant et après 1787. *La Méthode de nomenclature* introduit une telle rupture dans l'histoire de la chimie, que la langue naturelle des anciens chimistes est devenue, pour nous, langue étrangère. Mais à travers ces changements d'habitudes de langage, c'est toute la chimie qui a été profondément modifée. Ces noms sont l'acte de baptème d'une nouvelle science rationelle, expérimentale, quantitative et rigoureuse. L'exemple typique d'une revolution scientifique.[33]

We will discuss later, in chapter 10, the question of whether the reform of nomenclature actually brought about a true Bachelardian "rupture" or revolution in the culture of chemistry, as Bensaude-Vincent proposed. But it seems worthwhile to consider briefly what fell victim to this nomenclatural reform. Overcoming the chaos of the old chemical names implied dispensing with the practice of using names to indicate the substances' provenance, creator or discoverer, properties and uses, and methods of preparation. Constructing names solely according to composition resulted in names that were silent on the mundane contexts of these substances. Thus one may ask whether we encounter in the new nomenclature the artificial language of a science that shed its traditional links to artisanal practice, and emerged as a science proper through just this separation. Does the new nomenclature indicate the emancipation of academic chemistry from the worldly sphere of artisanal chemical practices? There is no simple answer to this question. If "emancipation" means cutting ties to and abandoning engagement in mundane chemical practices, the answer is clearly no. If it means claiming the superiority of academic chemistry over the chemical practices of miners, metallurgists, dyers, pharmacists, and so on, the answer is probably yes — at least as regards the understanding of our four authors. In any case, for them, the change they wanted to initiate with the new nomenclature did not entail any separation of academic chemistry from the chemical technology.

The entire quartet was almost uninterruptedly engaged with practical chemistry, be it as promoters, inventors, commissioners, or in other capacities.[34] This is well documented for the period of the French Revolution, the Directorate, and Napoleonic rule, when all of them — even Lavoisier up to his arrest in November 1793 — served on many governmental committees in charge of controlling and improving both economic-technological affairs, often with a military background, and educational-scientific institutions. Among the former were commissions such as the *Bureau de Consultation des Arts et Métiers*[35] (Berthollet, Fourcroy, and Lavoisier), the *Commis-*

---

32 On the spread of the new nomenclature in Europe, see Crosland [1962] pp. 177–214; Bensaude-Vincent and Abbri [1995]; and Crosland [1962] p. 212ff..

33 Bensaude-Vincent [1983] p. 1. In Poirier [1993], R. Balinski translates: "There was a before and an after 1787. *La Méthode de nomenclature* introduced such a rupture in the history of chemistry that the natural language of the old chemists became a foreign language for us. All of chemistry was profoundly modified. These names are the baptismal certificates of a new rational, experimental, quantitative, and rigorous science. The typical example of a revolution."

34 See Guerlac [1977].

35 See Gillispie [2004] p. 201f.; and Sadoun-Goupil [1977] p. 32.

sion des Poids et Mesures[36] (Berthollet, Fourcroy, Guyton de Morveau, and Lavoisier), the *Commission d'Agriculture et des Arts*[37] (Berthollet), and the *Régie des Poudres*[38] (Berthollet, Fourcroy,[39] and Guyton de Morveau). This engagement was not a novelty of the French Revolution, which mobilized savants for its survival, but predated the revolutionary age as indicated by a few examples. Since 1784, Berthollet had served as appointed inspector of dye works and director of the *Manufacture Nationale des Gobelins*.[40] Fourcroy, who always was interested in applications of chemistry in medicine, had been particularly engaged in improvements to the methods by which mineral waters were assessed with respect to their medical value.[41] Likewise Lavoisier's occupations in various fields of practical chemistry before 1789 are well-known.[42] Lavoisier was further an academic chemist who laid a direct claim to guiding and controlling practical chemistry. This becomes particularly apparent in the fact that he acted as commissioner of the old royal Régie des Poudres et Salpêtres.[43] Guyton de Morveau, finally, had been the driving force in promoting the chemical crafts of his native town Dijon. He was personally involved in industrial enterprises and was engaged particularly in color preparation, saltpeter and soda production, and, as a curiosity, in the production of a suitable gas for hot air balloons, a *dernier cri* of the 1780s.[44]

4.4 Classification in the *Méthode*

As its title states, the *Méthode de nomenclature chimique* was a treatise on chemical nomenclature, or, more precisely, on the methods of constructing names of chemical substances. As Lavoisier stressed, the aim of the work was to present a method for systematic naming—*"une méthode de nommer."*[45] Classification was one prerequisite of this method, and linguistic rules were another. Whereas the authors elaborated broadly on the latter—not only in the opening *Mémoire* by Lavoisier, but also on several occasions in Guyton's *Mémoire*—, they said almost nothing about their mode of classifying, at least not verbally, in the essays of the *Méthode*. Yet their classification is presented in form of a table inserted in the *Méthode*, the *Tableau de la nomenclature chimique*.

The table of chemical nomenclature was posted in the *Académie des Sciences* in the second half of April 1787—Lavoisier read his fundamental *Mémoire* on April 18, Guyton his *Mémoire* on May 2—to give the members of the Academy the opportu-

---

36 See Gillispie [2004] p. 277f.; and Sadoun-Goupil [1977] p. 32.
37 See Sadoun-Goupil [1977] p. 32f.
38 See Gillispie [2004] pp. 29ff., 395; and Sadoun-Goupil [1977] p. 34.
39 On the somewhat contradictory evidence of Fourcroy's service or non-service on the commission, see Smeaton [1962] p. 41ff.
40 See Nieto-Galan 2001; Sadoun-Goupil [1977] p. 21f.
41 See Smeaton [1962] p. 19.
42 On these occupations, see, for instance, Donovan [1996] p. 188ff.; and Poirier [1993] pp. 84ff., 115ff., 154ff., and 198ff.
43 On Lavoisier and the *Régie Royale des Poudres et Salpêtres*, see Bret [1994]; and Mauskopf [1995].
44 On Guyton's practical activities, see Gillispie [1970–1980] vol. V p. 602.
45 Guyton de Morveau et al. [1787] p. 17.

nity to get acquainted with the principles of the proposed nomenclature. It would have been hard to derive these principles from the *Méthode*'s two synonymies of old and new names, which followed an alphabetical order.[46] They were, however, easily recognizable in the names' arrangement in the classificatory scheme of the table. First, it could be seen in this table how the names of substances pertaining to the same classificatory unit were constructed, say, the class of salt. Second, the rules of how the names of different related classes—say, those of salts and acids—were linked also became transparent in this table. Thus it comes as no surprise that the essays by Guyton and Fourcroy merely explicated the table to explain the new nomenclature in detail. This mode of clarifying the proposed chemical nomenclature presupposed that their fellow academicians and, later, the readers of the *Méthode*, would be able to recognize the table as a classificatory scheme and, further, to understand the concrete way the chemical substances were classified by the four authors.

The *Tableau de la nomenclature chimique* played a central role in both the construction and the understanding of the methodical nomenclature, precisely because it was simultaneously a table both of nomenclature and of classification of chemical substances. The methodical character of the new nomenclature was essentially based on the classification of the substances to be named. At this point we must leave unanswered the question as to whether this classification was a fruit of the new methodical nomenclature or rather one of its prerequisite conditions, or whether a co-evolution of nomenclature and classification took place. We will come back to this question in chapter 10. What must be emphasized here is the interconnectedness of nomenclature and classification. According to Condillac, Lavoisier's authority in linguistic matters, classifying consisted of nothing other than naming.[47] True, classification is always involved where generic names are given to things. But classifying does not always consist of naming. As we will see, it often employs means other than words. In any case, in order to devise a methodical nomenclature like that of the *Méthode*, the prerequisite classification must have achieved a highly systematic character. It is precisely this systematic classification of chemical substances that underlies the proposed nomenclature of the *Méthode,* which part II subjects to scrutiny—the structure of this classification, its principles, historical background, and context.

---

46 One in the alphabetical order of the traditional names, the other in that of the new names.
47 See, for instance, part I, chapter 4 of Condillac's *Logique* (Condillac [1948] p. 379).

5

The *Tableau de la Nomenclature Chimique*

Much has been written on the *Méthode de nomenclature chimique*, and much has been said in praise of the *Tableau* of the *Méthode* and its outstanding significance in the history of early modern chemistry.[1] But despite this unanimously shared assessment and appreciation, no thorough and exhaustive analysis of the *Tableau's* classification of chemical substances has been presented by any historian of science up to now. Only the first column of the table, the column for simple substances, was hitherto subjected to closer inspection.[2] Thus our analysis of the table's classification cannot build on the results of earlier investigations, but has to start from scratch. This means we first have to provide an overview on the table to which we can refer and return whenever desirable in the course of the subsequent investigations. In what follows we will first describe the arrangement of the table and then analyze its classification's formal features.

5.1 Description of the *Tableau*

A first glance at the table (figure 5.1) shows that the classificatory system it renders cannot be grasped immediately.[3] The table is crowded with such an enormous number of names of chemical substances that it takes some time to recognize structures and patterns. The first task is, therefore, to pave a way to such structures and patterns. For this purpose, it might be useful to present some diagrams that approach the main structure of the table step by step.

Our first diagram (figure 5.2) aims to make the table more transparent by merely reducing the huge number of names. In many fields, it names only a few exemplary substances and indicates those left out by "etc." Take for instance the field III/b, the field with acids: eight acids (some with different degrees of saturation) are listed; that is, eight out of 31 acids in the table. Likewise, the table's field with the acidifiable bases of the different acids, field I/b, contains 26 such bases, not only five as in the diagram. Another example: the field I/c, the field with metals, contains seventeen metals or semi-metals and not only four as in this diagram. Most of the fields have been shortened in this manner.

This reduction is justified by the table. In the cases of the salts (field V/b) and of the metal oxides (field III/c), the authors of the table themselves listed only a selec-

---

1   See the literature referred to above in footnote 1 to chapter 4. This collection of secondary literature could easily be extended if we drew in literature on the four authors of the *Méthode* or on the Chemical Revolution in general, which usually addresses more or less elaborately the nomenclatural enterprise of 1787.

2   See Siegfried [1982] p. 36f.

3   Since the original of the *Méthode* is practically beyond the grasp of common readers, and because photo-mechanical reproductions of the original *Tableau* are usually unsatisfactory (as evinced by figure 5.1), one will appreciate that The MIT Press provides a link to a scan of the table on the book's website: <http//mitpress.mit.edu/0262113066>. Guyton de Morveau et al. [1994] included the *Tableau* newly typeset (pp. 100–107). However, it should be noted what was newly typeset there was not the *Tableau* of the original, but that of the *Encyclopédie méthodique* of 1786, which differs in some places from that of 1787. We will point to such differences where necessary. On the strange dating of the *Tableau* of the *Encyclopédie*, see Abbri and Beretta [1995] p. 280, item 5.

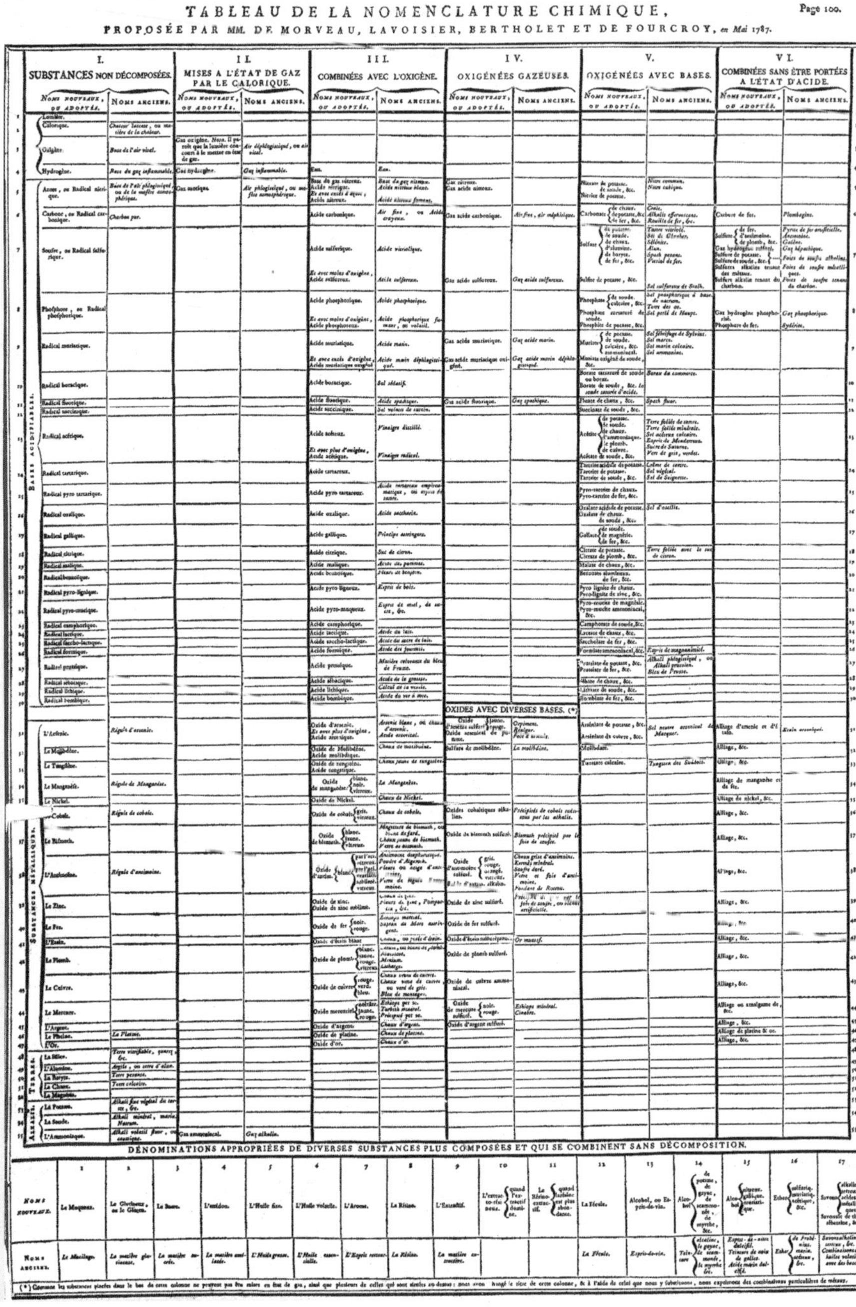

Figure 5.1: The *Tableau* of the *Méthode* from 1787.

tion of the substances pertaining to these fields. The reason our authors left out substances cannot have been simply brevity, in view of the fact that two exhaustive synonymies of old and new chemical names complete the *Méthode*. Rather, they could afford to leave out substances because the table is not an inventory of chemical substances, but a classificatory scheme. The completeness of such a scheme depends on whether or not it provides a place for every known kind of substance—classificatory completeness in a weak sense—or for every thinkable kind of substances as well—classificatory completeness in the strong sense.

The question of whether the table is complete in the strong sense or in the weak sense must take into account that the authors of the *Méthode* confined themselves to classifying and naming a rather small selection of the chemical substances known at the time. The wealth of substances extracted or distilled from plant or animal materials is almost entirely absent in the table, as is the majority of mineral raw materials known from mines or oryctological collections. In other words, the table does not try to classify and name all substances then known, but only a small part of them. Whether the table can claim a strong or weak classificatory completeness for this selection of substances depends, of course, on the criteria for this selection.

Since the four authors even did not mention the very fact that names were proposed for only a small selection of chemical substances, let alone argue for the criteria of their selection, these criteria become manifest only in analysis. Since no contemporary complained about this omission, we can conclude that the selection and its criteria must have been obvious to the historical actors. In hindsight, it is clear what was selected: the table provides slots only for pure chemical substances, that is, substances that are either simple substances (Lavoisierian elements), or chemical compounds that undergo reversible decompositions and recompositions.[4]

In our reduced table (figure 5.2), several fields can be identified immediately as slots for classes of substances: we already recognized a field with acids (III/b) and another with metals (I/c); field I/d is for the class of earths, field I/e for the class of alkalis, field III/c for the class of metal oxides, and field VI/c for the class of alloys and amalgams. In other cases, the classes can be identified by means of the headings above and beside the columns, or by resorting to the essays in the *Méthode*: field I/a is for the class of the most simple substances, field I/b for the class of acidifiable bases, field IV/b for the class of acids in a gaseous state, field IV/c for the class of combinations of actual or supposed metal oxides with several substances except for acids, and VI/b for certain compounds of sulfur, carbon, and phosphorus. However, there are also fields that are puzzling: two fields (V/b and V/c) seem to contain sub-

---

4   See the definition of our notion of "pure chemical substance" in chapter 2. As will become clear in the following, the table also included non-stoichiometric compounds, namely alloys and amalgams (field VI/c). Furthermore, in the appendix at the bottom of the table, one encounters classes of plant and animal substances that are not pure chemical substances in today's view. As regards the non-stoichiometric compounds, they point to a fundamental feature of Lavoisier's chemical system that is usually ignored and will be discussed in part III of this volume; namely, this system's lack of a stoichiometric underpinning. In part II, for the sake of adaptation to the historical actors' view as well as to reduce complexity, we include the non-stoichiometric compounds when speaking of "compounds." As regards the substances in the appendix, they will be discussed at length in part III.

		I	II	III	IV	V	VI
a	1	Lumière					
	2	Calorique					
	3	Oxygène	Gaz oxygène				
	4	Hydrogène	hydrogène	Eau			
b	5	Azote	azotique	Acide nitrique nitreux	Gaz acide nitreux	Nitrate de Nitrite de	
	6	Carbone		carbonique	carbonique	Carbonate de	Carbure de
	7	Soufre		sulfurique sulfureux	sulfureux	Sulfate de Sulfite de	Sulfure de
	8	Phosphore		phosphorique phosphoreux	.	Phosphate de Phosphite de	Phosphure de
	9	Radical muriatique		muriatique	muriatique	Muriate de	
	...	etc.		etc.	etc.	etc.	etc.
c	31	Arsenic		Oxide d'arsenic	Ox. arsenical de potasse	Arséniate de	Alliage d'arsenic & d'étain
	32	Molybdène		de molybdène	Sulfure de molybdène	Molibdate	de & de
	33	Tunstène		de tunstène		Tunstate	de & de
	34	Manganèse		de manganèse			
	...	etc.		etc.	ctc.		etc.
d	48	Silice					
	49	Alumine					
	50	Baryte					
	51	Chaux					
	52	Magnésie					
e	53	Potasse					
	54	Soude					
	55	Ammoniaque	ammonical				

Appendix

Le muqueux Le glutineux Le sucre L'amidon etc.

Figure 5.2: Reduction of the *Tableau de la nomenclature chimique*.

stances of one and the same class, namely salts. The same holds for the fields II/a, b, and e, all of which contain gases conceived as simple at the time. Thus, fields apparently do not always indicate classes of substances. This is confirmed by fields with only one substance—fields II/b and e, and III/a. The appendix, finally, contains substances that do not seem to form a class.[5] We can summarize this first description of the table by means of another diagram (figure 5.3). Using the original denominations of the different taxonomic units, which can be found either in the table or in the essays of the *Méthode*, this diagram renders the classes of substances and their

	I	II	III	IV	V	VI
	Substances non decomposées	*Mises à l'état de gaz par le calorique*	*Combinées avec l'oxygéne*	*Oxygénées gazeuses*	*Oxygénées avec bases*	*Combinées sans être portées à l'état d'acide*
a	*Substances qui se rappochent le plus de l'état de simplicité*	*Gaz*	*Eau*			
b	*Bases acidifiables*		*Acides*	*Oxygénées gazeuses*	*Sels neutres*	*Composés simples du Soufre, Carbone & Phosphore*
c	*Substances metalliques*		*Oxydes metalliques*	*Oxydes avec diverses bases* / *Oxydes avec diverses bases*		*Alliages & amalgames*
d	*Terres*					
e	*Alcalis*					

Appendix

Dénominations appropriées de diverses substances plus composées, & qui se combinent sans décomposition

Figure 5.3: The classes of the *Tableau* and their arrangement.

---

5   On the appendix, see chapter 14.

arrangement. Apart from the appendix at its bottom, the table consists of six vertical columns juxtaposed horizontally. Each of these columns is divided vertically into fields that represent actual or possible classes of chemical substances. Two columns (II and V) contain only one actual class, the other columns contain two, three (if field III/a is taken as a class), or even five classes. Thus, we find fourteen (thirteen) actual classes in the table, which are distributed into six higher taxonomic units.

Two of the columns, namely column II and the upper part of column IV, contain classes of gaseous substances.[6] The two classes of gaseous substances comprise compounds of the matter of heat, with either simple substances (II/a,b, and e) or with oxygenated substances (IV/b), also known as acids. In the following we will focus on columns I, III, V, and VI, where we find the classes of substances the understanding of which was at the core of the controversy between adherents of the phlogistic and the anti-phlogistic chemical systems, namely the classes of acidifiable and salifiable bases, acids, and salts, and those of metals and metal oxides or metal calces.

The column for simple substances

Turning now to the columns and their classes, the first feature of the classificatory structure that must be stressed is the exceptional position of column I. Whereas the five other columns, without exception, contain the classes of substances regarded by the authors of the *Méthode* as compounds, column I comprises only classes of substances that these men took to be simple substances—that is, non-compounds.

For two reasons, the term "element" is deliberately avoided here in favor of "simple substance," a phrase often used by the historical actors. Firstly, the authors of the *Méthode* themselves left open the question as to whether the substances in column I were true chemical elements or, alternatively, apparently simple substances that had so far resisted every attempt to decompose them, as the heading of this column—*Substances non decomposées*—highlights.[7] Actually, it was then already known that one of these supposedly simple substances, namely ammoniac, could be decomposed,[8] and the same was expected to happen to other simple substances in the near future.[9] When dealing with eighteenth-century chemistry, a relativistic understanding of the simplicity and compoundness of chemical substances seems much more appropriate than our present absolute distinction between elements and compounds. Secondly, the avoidance of the term "element" will allow us to trace fundamental structures of the table back to seventeenth- and early eighteenth-century chemistry, where the

---

6   A precise understanding of the specific arrangement of these two classes would require delving into Lavoisier's theory of gases, which is beyond the scope of this chapter.
7   See Guyton de Morveau et al. [1787] p. 28: "[…] les corps simples, c'est-à-dire ceux qui n'ont pu jusqu'à présent être décomposés […]." As to Lavoisier, see ibid. p. 17 and also the famous passage on the notion "element" in the "Discours préliminaire" of his *Traité élémentaire de Chimie*: Lavoisier [1789] p. XVIIf. and Lavoisier [1965] p. XXIV.
8   See Guyton de Morveau et al. [1787] p. 82.
9   See Guyton de Morveau et al. [1787] p. 17f. As will be discussed in chapter 10, Lavoisier subjected the classes of simple substances to substantial revisions only two years later, dismissing, among other things, the entire class of alkalis and, of the class of acidifiable bases, all organic bases—see the "Table of Simple Substances" in his *Traité élémentaire de Chimie*.

terms "element," or "simple principle," had entirely different meanings, and where the distinction between relatively simple substances and substances compounded out of simpler building blocks was made in a way similar to the later distinction made by the authors of the *Méthode*.[10]

Column I is for simple substances in a relativistic sense and consists of five classes.[11] The first class is the class of substances that are thought to come very close to the absolute state of simplicity—*Substances qui se rappochent le plus de l'état de simplicité*.[12] It contains four substances:[13] two substances which are unknown today—matter of light (*lumière*) and matter of heat (*calorique*), and two others with familiar names—*oxygène* and *hydrogène*—which are also unknown and must not be mistaken for our oxygen and hydrogen. Rather, the latter can be found in field II/a under the names oxygen gas (*gaz oxygène*) and hydrogen gas (*gaz hydrogène*).[14] For the authors of the table, these gases were compounds resulting from the combination of the simple substance calorique with the simple substances oxygène or hydrogène, respectively. Calorique, oxygène, and hydrogène, out of which these gases were thought to be composed, were postulated substances which were just as unproduceable then as they are today. Such postulated substances are characteristic of column I, as is shown most impressively by its second class.

This second class is the class of the acidifiable bases (*bases acidifiable*) or the radicals (*principes radicaux*) of acids. Acids were taken by the authors of the *Méthode* to be combinations of oxygen and such acidifiable bases.[15] Hence there were just as many acidifiable bases as there were acids—namely 26.[16] However, the decomposition of acids and the separation of acidifiable bases was accomplished in only three cases—those of sulfur, charcoal, and phosphorus. This means that 23 of the 26 acidifiable bases were postulated substances.[17] The understanding of the relationship between the acidifiable bases and their corresponding acids was one of the main issues of dissent between adherents and opponents of the phlogistic chemical system. For the latter, such a relationship existed because each of these acids was thought to be an oxide of one of the acidifiable bases; for adherents of the phlogistic system, the three known acidifiable bases—sulfur, carbon, and phosphorus—were thought to be compounds of phlogiston and one of the acids.[18] Thus, it was not controversial that sulfur, carbon, phosphorus, and the other substances in field I/b pertained to the same

---

10 This is not to deny the fact that Lavoisier and his collaborators aimed at a clear distinction of simple and compound pure chemical substances and, indeed, launched a development that eventually led to our unambiguous distinction of chemical elements from chemical compounds. We will come back to this issue below in chapter 10.

11 For a discussion of column I, see Siegfried [1982] p. 36f.

12 See Guyton de Morveau et al. [1787] p. 30.

13 In Guyton de Morveau's essay, *azote* is listed as fifth substance of this class; in the *Tableau*, however, this substance is put in the class of the acidifiable bases. See Guyton de Morveau et al. [1787] p. 30.

14 In order to reduce the abundant complexities given with the table, in the subsequent analysis the oxygen of field I/a is taken as if it were our oxygen.

15 This conception of acids was actually Lavoisier's theory, which Berthollet never accepted. As we will see later, much of the elegance of the table's classification rested on this strange understanding of acids.

16 Acids of the same acidifiable base but of different degree of saturation are counted as one acid.

17 This also holds for *azote*, as the *gaz azotique* in field II/b is our nitrogen, not the *azote* in field I/b.

18 In this controversy it was an advantage for the adherents of the phlogistic system that their system did not require them to assume a large number of hypothetical acidifiable bases.

class of chemical substances. Rather, what was controversial was whether they were relatively simple or more compound substances and, therefore, whether or not their class was correctly positioned in column I. As we will see, the important differences between adherents of the phlogistic chemical system and the authors of the table were mainly questions as to the arrangement of the classes rather than the classes themselves, and least of all the distribution of substances into these classes.

The third class of simple substances is that of the metals, semi-metals included. It is followed by the class of the earths, and the fifth and last class of column I is that of the alkalis. In eighteenth-century chemistry, these last three classes of column I constituted a particular taxonomic unit, that of the so-called bases of neutral salts, because the substances comprised were known to form salts with acids. This traditional unit is continued in the table.[19] As to metals, the authors of the *Méthode* followed the Swedish chemist Torbern Bergman and regarded, not the metals, but metal oxides (field III/c) as salifiable bases.[20]

The columns for compounds

Columns II through VI are for classes of compounds, the classes of compounds in one and the same column being analogous substances. Column III comprises three classes: first, a possible class that contains only one substance, water; second, two actual classes, those of acids and of metal oxides.[21] The authors of the table assumed that these classes of substances were analogous because they considered all of them to be oxides. The first class reflects the synthesis of water so recently accomplished for the first time by Henry Cavendish (1731–1810) and Lavoisier, and its interpretation by the latter. In the classes of acids and metal oxides, we encounter another main issue of dissent between adherents and opponents of the phlogistic system. These factions differed over the question of whether these classes were analogous because all of the pertinent substances resulted from the removal of phlogiston by combustion (so the adherents of the theory of phlogiston), or from the combination with oxygen (so the opponents). Accordingly, they differed with respect to the question of whether acids and metal calces were relatively simple substances (so the adherents) or chemical compounds (so the opponents). We will come back to this later.

---

19 The table seems to render this traditional unit implicitly by ordering these three classes *en suite*. Guyton de Morveau's remark on this arrangement is telling. He explains why, in column I, the metals come next after the acidifiable bases, but expresses it by saying that, for such and such reason, he will "begin" with the metals: "[…] c'est par celles-la que nous avons cru devoir commencer […]." (Guyton de Morveau et al. [1787] p. 53.) But he neglects to say what exactly he is going to "begin" with the metals. It seems very likely that every contemporary understood immediately that he was speaking about the group of classes of salifiable bases and that he argued for arranging the metals before the two other classes of this group. Indeed, this arrangement required justification, since the usual French arrangement of this group was first the earths, then the alkalis, and finally the metals.
20 The authors of the *Méthode* attached little importance to this question, for they clarified in a mere footnote that, in the heading of column five, "base" meant metal oxides in the case of the salts with a metallic base. The effect of this was that not even the appointed revisors of the table took notice of this distinction. See Guyton de Morveau et al. [1787] pp. 93 n. 1 and 242.
21 As stated above, we are skipping column II and the upper part of IV, the columns for gases, since we cannot go into Lavoisier's conception of these substances here.

In column V we encounter the realm of neutral or middle salts; that is, of salts that eighteenth-century chemists considered combinations of acids with one of the salifiable bases—that is, metals or metal oxides, respectively, earths, or alkalis.[22] Consequently, the column contains only one class that by and large corresponds to the class of acids in column III. The few salts formed with a metal oxide instead of an acid[23]— field IV/c, lines 31–33—obviously did not induce the authors of the table to introduce a separate class. [24]

	I	II	III	IV	V	VI
	Classes of Simple Substances	Classes of Compounds				
		I + Matter of Heat	I + Oxygen	III + Matter of Heat/Others	III + I/d-e and III/b + III/c	I + I
a	Most Simple Substances	Gases	*Eau*			
b	S, C, P ...		Acids	Gaseous Acids	Neutral Salts	Compounds of S, C, and P
c	Metals		Metal Oxides	Compounds of Metal Oxides		Alloys
d	Earths					
e	Alkalis					

Figure 5.4: Classificatory structure of the *Tableau de la nomenclature chimique.*

22 In the second half of the eighteenth century, the classificatory terms "neutral salts" and "middle salts" were used synonymously for the most part. Since the authors of the *Méthode* used the term neutral salts, this term will be used throughout part II, even in cases where the historical actors preferred the term middle salts.
23 The production of these salts from certain metal oxides may have been one of the reasons for replacing the name *"chaux metalliques"* (metal calces) with the name *"oxides metalliques."* (See Guyton de Morveau et al. [1787] p. 55f.) At least some contemporaries recognized in *oxides metalliques* not only a reference to the composition of the substances in question, but also an allusion to the acids: in the German edition of the *Méthode*, *oxide* is rendered with *Halbsäure* (semi-acid). See Guyton de Morveau et al. [1793] pp. 56 and 82.
24 See Guyton de Morveau et al. [1787] p. 94f.

Column VI, finally, comprises two classes of combinations of simple substances that are "in their natural state, without being oxidized or acidified;" that is, combinations of simple substances that do not involve oxygen.[25] The first class, which corresponds to the class of acidifiable bases in column I, contains such combinations of sulfur, carbon, and phosphorus with metals or alkalis. The second class, which corresponds to the class of metals in column I, contains alloys and amalgams.

So far, the table does not seem to provide a place for one class of compounds known to eighteenth-century chemists, namely the combinations of sulfur and alkalis with metal oxides, or what Lavoisier regarded as such oxides.[26] But we find these compounds placed in the lower half of column IV. However, this does not indicate that the authors of the *Méthode* regarded this class of compounds as analogous to that in field IV/b. Rather, this placement has the appearance of some embarrassment or compromise, as this column was dedicated to classes of gaseous compounds.[27] In any case, this first survey of the table's classificatory structure can end at this point and may again be summarized by a diagram (figure 5.4), which also will be of use for our further investigations.

5.2 The formal classificatory structure of the *Tableau*

We can begin the analysis of the formal structure of the *Tableau's* classification by distinguishing the hierarchical ranks of its taxonomic units. At the bottom we have the rank of chemical substances or species. The table renders this species rank using just the substances' names.[28] The next higher rank unites species—for instance, salt species formed out of the same acid species with different species of salifiable bases (field V/b, lines 5 through 30)—and could therefore be named the rank of genera. This rank is marked in the table by horizontal lines as well as by corresponding numbers in the margins. Very often, however, the fields demarcated by these lines contain the name of only one species and could therefore also be determined as indicating the species rank. This ambiguity of taxonomic units, which are rendered by graphical means alone, is characteristic of the table. We also encounter it at the next higher rank, which is signified in the table by braces on the right hand side. This taxonomic rank unites genera such as the salt genera (field V/b) or the genera of alloys and amalgams (field VI/c), and could therefore be called the rank of orders. In many cases, however, it unites fields with only one species and thus appears rather to represent the rank of genera. Finally, some of these genera/orders are united in taxonomic units of yet a higher rank, represented by the tableau's columns. This rank could be named

---

25 "[…] substances simples combinées dans leur état naturel, & sans être oxigénées ou acidifiées […]." Guyton de Morveau et al. [1787] p. 97.
26 The very first substances listed in this class expose the problem encountered here: the authors of the *Méthode* regarded *orpiment* or *yellow arsenic* (trisulfide of arsenic) and *realgar* or *red arsenic* (disulfide of arsenic) as combinations of sulfur with *white arsenic* (trioxide of arsenic) instead of arsenic proper.
27 See Guyton de Morveau et al. [1787] p. 90.
28 As said before, in some cases the names of a genus of species are not spelled out but indicated collectively by giving an example: "Alliage de nickel" (field VI/c, line 35), for instance, stands for all binary alloys of this metal.

that of classes. However, because of the ambivalence of the taxonomic units they unite, these six columns are inevitably ambiguous themselves with respect to their precise taxonomic rank.

It is in conformity with such ambiguities that, in contrast to the custom in botany, eighteenth-century chemists usually did not care very much for elaborated and consistent denominations of classificatory units and ranks. Lavoisier, for example, used the terms class and genus promiscuously.[29]

Among the graphical means by which the table represents the taxonomic units and their relations, there is one by which it transgresses the ordinary classificatory structure of hierarchically nested taxonomic units. The table arranges the taxonomic units in the manner of a matrix. Because of this layout, all of the genera and species are related in a twofold way to other units of the same rank. Vertically, they are related to those taxonomic units that are of the same class—either the class of simple substances (column I) or one of the five classes of compounds (columns II–VI). Horizontally, the species are related to their compounds when they are simple substances.[30] Or, if they are compound substances, they are related to one of their components and at the same time to other compounds of this component. Thus the *Tableau* contains two different, but coordinated classificatory systems: vertically, a system that distributes genera of substances into either the class of simple substances or into one of the five classes of compounds and horizontally, one that orders compounds according to one simple substance that these compounds have in common.[31]

The horizontal structure of the table deserves particular attention, for it is this particular structure to which the nomenclature of the table is immediately linked. The determination of the names for substances takes place along this horizontal coordinate in such a way that, apart from a few exceptions, each horizontal line of a simple substance and its compounds is rendered, according to certain rules, as a linguistic line of derivations developed from the simple substance's name, as shown in the following example: *Soufre—Acide sulfurique/sulfureux—Gaz acide sulfureux—Sulfate/Sulfite de potasse—Sulfur de fer* (line 7).[32]

However, the invention and application of such rules presupposed classes of substances that were already distributed into the five principal types of compounds (column II–VI). This distribution, along with the establishment of these five classes of compounds, was neither a matter of linguistic distinctions or derivations nor of formal classificatory rules, but of classificatory considerations that rested on specifically chemical assumptions and convictions held by the authors of the *Méthode*. By turning to these convictions, we close the description and analysis of the table's formal structure and open the investigation of its conceptual framework.

---

29 "[…] le nom de classe ou de genre […]." Guyton de Morveau et al. [1787] p. 19.

30 Not for formal but for functional reasons, the table could not render these relations for the last two classes of simple substances. Otherwise the tableau would have lost its unambiguity, as the same compounds would have occurred at different places. In chapter 10 we will come back to the problem of overlapping classes in non-encaptic classifications like our *Tableau*.

31 In chapter 7 we will encounter another superposition of different structures in this tableau; this time, however, structures of not a formal, but a specific chemical character.

32 We will come back to this in chapter 10.

6

Classifying According to Chemical Composition

As shown in the previous chapter, the *Tableau* provided six classes for the genera of pure chemical substances: one for the genera of simple substances and five for the genera of different types of compounds. This most principal feature of its classificatory structure relied on two premises. It was first assumed that the chemical substances in question could be divided into either simple substances or compounds and second, that all chemical compounds were classified according to their chemical composition. Classification according to composition is, therefore, the most fundamental chemical principle embodied in the classificatory structure of the table.

Classification according to chemical composition was not the most natural way of classifying chemical substances. Actually, eighteenth-century chemists employed several different modes of classifying substances in plant and animal chemistry as well as in various fields of technological chemistry, and continued to do so even after Lavoisier's Chemical Revolution. As we have shown in part I, it was common in the eighteenth century to classify substances according to their provenance from the three kingdoms of nature. Another ubiquitous criterion for classifying was in accordance with a substance's perceptible properties—color, taste, consistency, combustibility, solubility and so on.[1] Under technological aspects, chemical substances were ordered according to the techniques employed in their preparation and their practical purposes, such as medical applications.

Classification according to chemical composition thus cannot be regarded as a natural choice. Moreover, as we have shown in part I, this specific taxonomic option did not exist in the seventeenth century, as it rested on historical preconditions that emerged gradually by the end of the seventeenth and the early eighteenth centuries. Among these historical preconditions the following three were the most significant: first, the existence of pure chemical substances along with reversible chemical operations in chemical practice; second, on the basis of the former, the development of the modern conceptual system of chemical compound and affinity; and third, linked to this, experimental strategies for establishing the composition of compounds that became known as the analytical method. This chapter will shed further light on these three historical preconditions of classification according to composition.

6.1 Pure chemical substances

The premise that all of the substances comprised by the *Tableau's* classificatory structure are either simple substances or compounds is not only an elementary presupposition for classification according to composition, it is at the same time the basic criterion for the selection of chemical substances that can be classified within this framework. The substances assembled in the tableau of the *Méthode* represent a highly selected group of substances compared to the archipelago of materials—

---

1   As we will see in the next chapter, classification according to properties also occurred in the 1787 table, though only as a complementary feature.

including raw minerals, organized parts of plants and animals, vegetable and animal juices and extracts, and composite commodities such as glass, soap, and porcelain—which late eighteenth-century chemists studied in their laboratories and at their writing desks.

For a historical understanding of this selection, its tacit character is particularly telling. In the text of the *Méthode* it was not mentioned, let alone clarified, that the classificatory structure of the table implies a selection of chemical substances. Its authors argued neither for such a selection nor for the underlying distinction of the selected chemical substances from others. They evidently felt no need to justify this selection and, as a matter of fact, none of their contemporaries asked them to do so. From these circumstances one can conclude that both the tacit selection and its tacit criterion were not introduced by them for the first time. Rather, they proposed a new nomenclature and classification for a selection of substances that had been singled out and well established as a particular subject matter of chemical investigation independent of and long before their classificatory attempt.

The particular field of chemical theory and practice to which their selection of substances pertained is plainly indicated by the chemical operations connecting the substances of the table's classes. Focusing on the classes that were well-established before the *Méthode*—the classes of the traditional salifiable bases (I/c–e), the acids (III/b), the metal oxides or metal calces (III/c), the salts (V/b–c), and the alloys (VI/c)—we find three basic types of chemical processes involved: first, those in acid solutions, mainly in connection with the production or decomposition of salts; second, calcination of metals and reduction of metal calces; and third, the combination of metals into alloys.

These processes have one feature in common that was of utmost significance for the development of the modern concept of chemical composition. In all of these operations compounds were decomposed into stable building blocks, either simple or more proximate components, and recomposed out of these components, either directly or via some intermediate steps. We argue that this unity of chemical analysis and resynthesis, along with the observation of a recurrent, stable set of components, is what distinguished these processes from those applied by chemists in plant and animal chemistry as well as in most areas of mineralogy at this time. And it is this reversible decomposition and recomposition that allowed the distinction of chemical compounds from other composite materials, quite independent of any theory of chemical elements and chemical bonds. Chemical compounds in the modern sense were substances that could not only be decomposed, but also recombined by chemical synthesis out of tractable ingredients in the laboratory.[2] The group of substances tacitly selected by the authors of the *Méthode* were chemical compounds and their simpler building blocks, which can be designated as "pure chemical substances."[3]

---

2   See also our arguments in chapter 2.
3   Porter [1981] used the term "laboratory materials" or "laboratory compounds" to designate these substances (pp. 567–568). Focusing on mineralogy in eighteenth-century Germany and Sweden, however, he recorded the fact that the authors of the *Méthode* restricted their enterprise to such pure laboratory substances without accounting for it.

Our notion of pure chemical substances is an analytical tool that highlights patterns in the chemical behavior of substances in chemists' experimental practice (see chapter 2). Substances such as metals, alkalis, acids, and earths were traceable in chemical transformations. In the making of salts and alloys, these traceable substances behaved like stable building blocks that were preserved in chemical transformations and could be recovered by decomposing the salts and the alloys. Similarly, the alloys and salts made from these ingredients could be decomposed into the original ingredients and recomposed from them. And further, these decompositions and recompositions seemed to be governed by elective forces or chemical affinities. However, the bulk of raw materials and processed substances which seventeenth- and eighteenth-century chemists studied in their laboratories did not display this kind of recurrent pattern of chemical behavior. Their chemical transformations were for the most part much more difficult to trace. And these materials were far more complex, as they yielded many products in chemical decompositions, rather than just two as did the paradigmatic classes of salts and alloys. The modern chemical concept of pure chemical compounds in the sense of stoichiometric compounds did not yet exist in the eighteenth century, not even in Lavoisier's chemistry. Yet the materiality of "pure substances" allowed chemists to trace chemical transformations, whereas in most other cases this was impossible. We argue that, at a time when the modern chemical concept of purity did not yet exist, there nevertheless existed a material purity that may be compared to the absence of background noise in experimental physics. It is this absence of background noise, or material purity, connected with a distinct recurrent pattern of chemical behavior—analysis and resynthesis—that we have in mind when we speak of "pure chemical substances" in eighteenth-century chemistry.

A few additional clarifications concerning our notion of pure chemical substances are in order. First, inasmuch as "pure" also means "single," our notion of pure chemical substances highlights another feature distinguishing the pure chemical substances from the vast majority of plant and animal materials studied by eighteenth-century chemists: the mode of individuation. Whereas in eighteenth-century plant and animal chemistry chemists drew the boundaries of single substances by referring to observable properties, both chemical and physical, the individuation of pure chemical substances was detemined by experimentation and the tracing of substances in experiments.[4] The single substances were the traceable ones that were preserved in chemical analyses as building blocks of chemical compounds or bound together into a single compound by chemical affinities. Reversible decompositions and recompositions and replacement reactions were also the experimental means by which chemists distinguished single substances from a mixture of several substances. Second, the laboratories in which the pure substances were handled were not exclusively those at academic institutions. The operations in question were also performed at several sites of practical artisanal chemistry. As we will discuss in the next chapters, most of them had been operations in artisanal contexts long before they were performed in aca-

---

4   See chapter 12.

demic chemical laboratories, and they also left academic institutions to be applied at more mundane sites. Likewise, the adjective "chemical" does not imply that only academic chemists studied these substances; apothecaries, assayers, and other groups of artisans did so as well. Third, the notion of pure chemical substance indicates a distinction between natural raw materials and chemically processed substances. Pure chemical substances are only rarely found in nature, but were chemical preparations manufactured by both craftsmen and academic chemists.

Eighteenth-century chemists were well aware of the artificial nature of pure chemical substances.[5] In the second half of the seventeenth century, when chemical operations in acid solutions became an important focus of chemical activities, many chemists differentiated, for example, the copper sulfate obtained in mines from the copper sulfate composed in the laboratory or workshop. The latter was understood as a chemical combination of its components, the former was not. More precisely, the modern conceptual system of chemical compound and affinity, which actually originated in the reflection upon this practice of composing and decomposing substances, was not smoothly applied to substances found in nature. The distinction between artificially composed and natural substances gradually faded away, however, as the analysis of the latter was routinely supplemented by the attempt to recompose these substances from their components.[6] With each successful resynthesis of such natural substances, chemists became more and more accustomed to understanding them in the same way as pure laboratory substances—as chemical compounds in the modern sense. Disbanding the distinction between nature and art for these substances, early eighteenth-century chemists at the same time separated the domain of pure chemical substances from those materials that could not be resynthesized. The criterion of natural origin remained significant for substances that could only be subjected to chemical analysis but not to resynthesis, but it became irrelevant for those that could be resynthesized in the laboratory and for the recurrent simpler building blocks of these substances. Accordingly, pure chemical substances were this latter group of more compound and simpler substances, defined regardless of whether a particular sample was actually produced in the laboratory or found in nature.

6.2 Composition

Classification according to chemical composition, as we find it in the *Méthode*, relied on a chemical concept of "composition" that was far from universal. The modern notion of chemical composition is a historical one. It was not familiar to seventeenth-century chemists, nor did it remain unchanged after its introduction around 1700. We argued in part I that at the turn of the seventeenth century, ongoing experiential investigations and changes in chemists' understanding of both chemical transformations and the constitution of substances coalesced in the modern conceptual network of

---

5   On the following, see Klein [1994a] part VI.
6   On an early attempt, see Homberg's and Geoffroy's analysis and resynthesis of sulfur (Klein [1994a] 25f.).

chemical compound and affinity.[7] In the latter conceptual system, the notion of chemical composition obtained a distinctive meaning: chemical compounds were composed of building blocks tied together by elective chemical affinities. Rather than philosophically defined entities—such as atoms, corpuscles, Aristotelian elements, and Paracelsian principles—chemists identified these building blocks as chemical substances, which in experiments could be both isolated and manipulated to recompose the respective chemical compound. There was no ontological difference between a compound and its components: both of them were ordinary chemical substances. Moreover, the simplicity of the building blocks and the compounded nature of chemical compounds was not an absolute, but rather a relative difference. For example, metals, the simple building blocks of alloys, could be decomposed by calcination into metal calces and phlogiston. Likewise, vitriolic acid was a simple building block of the neutral salt vitriol; nevertheless, by the mid-eighteenth century most chemists agreed that vitriolic acid was not an absolutely simple element. The adherents of the phlogistic system regarded vitriolic acid—along with other mineral acids, alkalis, earths, and metal calces—as a chemical substance that could not be decomposed in experiments. But they did not settle the question of how such stable experimental building blocks related to the hidden entities postulated by the chemical philosophy of elements and principles—or even left this question deliberately open. What counted most was the fact that substances such as vitriolic acid could be isolated out of chemical compounds and used as building blocks to resynthesize the original compound or to form new ones. And it was this very separability as isolated substances and their tractability as building blocks that rendered them chemical components in the modern sense. The development of the philosophical concept of principle or element, vis-à-vis "simple substance" in the sense of a chemical building block preserved in chemical reactions, deserves particular attention with respect to the changes in the mode of classification and ontological shifts in the *Méthode* and the Chemical Revolution more broadly.

As shown in part I,[8] elements and principles in the Paracelsian tradition were conceptualized as generators rather than physical components of *mixta*, the perceptible substances chemists dealt with.[9] In most sixteenth- and seventeenth-century chemical philosophies—especially those not combined with corpuscular theories—principles were defined as *semina* invested with a set of qualities that generated perceptible substances in co-action with the elemental matrix. Both elements and principles were constituting agents, which engendered corporeality, perceptible properties, and virtues of natural bodies. In the seventeenth century, French chemists such as Etienne de Clave, Nicaise Le Febvre, Christopher Glaser, and Nicolas Lemery supposed five elemental constituents or principles of natural bodies, namely water or *phlegma*, an acid

---

7   See chapter 2; see also Klein [1994a] part VI.
8   See chapter 2, where we not only address exemplary seventeenth-century theories of chemical principles, such as Le Febvre's, Boyle's, and van Helmont's, but also enter into a critical discussion of recent secondary literature on this topic.
9   It is important not to mistake these *mixta* with our mixtures. *Mixta* were homogeneous substances and were strictly distinguished from mere aggregations of materials.

spirit or mercury, an inflammable sulfur or oil, a fixed salt, and an earth.[10] *Mixta*, the perceptible substances chemists dealt with, were understood as marked by but not consisting of such principles, if "to consist of" means to be composed of pre-formed building blocks that can be separated and isolated as manipulable pure chemical substances.[11] Furthermore, Parcelsians understood all mixts as being constituted by all three or five principles. They differed, therefore, not in terms of the set of principles that form them, but in terms of the proportions or different states of equilibrium or domination between the constituents.[12]

As a result of a general decline in organicistic and Aristotelian theories of matter in the course of the seventeenth century and of a coupling of corpuscular theories with the philosophy of principles, the philosophy of elements underwent fundamental changes, though only gradually and in the form of slight modifications. Because of this gradual character of the conceptual revisions, it is almost impossible to tell without thorough analysis of the specific case at hand what exactly an author of the late seventeenth or the early eighteenth century actually meant when speaking of principles and mixts. By the middle of the eighteenth century, most chemists questioned the assumption that every natural mixt was constituted by all of the five principles. Rather, classes of substances were thought to differ with regard to the principles by which they were constituted. The very notion of natural mixts was affected more and more by these conceptual changes. Around 1750 the majority of chemists regarded the ultimate principles as a kind of physical component of mixts, the fact that they still associated them with features of the old Paracelsian elements and principles notwithstanding.[13]

Alongside these changes, the qualitative notion of the dominance of constituting principles slowly shifted to a more quantitative one, which derived the differences among mixts from the different proportions of their constituents. Eighteenth-century chemists became more and more accustomed to the idea that perceptible substances contained different amounts of principles. In their theories they began to distinguish between natural mixts according to estimations of the particular ratios of the constituent principles each of them presumably contained. As a consequence, classifications constructed in the framework of eighteenth-century philosophies of elements and principles may superficially resemble classifications constructed using resources

10 Eighteenth-century chemists often enlarged this list in order to adjust the chemical philosophy of principles with accumulated experimental experience. Boerhaave, for example, listed nine principles — see Klein [2003b]. On the French chemical textbook tradition and the theory of five chemical principles presented in them, see Debus [1991]; Kim [2001], and [2003] pp. 17–62; Klein [1994a] pp. 36–56, and [1998b]; and Metzger [1923].

11 It is important to note that in the Paracelsian tradition, the term "separation" meant a creative act of reconstituting the original seminal principles. See Le Febvre [1664] pp. 17–20.

12 The term "proportion" is not quite appropriate in this context, as Paracelsian chemists did not regard principles as pre-formed physical parts.

13 Even the group of the most simple substances in the table from 1787 (field I/a) may be reminiscent of the elements and principles of the old understanding. (We will come back to this group in chapter 8 in connection with phlogiston.) However, the difference between the simple substances in the fields I/b–e and elements and principles in the traditional sense is not "only a matter of degree" (Cassebaum and Kauffman [1976] p. 456) but one "of principle." The point missed by Cassebaum and Kauffman is that the simple substances in the table are pure chemical substances, which cannot be said of the former elements and principles.

derived from the modern conceptual network of chemical compound and affinity. As it is easy to confuse these two modes of classification, the excursus included in the next section will analyze three examples of eighteenth-century classification according to elements and principles.

6.3 The arduous career of the analytical method

The ambiguities in mid-eighteenth-century chemists' understanding of principles also concerned the crucial question of whether or not such principles could be separated from natural mixts as isolated laboratory substances; that is, whether or not they could be regarded as chemical components in the modern understanding. As we will see in the next section, eighteenth-century chemists were not sure whether their inability to separate the ultimate principles in a pure state was due to the nature of ultimate principles or to the still-imperfect state of chemical art. With this, we arrive at the third historical precondition of the modern concept of chemical composition and classification according to composition; namely, suitable experimental procedures, methods, and standards for the acquisition of certain knowledge of a compound's composition.

Back in the early eighteenth century, chemists were already able to identify and classify several groups of substances—namely salts, alloys, and watery solutions—based on their knowledge of composition. This is apparent in the famous affinity tables, the first of which appeared in 1718. In these cases, chemists' assumptions about substances' composition rested on the analytical method, meaning the experimental separation of a compound's components, complemented by its recomposition from these components. Chemical operations in practical fields such as salt production and metallurgy, which comprised separation and recombination of components, constituted the background of these achievements, as will be discussed in detail in chapter 8. These achievements appear paradigmatic for the acquisition of certain knowledge of a compound's composition, not only in hindsigh but also in the eyes of the historical actors.

There is clear evidence that the method of analysis complemented by resynthesis was not only applied early in the eighteenth century, but also regarded as the most reliable method for the acquisition of knowledge of composition. It may suffice to quote a few prominent chemists. In 1732 the famous Dutch chemist Herman Boerhaave (1668–1738) prescribed the method for the acquisition of experimental knowledge about the composition of bodies and materials as follows:

> The effect produced should rather consist in a *separation of the parts, such as they were originally,* than be changed by the operation: for as bodies, by this art, are resolved either into their natural parts, or parts acquired in the operation, it is plain that the first alone should here be obtained; *so that by re-adjusting the same parts, we may be certain of producing the original subject again.*[14]

---

14 Boerhaave [1753] vol. II p. 3; our emphasis.

Knowledge of chemical composition, Boerhaave stated here, was acquired first through the experimental separation of the parts. Second, the chemical parts had to be separated "as they originally were," without being simultaneously transformed. Third, the original substance also had to be reproduced from the separated parts in order to confirm that the parts had not been altered during the process of separation. Such alterations took place easily, as the fire or reagents used in chemical separation sometimes interacted with the parts and thus transformed them into new substances.

Similar voices were heard at the middle of the eighteenth century. In 1749, in the introduction of his *Elemens de Chymie théorique*, Pierre Joseph Macquer highlighted the analytical method when he defined a principal goal of chemistry as follows:

> to separate the different substances which enter into the composition of a body [...] and again rejoin them to make the first mixt reappear with all its properties.[15]

And in 1753, Gabriel François Venel proclaimed in his article *Chymie*, contained in Diderot's *Encyclopédie*:

> If the chemist succeeds in combining one by one all the principles that he has isolated and in recomposing the body that he has analyzed, then he achieves the equivalent of chemical demonstration.[16]

However, despite unmistakable statements like these, the ideal analytical method proved to be less successful or even inapplicable with respect to substances other than salts, alloys, or watery solutions. For a historical understanding of the development of the analytical method these less successful applications are of particular interest. They illuminate the delicate context in which this method became accepted as the standard procedure for procuring certain knowledge of composition.

Plant and animal substances in particular resisted the analytical ideal of analysis and resynthesis. At the beginning of the nineteenth century, Louis Jacques Thenard reported an episode which demonstrates that analysis followed by resynthesis was a broadly accepted chemical proof in the French Enlightenment discourse, which, however, did not work in the case of substances extracted from plants: "While taking a course in chemistry with G. F. Rouelle, Jean-Jacques Rousseau said that he did not believe in the analysis of flour, so long as he did not see that the chemists can remake it."[17] Of course, neither Guillaume François Rouelle nor Thenard were able to remake flour from its components.[18] As we will discuss in depth in part III, in the second half of the eighteenth century chemists analyzed entire plants and the roots, leaves, flowers, and other organs of plants, no longer by dry distillation but by solvent extractions and other techniques allowing them to separate the more compound or

---

15   Macquer [1749] p. 1.
16   Diderot and d'Alembert [1966] vol. 3 p. 418.
17   Thenard [1813–1816] vol. 3 p. 3f.
18   Thenard reminded his audience of this episode only to immediately rebuff Rousseau. "This philosopher [Rousseau] would undoubtedly speak differently today," he remarked, "since we can explain the causes of the failure to re-synthesize flour and other organic materials" (ibid.). In the second decade of the nineteenth century, chemists had broadly accepted the Lavoisierian theory of the composition of organic compounds, which also offered an explanation for why it was impossible to resynthesize plant materials from their analytical components (see part III).

"proximate principles" of plants rather than their most simple elements or principles. Although chemists were unable to resynthesize plants and their organs from their analytical components, the new techniques allowed them to repeat analyses of the same kind of plant in such a way that the experiments yielded more or less identical analytical products. By the end of the century, the repeatability of analysis reassured chemists of their analytical success in plant analysis.[19]

However, chemists were far less successful when they attempted to further analyze the extracted plant substances in order to establish knowledge about their elemental composition. Not only flour, which aroused Rousseau's objections, but all kinds of substances separated from plants or animals posed difficulties to elemental analysis. The pre-Lavoisierian chemists were well aware of these difficulties and of the uncertainty of their knowledge about the elemental composition of plant and animal materials. When they submitted different kinds of plant and animal materials to dry distillation—which was the standard analytical method for elemental analysis prior to Lavoisier's introduction of the combustion method—they always obtained very similar analytical products, such as empyreumatic oils, volatile acids, insipid waters, fixed salts, and earths. In 1749, Pierre Joseph Macquer described the problems with respect to the vegetable and animal oils. Based on elemental analysis by dry distillation, Macquer claimed that oils were composed of phlogiston, water, an acid, and a certain quantity of earth. Yet he also added that there may have been other principles contained in oils that had escaped his attention.[20] The sole means that would have allowed chemists to be reassured about their analytical results—the resynthesis of the original compound from the analytical products—was barred in plant and animal chemistry as Macquer conceded:

> there is no invariable (*bien constante*) and well-known (*bien avérée*) experiment which proves that one can produce oil by combining together solely these [four] principles; these sorts of re-compositions are nonetheless the unique means we possess to assure ourselves that we know all the principles which enter into the composition of a body.[21]

According to Macquer, not only the quantity of the components isolated in the chemical analysis of plant and animal materials varied when the analysis was repeated, but even their quality and number.

As we will see in the following section, plant and animal materials were not the only troublesome objects in eighteenth-century chemists' investigations of chemical composition. The ultimate elements or principles postulated by chemical philosophy were also problematic. When chemists attempted to study the composition of substances from their most simple ultimate principles or elements, they encountered obstacles to the actual separation of these hidden entities. Although by the middle of the eighteenth century most chemists no longer regarded the simple principles of mixts as incorporeal semina, but rather as bodily component parts of compound chemicals, the ultimate principles often appeared to be as inseparable as the seminal

---

19 Largely the same can be stated with respect to the analysis of animal tissues by means of extraction.
20 Macquer [1749] p. 179.
21 Ibid. p. 179f.

principles of the Paracelsians. Despite such experimental difficulties, not all chemists refrained from attempts to classify substances according to composition from the ultimate principles. These attempts deserve attention in the context of our investigation, as they may be easily confused with classifications based on that modern conceptual network of compound and affinity and the modern analytical method embodied in affinity tables and the *Méthode*. Moreover, these exceptional classifications, the three most important of which we will analyze in the following excursus, seem characteristic of the precarious state of the analytical method in the first half of the eighteenth century.

Excursus: Ambiguous classifications by Homberg, Stahl, and Macquer

In the first of a series of articles published between 1702 and 1709,[22] Wilhelm *Homberg* (1652–1715), then a leading figure among the chemists at the Parisian Academy of Sciences,[23] raised doubts about the possibility of isolating all of the ultimate chemical principles. Like his French predecessors, Homberg assumed five principles—salt, sulfur, mercury, water, and earth—to be the ultimate constituents of substances. Yet he stated that two of these principles, sulfur and salt, were always joined with one of the three other principles: "The principles of sulfur and salt are perceptible by us only when they are joined with some of the other three principles which serve as their vehicle."[24] In other words, Homberg believed it impossible to separate the ultimate principles sulfur and salt in a pure form and to study their properties in an experimental way.

In the same article, Homberg provided a short overview of the substances known at the time that could be analyzed by means of the fire (*par le moyen du feu*); that is, through distillation or combustion. This survey represents a rough classification of chemical substances, whose first paragraph at first glance even seems to suggest a classification according to composition.[25] Like most chemists of his time, at the highest taxonomic rank Homberg distinguished between mineral and vegetable substances, including animal materials in the latter. As a kind of corroboration of this traditional, naturalistic distinction, he added considerations about composition from the ultimate principles. Not all substances, he asserted, can be reduced into the same

---

22 Homberg [1702] pp. 33–36. See also Di Meo [1984]. Although Di Meo provided an elaborate analysis of Homberg's essay, emphasizing that scientific theory and practice form the *intermédiaire* between *le nom* and *la chose* (p. 160), he neglected Homberg's practice of determining principles. He thus misses the character of Homberg's classification. Kim [2003] p. 84ff. and Principe [2001] p. 538ff. discuss Homberg's important 1702 essay under different aspects.

23 For life and work of Homberg, see Partington [1961–1970] vol. III p. 42ff.; Gillispie [1970–1980] vol. VI p. 477ff.

24 "Le souffre & le sel principes ne sçauroient paroître à nos yeux sans être joints à quelques unes des trois autres principes qui leur servent de vehicule." Homberg [1702] p. 34.

25 In this paragraph, Homberg defined chemistry as the art that reduces composed bodies to their principles by means of fire and composes anew bodies in fire through the mixture of different matters: "l'art de reduire les corps composez en leurs principes par le moyen du feu, & de composer de nouveaux corps dans le feu par le mêlange de differentes matieres" (Homberg [1702] p. 33). It would, however, be mistaken to read this definition as indicating the unity of analysis and resynthesis. The bodies "composed anew by fire" are not necessarily the same ones that were "reduced" in the first place.

principles.[26] Whereas in mineral substances, one finds all five principles—salt (*le sel*), sulfur (*le soufre*), mercury (*le mercure*), water, and earth (*l'eau & la terre*), in vegetable substances only four principles occur since the mercury principle is never found in them.[27]

With respect to the next lower rank, Homberg claimed to classify the substances according to different combinations of their ultimate principles: "the different combinations of these five matters among each, or some of them, make up the great variety of all bodies."[28] Accordingly, he distinguished four classes of mineral substances: 1) metals containing mercury (*du mercure*), a sulfurous matter (*une matière sulphureuse*), an earthy matter (*une matière terreuse*), and sometimes also a saline matter (*une matière saline*); 2) fossil salts containing acids that contain some sulfurous matter themselves (*quelque matière sulphureuse*), a little fixed salt (*peu de sel fixe*), and a little earth (*un peu de terre*); 3) stones containing only a lot of earth (*beaucoup de terre*) combined with a little sulfurous vapor (*un peu de vapeur sulfureuse*); and 4) earths (*terres*) containing an acid salt (*du sel acide*), sometimes a little fixed salt (*un peu de sel fixe*), and a little sulfurous matter (*un peu de matière sulfureuse*). The vegetable substances, however, could not be divided in this way since Homberg assumed that all of them contained the same principles: salt, water, earth, and a sulfurous matter.[29]

This is a puzzling classification, indeed. On the one hand, Homberg had explicitly stated that it was impossible to isolate in a pure form all of the five ultimate principles. But on the other hand, he also seems to claim that knowledge about composition from the ultimate principles informed his classification. How could these two obviously conflicting claims be reconciled? It is necessary to read carefully what Homberg really stated. He did not state that his knowledge of the substances' composition from ultimate principles was based on the actual isolations of these principles. Rather, he spoke of sulfuric matter (*matière sulfureuse*), salin matter (*matière saline*), and so on. These "matières" were indeed separated experimentally. But they were not identical to the pure principles. Rather, they actually were mixts supposed to consist *mainly* of the principle at stake and hence representing it in the purest form that chemical art could achieve. Obviously Homberg thought it justified to draw conclusions from the actually separated substances to the pure, but still hidden, ultimate principles.

In the context of our investigation it must be emphasized that Homberg's coupling of classification and determination of composition from ultimate principles did not actually engender a new mode of classification. Rather, traditional classes were pre-

---

26 *"Tous les corps* [...] *ne se reduisent pas dans les mêmes principles."* Homberg [1702] p. 34.

27 On closer inspection, the explicit criterion of this distinction proves to be questionable since the mercury principle was also lacking in three of the four classes of mineral substances that Homberg distinguished next. Thus, one may wonder why he did not divide the realm of chemical substances into those that contain the mercury principle—the metals—, and those that do not—the other minerals as well as the vegetable substances. One may conclude from this that classification according to provenance formed the unshakable basis for this division.

28 "[...] les differentes combinaisons de ces cinq matieres, ou de quelques unes d'entr'elles, font la grande variete de tous les corps." Homberg [1702] p. 34.

29 Ibid. p. 35.

served and merely superimposed upon the philosophy of principles. These traditional classes were ordered according to the natural origin and observable properties of the substances rather than to composition. All of Homberg's classes of mineral substances actually rested on their natural origin and observable physical and chemical properties. The upshot is that he followed and preserved the naturalistic method of ordering materials but presented it in a new, philosophical way.

The famous German chemist Georg Ernst Stahl (1660–1734) had even more qualms than Homberg about the experimental separation of ultimate principles.[30] A "pure, natural resolution" of bodies into principles, he proclaimed in his *Fundamenta Chymiae* (1723), is "not easily obtainable from the Chemistry of these days, and so can hardly be come at by Art."[31] Only "very rarely," he added, are principles found in a state of purity.[32] Nevertheless, in the same textbook Stahl implemented assumptions about composition from the ultimate principles when he distinguished four main classes of substances, which structured the practical part of this work: the classes of salts, waters, sulfurs, and earths.[33] Stahl described and classified only "mixts" and simple "compounds," substances defined in the context of his theory of a graduated order of composition as comparatively simple compounds, but not the more compounded raw minerals, plants, and animals.[34] But even if his classification was not comprehensive, it aimed at a more philosophical mode of ordering materials than the common naturalistic classifications.[35] Stahl claimed explicitly that his classification followed an "analytical method," and that each chemical species must "be considered according to its essence, and component parts."[36] Here "essence" meant essential or chemical properties, and "component parts" meant the most simple ultimate principles.

It may suffice to scrutinize this taxonomic approach in one case, that of the class of salts. Stahl's class of salts comprised the acids, alkalis, neutral salts, and sugar.[37] From his descriptions of these groups of salts, it can be taken that he grouped together substances that had an acid, alkaline, sweet, or salty taste; were corrosive; were soluble in earths; and could be used as a solvent.[38] Accordingly, right at the beginning of his discussion his definition of the class of salts was based on percepti-

30 For biographical information on Stahl, see Partington [1961–1970] vol. II p. 653ff. and Gillispie [1970–1980] vol. XII p. 599ff.

31 We quote from the English translation Stahl [1730] p. 4.

32 Ibid. p. 12. See also Stahl's *Zymotechnia fundamentalis*: "[es ist] bißhero unmöglich gewesen, die anfänglichen Grundwesenheiten der natürlichen Dinge, eintzeln und einfach, entweder irgendwo zu finden, oder aus ihren Mischungen zu scheiden, und dieselbe gantz bloß und rein, ausser den Stand ihrer würklichen Verbindung mit andern Dingen auf einigerley Weise darzustellen." We quote from the 1748 German edition of Stahl's *Zymotechnia fundamentalis*—Stahl [1748] p. 66.

33 As the title, *Fundamenta Chymiae*, expresses, theoretical points of view dominated in this chemical textbook. In other writings, Stahl classified according to natural origin and properties. For example, pharmaceutical properties are a classificatory criterion in Stahl's *Fundemantum pharmaciae chymicae* of 1728.

34 For this theory, see chapter 15.

35 But within the overarching frame of these four classes, Stahl also considered the natural origin of the substances from the three natural kingdoms.

36 Stahl [1730] p. 76.

37 With the exception of sugar, Stahl followed the understanding of salts prevalent at the time, which will be discussed more comprehensively in chapter 9.

ble properties: "Salts in their own essence are of a fluid consistence, and of a mean degree of volatility; corrosive, and soluble in earths."[39] But at the same time he also claimed that all salts were mixts "consisting of Earth and Water, both of them in their simple, genuine state, intimately combined together."[40] How did Stahl evince the latter claim, which highlighted composition from the ultimate principles as a taxonomic criterion? "In order to discover the nature and essence of Salts,"[41] Stahl described a series of experiments with concentrated nitric acid, or *aqua fortis,* and several kinds of metals, lime or crab's eyes (consisting of calcareous earth), and various kinds of alkalis, which formed salts toghether with acids. The experiments were well-known as an experiential base for the concept of elective chemical affinities, and hence had entered E. F. Geoffroy's 1718 table of chemical rapports.[42] Stahl concluded from his experiments that metals, earths, and alkalis rendered the acid "more gross and dense, by the admixture of terrestrial Aggregates," thereby creating the solid consistence of salts.[43] As all of these terrestrial aggregates contained the ultimate principle earth, and as the liquid acid contained the principle of water, which caused its liquidity, therefore salts consisted of earth and water.[44] The different species of salts of one acid differed in their content of earth.

The example shows that Stahl's "analytical method" differed greatly from the analytical ideal of analysis followed by resynthesis. He deduced the presence of ultimate principles from his observations of changes of properties of compound substances in chemical reactions. Like Homberg's classification, his classification according to "component parts" rested not on actual investigations of isolated principles, but on interpretations of experiments with ordinary substances in light of the philosophy of principles.

Stahl's "analytical method" of classification according to ultimate principles did not meet with much acclamation by German or European chemists. Only few chemists actually adopted his method of classifying substances according to their ultimate principles, and they did so only in very specific theoretical contexts. Among these chemists was Pierre Joseph Macquer (1718–1784), who became a influential chemist in France in the 1750s and later in all of Europe through his *Dictionnaire de chimie* of 1766.[45] In the context of his theoretical chemistry, Macquer largely accepted Stahl's four major classes, with modifications that resulted in the four main classes of salts,

38 What was shown for the salts can also be observed in Stahl's treatment of the other three main classes. He ordered under the label "sulfurs" inflammable and colored materials such as ordinary sulfur, compounds of sulfur, vegetable and animal oils, pitch, rosins, and gums. "Waters" were fluid and insipid substances such as spring water, rain, dew, mineral and other subterranean waters, and distilled waters, as well as mercury. The class of "earths" assembled solid substances that were not soluble in water and were "fixed" (that is, resistant to fire), such as chemically isolated earths and metals, as well as vegetable and animal gums and mucilage. All of these distinctions followed by and large the prevailing demarcations of chemicals by eighteenth-century chemists, as we will see in chapter 9.
39 Stahl [1730] p. 77. The phrase "in their own essence" meant in their purest form. As becomes clear from Stahl's subsequent discussion, in this he referred to liquid acids, which he considered to be the purest group of salts.
40 Ibid. p. 77.
41 Ibid. p. 78.
42 See columns 1 and 3 in Geoffroy's table, in Geoffroy [1718].
43 Stahl [1730] p. 78.
44 Stahl did not present an argument for why the purest kind of salts, liquid acids, contained earth.

calces, metals, and oils.[46] Like Stahl, Macquer combined hypotheses taken from his philosophy of ultimate principles with observations of perceptible properties of ordinary substances. For instance, Macquer, too, claimed that the substances belonging to the class of salts consisted of the two simple principles water and earth, which differed quantitatively in the three different genera of salts—that is, the mineral acids,[47] fixed alkalis, and neutral salts.[48] And like Stahl, Macquer also corroborated this view by referring to the chemical properties of salts. All salts dissolved in water, and several salts combined with lime, sand, or other kinds of perceptible materials designated as "earths."

However, the experiments to which Macquer referred were not unambiguous. First, they demonstrated that salts did not always combine with all kinds of earths.[49] Second, and even more importantly, Macquer was well aware that the perceptible materials designated "water" and "earth" and employed in experiments were not identical with the simple principles designated "water" and "earth." The connection between the perceptible experimental materials and the ultimate principles was established only by theoretical considerations, which connected the philosophy of principles with the concept of chemical affinity and the observation of chemical properties of substances. Macquer assumed that on the deeper, hidden level of ultimate principles, the chemical affinities displayed between two different perceptible substances were affinities between the same kind of principles. He postulated the "general law" that "all similar substances have mutual affinity among each other and consequently are disposed to unite, like water with water and earth with earth."[50] Since according to his philosophy of principles salts consisted of the ultimate principles water and earth, and since the perceptible substances water and earth also contained these ultimate principles in abundance, it followed from his "general law" that "all saline substances must have affinity with earth and with water, and be able to combine and unify with one or the other of these principles."[51] Based on this theory, Macquer then presented the following tautological argument: the reason why salts dissolved in water and combined with earths to form more compounded substances was that salts consisted of the ultimate principles water and earth. Macquer was fully aware of the precarious state of his reasoning. At the end of his chapter on the class of salts he conceded: "To tell the truth, chemists have not yet succeeded in producing a saline material by combining an earth and water."[52]

---

45 For life and work of Macquer, see Partington [1961–1970] vol. III p. 80ff. and Gillispie [1970–1980] vol. VIII p. 618ff.

46 In this classification Stahl's class of waters is entirely dismissed; calces and metals, which in Stahl's system belonged to one and the same class, that of earths, constitute two different classes, and the classes of salts and oils overlap with Stahl's classes of salts and sulfurs.

47 Vegetable acids and salts were not included as they were considered more compound mixts.

48 Macquer [1749] pp. 23–24.

49 Acids combined only with absorbent earths, such as lime, but not with vitrifiable earths such as sand, whereas alkalis combined only with vitrifiable earths into glass. Ibid. pp. 28–32.

50 "Toutes les substances semblables ont d l'affinité ensemble, & sont par conséquent disposées à se joindre, comme l'eau à l'eau, la terre à la terre." Ibid. p. 20.

51 Ibid. p. 23.

52 Ibid. p. 26: "à la vérité, les Chymistes n'ont pu jusqu'à présent parvenir à produire une matiere saline en combinant ensemble la terre & l'eau."

By means of his theory of affinity, Macquer established a link between experiments and observations concerning the elective combination and solubility of substances and the philosophy of ultimate principles. This sophisticated move accords with the fact that Macquer was one of the chemists who made E. F. Geoffroy's table of affinities prominent in the European chemical community. Like his predecessors Homberg and Stahl, it enabled Macquer to classify substances according to their composition from simple principles. But as our analysis has shown, we must be cautious about the methods upon which such classification rested.

Subjecting ultimate principles to the analytical method

We thus conclude that in the first half of the eighteenth century elemental analysis and classification based on ultimate elements or principles of substances did not correspond with chemists' methodological imperative to prove knowledge of a substance's composition by resynthesizing it from its analytical parts. Moreover, as many leading chemists conceded that chemical art was not actually able to separate the ultimate chemical principles in a pure form, statements about composition from ultimate principles could not rely on the experimental examination of isolated simple principles. Rather, when chemists undertook experiments in this context, they observed physical and chemical properties of substances that they held to be compounds and not simple substances. Then, from properties such as color, solubility in water or combustibility, they inferred the existence of certain kinds of ultimate principles that caused these properties. Thus, knowledge claims about composition from ultimate principles and, consequently, classification according to such principles were embedded in experiments, but did not meet the collectively accepted standards of experimental proofs developed in those fields of investigation mapped by the tables of chemical affinity.

Therefore, it is of extraordinary importance to realize the difference between statements about composition from ultimate principles and statements about chemical compounds and simple components listed in chemical affinity tables and the *Tableau* of 1787. Ultimate principles differed from perceptible chemical substances chiefly with respect to two peculiarities betraying their descent from the ancient Paracelsian principles: first, they were thought to be recognizable only indirectly by the properties of the compounds in which they were contained; second, they could not actually be separated as isolated perceptible substances. In other words, they were not proven components according to the standards of the analytical method. It was not until the last quarter of the eighteenth century that chemists drew consequences and rejected this kind of chemical analysis. In 1782, Fourcroy explicitly dismissed the claim that components can be recognized by a compound's properties:

> Stahl and his partisans have alleged that compounds always participate in the properties of the bodies that compose them, and that the compounds bear the mean of the properties of their principles. They have even pushed this idea to the point of believing that it is possible to determine from the properties of the compound the nature of those substances that entered into its composition.[53]

At about the same time, Torbern Bergman proclaimed that the analytical method was
the only method that could be trusted when establishing components of compounds:

> In investigating the principles of a body, we must not judge of them by a slight agree-
> ment with other known bodies, but they must be separated directly by analysis, and that
> analysis confirmed by synthesis. It is well known that bodies by composition acquire
> new properties which did not appear in any of the component parts separately, and on the
> contrary, some of the original properties decay, or even disappear; hence, it appears how
> little analogy is to be trusted to.[54]

In closing our investigation of the arduous career of the analytical method, a con-
curring theoretical development should be addressed briefly. In his *Fundamenta
Chymia* of 1723, Stahl had proposed another theory that was much better received in
the European chemical community than his "analytical method," namely the theory
of a graduated order of composition.[55] Stahl distinguished between "mixts" or "primary mixts," which were composed immediately of ultimate principles, and "secondary mixts" or "compounds," which were composed of mixts and were resolved in
chemical analysis not into the ultimate principles, but into compound parts or mixts.
"All the darkness and disputes about Principles," Stahl stated, "arise from a neglect of
that real distinction between original and secondary Mixts, or Mixts consisting of
Principles and Bodies compounded of Mixts."[56]

Stahl's theory of a graduated order of composition was a major theoretical transition in eighteenth-century chemistry. It proved to be particularly significant for plant
and animal chemistry, as we will discuss in detail in part III. With respect to the issue
of composition, this theory was congruent with experiments in both mineral, plant
and animal chemistry. Moreover, it allowed a partial employment of the analytical
method in cases where determinations of ultimate principles failed. It should be
noted, however, that the spread of this theory did not imply that eighteenth-century
chemists unambiguously discarded the search for ultimate principles in favor of
explorations of the more compound proximate components. The adoption of this theory was perhaps indicative of a general pragmatic shift in focus from philosophical
principles to components that could actually be isolated as pure chemical substances;[57] but if so, this was not a simple linear process.

---

53 Fourcroy [1782] vol. I p. xlix; English translation by Porter [1981] p. 546.
54 Bergman [1784] vol. I p. xxvf.
55 On this theory, see chapter 13.
56 Stahl [1730] p. 5; Stahl also distinguished "mixts" and "compounds" from mere "aggregates" made
    from several matters, mixts, or principles, by drawing on atomistic ideas. In mixts and compounds, the
    atoms of the different components unite into coherent or individual "compound atoms," whereas, in
    aggregates, the atoms of the different materials do not form unified atoms. "An *Aggregate* is
    distinguish'd from an *Atom*, in that an *Atom* is one numerical individual; but an *Aggregate* several *Atoms*
    combined together by contiguity" (ibid. p. 11).
57 Holmes argued that an increase in chemists' pragmatic attitude may have been the background of the
    transformation process from the Paracelsian chemistry of principles to the modern chemistry of
    compounds. Holmes [1989] pp. 36f. and 40f.

The rigorous application of the analytical method to principles

Indeed, the search for ultimate principles did not cease in the last decades of the eighteenth century. As is shown by the 1787 *Tableau's* class of substances that come closest to the state of simplicity (field I/a), the issue of such principles still occupied the four authors of the *Méthode*. The real change that took place during these last decades was the rigorous application of the analytical method to such principles as well. The "Discours préliminaire" of Lavoisier's *Traité* of 1789 is doubtlessly the most famous document on the subject:

> All that can be said upon the number and nature of elements is, in my opinion, confined to discussions entirely of a metaphysical nature. The subject only furnishes us with indefinite problems, which may be solved in a thousand different ways, not one of which, in all probability, is consistent with nature. I shall therefore only add upon this subject, that if, by the term *elements*, we mean to express those simple and indivisible atoms of which matter is composed, it is extremely probable we know nothing at all about them; but, if we apply the term *elements*, or *principles of bodies*, to express our idea of the last point which analysis is capable of reaching, we must admit, as elements, all the substances into which we are capable, by any means, to reduce bodies by decomposition. Not that we are entitled to affirm that these substances we consider as simple may not be compounded of two, or even a greater number of principles; but, since these principles cannot be separated, or rather since we have not hitherto discovered the means of separating them, they act with regard to us as simple substances, and we ought never to suppose them compounded until experiment and observation proved them to be so. [58]

In this statement, Lavoisier cut the bond between the old search for ultimate elements or principles and the chemical analysis that had been developing alongside that search for many decades. As we have seen above, that bond had been further elaborated and refined, especially by G. E. Stahl and P. J. Macquer. By contrast, Lavoiser proclaimed that it was metaphysical ballast, which caused endless problems. One of his main achievements, which may justify to some extent the claim that his chemistry was revolutionary, was the rigid destruction of the many sophisticated links his predecessors had created between experimental analysis and its perceptible analytical products, on the one hand, and theories of matter such as the philosophy of principles and atomism, on the other. Lavoisier's definition of elements or principles as substances which cannot be further decomposed by chemical analysis came as a postulate: we must not take elements to be more than substances that can actually be isolated from more compound substances in the laboratory; and we must not speculate about the possibility of further decomposing substances as long as we cannot achieve that decomposition in practice. This definition of "element" was relative, that is, it depended on the available tools and techniques of chemical analysis. Lavoisier did not argue theoretically for his notion of element, and he did not exclude the idea that simpler elements existed than the ones hitherto isolated by chemical art. Therefore he substituted the term "simple substance" for the ancient term "element." In so doing he left open some space for theoretical speculation about the proper ultimate

---

58   See Lavoisier [1965] p. xxiv.

elements The four simple substances which "come near to the state of simplicity," and which therefore are demarcated in the tableau of the *Méthode*—the matter of light, calorique, oxygen, and hydrogen—make manifest that open theoretical space

7

Simple Substances and Paradigmatic Syntheses

The principle of classifying according to composition we have discussed so far was the most fundamental and, in view of further developments of chemistry, without a doubt the most significant chemical principle of the *Tableau's* classification. But it was at the same time too general a principle to provide all of the criteria needed to construct a concrete classification of the pure chemical substances of the age. Additional criteria of independent origin were actually applied by the authors of the *Méthode*. This is obvious in the case of simple substances. By implication, such substances could not be distinguished and classified according to composition. Although the notion of simple substances was inherent in the *Tableau's* classification according to composition—which highlighted composition from the simple substances rather than from compound proximate components—the ordering of simple substances required further taxonomic principles. However, such additional criteria were needed not only for the subdivision of simple substances. For the construction of taxonomic units of compounds, too, the authors of the *Tableau* employed chemical principles and conceptions that could not be derived from the fundamental principle of classifying according to composition. In this short chapter we will investigate these additional chemical principles and conceptions that shaped the classification of the *Méthode*.[1] We begin with those used for the table's classification of simple substances.

7.1 Classification of simple substances

When looking for the criteria that governed the table's distinction of different classes of simple substances, we encounter one criterion that can be regarded as a derivative of the general principle of classifying according to composition, for the distinction of the class of most simple substances (field I/a) from the other classes of simple substances (I/b–e) follows this principle in a way. Apart from the separation of the four simplest substances, the authors of the tableau classified the simple substances according to their perceptible, mostly chemical, properties. What was said earlier in chapter 6 with respect to composition as a classificatory criterion can also be said with respect to properties: the choice of this criterion was in no way natural. A division according to provenance—that is, into mineral, vegetable, and animal substances—was another option. But this choice would have split not only the new class of acidifiable bases (field I/b), but also the traditional class of alkalis (field I/e). It thus seems very likely that the four authors chose a classificatory criterion that ensured the preservation of the classes of metals, alkalis, and earths (fields I/c–e), which were established long before 1787.

Since the *Méthode* is silent on the specific kinds of properties according to which these traditional classes were distinguished, we can be fairly sure that the definitions

---

1   In the following, we presuppose the description of the table given in chapter 5. In referring to the table's fields—field I/a, etc.—we refer to the diagrams in that chapter, particularly figure 5.4.

applied were common among eighteenth-century chemists. What were these common definitions?[2] To begin with the earths (in present-day terms, oxides and carbonates of light metals), there were three chief defining properties. To be taken as an earth, a substance had to be incombustible, insoluble in water, and not volatile in distillation. The defining properties of metals were chiefly metallic lustre, fusibility, and malleability. The table's class of metals also includes semi-metals; that is, in the eighteenth-century understanding, substances that lack malleability but fulfill the first two criteria of metals. Regarding semi-metals in particular, there were many disputes among eighteenth-century chemists.[3] Yet, focusing here on shared opinions, we can abstract from such disagreements. Alkalis, finally, were defined chiefly by two chemical properties: by the effervescence of their solution in acids and by the neutralizing effect of their reaction with acids.

A definition according to chemical properties also occurs in the case of the class of the acidifiable bases (field I/b)—a class newly created by the authors of the *Méthode,* which presupposed Lavoisier's anti-phlogistic chemical system. This class comprised simple substances that form acids when combined with oxygen. As discussed earlier in the description of the table, 23 of the 26 substances of this class were hypothetical substances, meaning substances not yet separated through chemical analysis of the respective acids. These hypothetical substances were thus defined exclusively as substances forming acids with oxygen. But this does not hold equally for the three actually isolated substances in this class, sulfur, carbon, and phosphorus. True, these substances formed acids with oxygen, too, and that's why they were put into this class. But, in contrast to the hypothetical acidifiable bases, they were not defined exclusively by this one chemical reaction. Rather, they were involved in several further chemical syntheses—even in those without oxygen, as is clearly shown by field VI/b, which is dedicated to the products of precisely such syntheses. Nevertheless, their quality of forming acids with oxygen was decisive in their being grouped together in the same class.

7.2 Paradigmatic syntheses

The remarkable fact that certain simple substances were classified, exclusively or mainly, by virtue of their ability to synthesize a specific compound like acids deserves further attention. In a classification according to composition, simple substances were by implication identified as potential components of compounds. A classification of simple substances as actual building blocks of a specific class of compounds, however, goes beyond what is given with the general component-compound structure of a classification according to composition. In the tableau, simple substances—potential components of an open horizon of compounds in common understanding—appear as predetermined building blocks of a particular class of

---

2   The fact that we confine our discussion to undisputed main features of these definitions implies no denial of controversial features with respect to their other aspects.
3   See, for instance, Kopp [1966] vol. III p. 94ff.

compounds. Such classes of predetermined building blocks (that is, of substances classified with respect to particular chemical syntheses) indicate a stratum of the chemical framework underlying the table's classification that has yet to be discussed. Superimposed upon its general component-compound structure, we find an additional structure of classification, constructed with respect to selected chemical syntheses.

Coming across such an additional classificatory structure, it seems natural to ask, is only the class of acidifiable bases indicative of this stratum, or were other taxonomic units of the table also formed with respect to selected chemical syntheses? And if so, with respect to which ones? Which chemical syntheses structured the classification of the table? Or, more precisely, the interpretation of which specific syntheses provided an additional chemical framework of the *Méthode*'s classification, which governs its more concrete features?

Upon closer inspection, metal oxides, earths, and alkalis show the same ambivalence as the three pure chemical substances among the acidifiable bases. Like sulfur, carbon, and phosphorus, they too are rendered by the table as substances with an open horizon of possible combinations, and, at the same time, as predetermined building blocks of particular syntheses. As to the possible combinations, the table covers a lot of combinations of metal oxides, earths, and alkalis with several other substances. These combinations belong to a variety of classes. Compounds of metal oxides can be found in the fields IV/c and V/b–c, compounds of earths in fields IV/c and V/b–c, and compounds of alkalis in the fields IV/c, V/b–c, and VI/b. However, for the authors of the *Méthode*, metal oxides, earths, and alkalis formed the group of salifiable bases, which, as the name tells us, were defined as ingredients of salt syntheses. When grouped together as salifiable bases, the unifying definition of these substances was obviously analogous to that of the acidifiable bases, which were defined as the class of predetermined ingredients for the chemical synthesis of acids. More precisely, metal oxides, earths, and alkalis were united in the class of predetermined ingredients for the chemical synthesis of neutral salts. Thus in the group of salifiable bases we encounter another selected synthesis, the understanding of which structures the classification of the *Tableau*, namely the chemical synthesis and conception of neutral salts.

The conception of neutral salts was indeed of extraordinary significance for the classificatory structure of the *Tableau*, as becomes immediately clear when we compare its extension with that of the conception of acids. The latter is included in the former conception. As already mentioned, in eighteenth-century chemistry the name neutral, or middle, salts was given to compounds synthesized by the combination of one of the salifiable bases with an acid.[4] Yet, for the authors of the *Méthode*, acids were not simple substances as assumed in the phlogistic system. Rather, according to Lavoisier's anti-phlogistic chemical system, acids were compounds of an acidifiable base and oxygen, and thus represented a conception that connected substances and classes of substances. Therefore the conception of neutral salts linked the classes of

---

4   See also our more elaborate discussion in chapter 9.

salifiable bases not simply with the class of acids, but with a class that, for its part, was linked with the class of acidifiable bases and, furthermore, with oxygen. The concept of neutral salts thus immediately connected the three classes of salifiable bases (I/d,e and III/c) with the class of acids (III/b). Furthermore, mediated by the latter, it also connected all of these classes with the class of acidifiable bases and with oxygen. Again, a diagram can depict the web of classes organized by the conception of neutral salts (figure 7.1).

	I	II	III I/b + Oxygen	IV	V III + I/d-e and III/b + III/c	VI
a	*Oxygen*					
b	Acidifiable Bases		Acids		Neutral Salts	
c			Metal Oxides			
d	Earths					
e	Alkalis					

Figure 7.1: Web of classes linked and organized by the conception of neutral salts.

The diagram shows the great extent to which the classificatory structure of the *Tableau* depended on the synthesis and conception of neutral salts. The conception of neutral salts was indeed a backbone of the table's classificatory structure in terms of both quantity and quality. But the web of classes linked and organized by this conception represented only one part of the structure. Two further webs of classes can be recognized. There is first a web of two classes of gaseous substances (fields II/a,b,e being taken as one class, and field IV/b) linked by the conception of compounds of matter of heat, or calorique. This web, however, provides order to only eleven substances. There is secondly a web of classes of chemical substances, the significance of which is hardly inferior to that woven by the conception of neutral salts. Although less comprehensive than the latter, its classificatory weight is unmistakable solely by the large amount of substances arranged by its three (four) classes—namely, the classes of metals, of metal oxides, and of alloys (fields I/c, III/c, VI/c), with which the class of compounds of metal oxides is associated (field IV/c). The classificatory weight of this web of classes resulted from the domain of chemistry it represents, the field of metallurgy. As will be shown in chapter 8, this segment of chemical practice with pure substances was as important throughout the eighteenth century as the production of salts. We argue that the combination of the web formed by the conception of metal oxides and alloys with that formed by the conception of neutral salts constitutes the backbone of the table's classification.

So far it might appear that the role of the conceptions of neutral salts and of compounds of metals for the classificatory structure of the tableau consisted merely of connecting a couple of classes of substances that also existed independently of these conceptions. But, rather than simply connecting given classes, the selected syntheses introduced structured clusters of classes into the classification. In other words, each selection of a kind of synthesis was also a selection of classes, which then had to be introduced and arranged in the classificatory scheme. By implication, each such selection is at the same time an exclusion of other possible classes pertaining to different kinds of syntheses and, thus, different ways of grouping the substances into classes. Finally, when what is chosen is not one single kind of synthesis, but a set of of different kinds of syntheses, there is no guarantee that the resulting mixture of different clusters of classes will not display inconsistencies. In practice, there was not much liberty as to which syntheses and which structured clusters of classes should be chosen. If the point was to achieve a meaningful classification of the substances of a certain field of chemical investigations, the significance of a synthesis for this field became the decisive criterion which nobody could afford to ignore, including the authors of the *Méthode*.

Obviously convinced that the three kinds of syntheses must be employed—along with the conception of neutral salts including that of acids, the conception of compounds of calorique, and the conception of metallic oxides and alloys—the authors of the *Méthode* had to find a unifying and consistent taxonomic arrangement for the different clusters of classes given by the selected syntheses. What type of taxonomy did they choose? At this point, it seems appropriate to relate the classification of the *Méthode* to contemporary classificatory enterprises in botany and the dispute about "natural" and "artificial" classificatory systems.[5] Almost simultaneously with the *Méthode*, in 1789, the French botanist Antoine-Laurent de Jussieu reduced the dispute to the methodological alternative of "method" versus "system." Using present-day terms, one can roughly say that the definition of "method" advocated by Jussieu was "upward classification by empirical grouping" and "system"—being, of course, an allusion to the famous (or notorious) sexual system of Linnaeus—meant "downward classification by logical division."[6] Does this alternative have any bearing on our chemical classification? Is it possible to decide which of the two methodical ways its authors followed? This indeed seems possible: the *Méthode* followed the "method." Our four authors obviously did not obtain the clusters of classes discussed through a "logical division" of higher taxonomic units. On the contrary, having introduced these classes with the conceptions of the selected kinds of syntheses, they faced the task of finding overarching taxonomic units into which these classes could be distributed.

---

5   On the following see Lefèvre [2001].
6   See Mayr [1982] pp. 158, 190.

The general principle of classifying according to composition provided one feature of this overarching arrangement of selected clusters of classes. By implication, it allowed the grouping together of all classes of simple substances, regardless of the cluster of classes to which they pertained. For the ordering of the different classes of compounds, however, the only arrangement suggested by this general principle would have been that according to degrees of aggregation: first, an order of classes of combinations of two simple substances; next, one of combinations of such a first-order combination with a further simple substance, followed by one of the combinations of two first-order combinations, and so forth.

For evident reasons, the authors of the *Méthode* did not follow this mechanistic way, which would have led to two instead of five columns of compounds—one fusing together columns II, III, and VI, the other columns IV and V. In this scheme the horizontal arrangement of compounds of one and the same simple substance discussed earlier would have been impossible. Rather, they stuck with their concrete understanding of the chosen classes' composition and singled out the classes of compounds that pertain to different clusters of classes, but are analogous with respect to composition. They found two such classes: the acids and the metal oxides. Both classes comprised substances that were thought of as combinations of simple substances with oxygen, but pertained to different clusters of classes. More precisely, both classes pertained to the cluster around the neutral salts; the class of metal oxides was at the same time also part of the cluster built around the metals. As a consequence, the authors of the *Méthode* were able to link the two most important webs of classes by uniting these webs' classes of simple substances in column I and their two analogous classes of compounds in the column combining simple substances with oxygen (column III).[7] The result of this linkage was the classificatory system given by columns I, III, V, and VI—except for the field VI/b (figure 7.2).

How did the classes of gases or compounds of calorique fit into this taxonomic scheme? Within the cluster of classes resulting from the conception of compounds of calorique, there was no class whose composition would have been analogous to that of any class of the two other clusters. Thus, the classes of gases had to be ordered into separate columns, namely the columns II and IV, which are not linked with but simply juxtaposed with and inserted between the columns formed by the classes of the two other webs.[8] Finally, the class of the compounds of sulfur, carbon, and phosphorus (field VI/b), which does not pertain to any of the three clusters of classes, was properly inserted in column VI. This class was indeed regarded as analogous to the class of the alloys.

Our investigation of the framework of chemical conceptions informing and structuring the classification of the table from 1787 can end here, and it seems natural to proceed to a historical assessment of the achievements of this important classificatory

---

7   The water in field III/a could be and actually was regarded as an acid or, more precisely, as a semi-acid—see Guyton de Morveau et al. [1787] p. 88.

8   As mentioned earlier in chapter 5, the two classes in column IV were not regarded as classes of analogous compounds by the authors of the *Méthode*. Rather, they simply used the free space in field IV/c for the class of compounds of metal oxides, thus avoiding the introduction of a further column.

	I	II	III I/b-c + Oxygen	IV	V III + I/d-e and III/b + III/c	VI I + I
a	*Oxygen*					
b	Acidifiable Bases		Acids		Neutral Salts	
c	Metals		Metal Oxides			Alloys
d	Earths					
e	Alkalis					

Figure 7.2: The classification according to the conception of neutral salts combined with that according to the conception of metal oxides and alloys.

enterprise. An essential prerequisite condition of such an assessment, however, is knowledge of the context of this classification. In particular, we need to know more about eighteenth-century chemists' modes of classification. Which features of the classification were actually original, and which were more or less traditional? How were the substances in question classified earlier in the eighteenth century and what was the relation of our classification to these earlier ones? Can the latter be regarded as forerunners of the table from 1787, or does the table mark a conceptual break with such earlier attempts? Moreover, for an assessment of the table's classification, we need to know more about even more elementary aspects of its context. What was the background of the pure chemical substances classified? Where did the webs of classes that structure the classification come from? Are these webs genuine constructions of academic chemistry, or are they also based on certain types of chemical operations pertaining to technological chemistry? Which operations and which fields of chemical practice are reflected in the table? In other words, we need to know more about this world of pure chemical substances and about chemists' modes of classifying before we can enter into an assessment of the *Tableau* from 1787. The following chapters 8 and 9 will study these aspects of its context.

8

Operations with Pure Chemical Substances

Almost all of the classification's main characteristics discussed in the previous chapters were given with or rested on its confinement to pure chemical substances. This confinement must be seen as a confinement to a particular domain of theory and practice, one which constituted the historical context of the *Méthode*. As a manifold domain, which, as we will see, comprised different fields of chemical practice as well as theory, a dense description would be required for any portrayal to do justice to its complexity. Such a portrayal is, of course, beyond the scope of this chapter. However, an outline of this domain is indispensable, as its very existence as a distinguishable domain might legitimately be questioned due to its complexity.

The different clusters of classes that we were able to distinguish in the table's classification are indicative of the chemical domain's extension in space and time, that is, its main fields of inquiry and its historical depth. We distinguished three such clusters, one around the conception of neutral salts including that of acids, another around the conception of gases, and a third around the conception of metal oxides and alloys.

The cluster of classes of gaseous substances indicates a field of chemical investigations that was still a novelty in the 1780s. Although thorough investigations of these substances can be traced back to the research of Joan Baptista van Helmont (1579–1644) in the first half of the seventeenth century and of Stephen Hales (1677–1761) in the first half of the eighteenth century, the beginning of the modern chemistry of gases is usually dated to Joseph Black's (1728–1799) research in the 1750s.[1] And the first climax of this field of chemical research is generally seen in Lavoisier's chemical theory. However, considering how disputed almost every conception of gaseous substances was at the time, one may hesitate to speak of the existence of a chemistry of gases as early as the 1780s, at least if this notion is understood to designate an established segment of chemical research with a core of shared conceptions.[2] There was certainly no research tradition on these particular pure chemical substances upon which the authors of the *Méthode* could base their classification of gaseous substances.

In contrast to this newcomer in the fields of chemical occupations, an established field of chemical theory and practice is indicated by the cluster of classes around the conception of metal oxides and alloys, which could be traced back to the dawn of human civilization. Even regarding only the early modern period, one finds a highly developed metallurgy as far back as the sixteenth century. Also, the cluster of classes around the conception of neutral salts originated from a particular technique of manufacturing drugs that developed gradually starting in the late sixteenth century, and which became an established field of chemical practice and an object of chemical

---

1   See, for instance, Siegfried [2002] pp. 114ff. and 152ff. For life and work of Hales, see Gillispie [1970–1980] vol. VI p. 35ff.; for life and work of Black, see Gillispie [1970–1980] vol. II p. 173ff; for Black's delayed reception in France, see Guerlac [1961] p. 10ff.

2   The most prominent figure in the field of "pneumatic chemistry" in the 1780s besides Lavoisier was Joseph Priestley, a well-known advocate of the phlogistic chemical system.

theory in the second half of the seventeenth century. Therefore these are the two fields of practical chemistry in which we can expect to find the chemical operations by which pure chemical substances were singled out independent of and prior to the modern conceptual system of compounds and affinity, namely the reversible operations of chemical analysis and resynthesis. These fields deserve attention as practical and historical contexts of the modern concepts of chemical compound and composition, and thus for the classification of pure chemical substances according to chemical composition.

8.1 Reversible operations in metallurgy

In the early modern period, metallurgy was not merely a craft in the sense of a practice of unlearned men based entirely on inherited rules and tacit knowledge.[3] On the contrary, we find metallurgical literature in the sixteenth century, such as the booklets written by assayers (*Probierbücher*), providing evidence of reflection upon the processes of assaying and separating metals. This reflection can be regarded as one of the germs of the modern conceptual system of chemical compounds and affinity. Georgius Agricola's *De re metallica* of 1556, doubtless the most important metallurgical book of that century, was still read in the eighteenth century.[4] The context of this sustained interest in the processes of combination and separation of metals was, first of all, the industry of mining, smelting, and processing metals for several economic purposes, which had begun to flourish at the turn of the fifteenth century and enjoyed enhanced favor in the mercantilist eighteenth century. Yet in the sixteenth and seventeenth centuries metals also attracted the attention of the followers of Paracelsus, and thus became the subject of chemical investigations in pursuit of medical and pharmaceutical purposes. In this new context, the metallic calces became a topic of special interest.

In sixteenth- and seventeenth-century metallurgy, the reversible operations of chemical analysis and synthesis were performed in the context of the precious metals' extraction from their ores and the separation of metals from each other.[5] It may suffice to expound on one example of such operations, the extraction and separation of gold and silver. Gold and silver were extracted from their ores by adding lead. The method consisted of melting the prepared ores, which had been liquified by means of fluxes, with lead.[6] This resulted in gold-lead and silver-lead alloys, which were separated mechanically from the lighter slag floating on the surface. The decomposition of the alloys took place in special furnaces where air and charcoal were used to oxi-

---

3   On the following, see Klein [1994a] p. 122ff., and [1994b] pp. 173f. and 180ff.
4   See Beretta [1997]. For life and work of Agricola, see Darmstaedter [1998]; Wilsdorf [1956]; and Gillispie [1970–1980] vol. I p. 77ff.
5   We here use "analysis" and "synthesis" in the contemporary meaning that does not distinguish between stoichiometric and non-stoichiometric compounds. The extraction of non-precious metals, usually available as oxides or sulfides, through reduction was not recognized as a chemical procedure prior to Stahl's phlogiston theory, but considered to be a mere smelting of metals from the ore.
6   For this process, see book IX of Agricola's *De re metallica* (Agricola [1556]).

Figure 8.1: Separation of gold and silver in a sixteenth-century metallurgical laboratory. From book X of Agricola [1556].

dize the lead to litharge (lead oxide), leaving behind the gold or silver. Whenever feasible, the lead was re-extracted from the litharge.[7] A common method of separating gold and silver employed *aqua fortis* (nitric acid).[8] The acid was poured over the gold-silver alloy in a glass flask and heated until the silver dissolved and the gold was deposited on the bottom. The silver was recovered by pouring the solution into copper vessels, where the silver precipitated.[9] This "wet" separation procedure not only lent technological prominence to a mineral acid, but was also the prototype for the

---

7   In place of lead, mercury was also used as a means of extraction, first for gold and later for silver as well. In this case a silver or gold amalgam is produced, which is then heated to separate the mercury from the silver or gold.

8   This method was developed in the Venetian metal processing trade starting at the end of the fifteenth century and assumed technological importance in sixteenth-century metallurgical engineering.

9   On this process, see book X of Agricola's *De re metallica* (Agricola [1950] p. 443ff.).

separation of metallic solutions through precipitation, a practice which spread extensively in seventeenth-century pharmacy, as we will see.[10] If one also considers the many metallic alloys created for various purposes in this period, a considerable number of chemically created compounds can be established that could be reversibly separated in metallurgical operations, either through assaying and extracting metals from ores or by re-extracting metals from alloys.[11]

How did assayers and other authors of metallurgical books conceive of these reversible operations? The authors of the sixteenth- and early seventeenth-century metallurgical writings understood them in a way that conformed to neither Paracelsian nor traditional Peripatetic notions. The chemical transformations were instead seen as assembling and disassembling substances that were preserved in these processes. When describing the metallurgical operations of separating the metals contained in alloys and of assaying ores, these writings represented both alloys and ores as substances containing certain measurable amounts of different metals. In the introduction to book X of his *De re metallica*, where "separation" (the German *Scheidung*) of metals is the topic, Agricola explained the necessity of this chemical operation in the following terms:

> Frequently two metals, occasionally more than two, are melted out of one ore, because in nature generally there is some amount of gold in silver and in copper, and some silver in gold, copper, lead, and iron; likewise some copper in gold, silver, lead, and iron, and some lead in silver.[12]

The ideas related to this talk of an amount of one metal in another one become especially clear in descriptions of the separation of metals in connection with methods of assaying. The methods of assaying were supposed to determine the quantity of a metal in an alloy or a ore containing one or more other metals. Just as the assaying of ores was to determine "whether ores contain any metal in them or not [...], whether it is much or little," the test carried out on an alloy, such as a coin, was to show "whether coins are good or are debased," and to see "if the coiners have mixed more [silver] than is lawful with the gold; or copper, if the coiners have alloyed with the gold or silver more of it than is allowable."[13] Furthermore, particular care was taken not to lose any metal, especially in separating and assaying precious metals.[14]

The metallurgical writings' talk of an amount of a certain metal in ores or alloys implies the idea of physical parts with a certain measurable weight. Alloys, seen by some contemporary authors of chymical writings (for example, Andreas Libavius) as perfectly homogeneous substances, the same in all their parts, were viewed by

---

10 Another common separating procedure for gold and silver was melting with sulfur or stibnite (antimony sulfide). Here silver sulfide was produced and the solid gold fell to the bottom. The silver was later recovered from the silver sulfide, for example, by adding lead, a process in which silver sulfide was reduced and lead sulfide was formed.
11 As stated in chapter 5, here we adopt the view of the historical actors to include non-stoichiometric combinations among the compounds as well.
12 Agricola [1950] p. 439. We quote the English translation by H. C. and L. H. Hoover.
13 Ibid. p. 219.
14 Agricola demands that the silver and gold should be separated so that "neither of them is lost" (ibid. p. 248).

authors of metallurgical writings as aggregates of various corporeal parts lying next to one another. Separating then meant disassembling such parts of what seemed to be an homogeneous whole.

This interpretation is corroborated when we compare descriptions of chemical operations with those of working processes performed with mechanical tools and procedures. With respect to the term "separation," it then becomes clear that this operation was by no means limited to the field of chemistry. When describing the mechanical preparation of ores for the extraction of metals by crushing with iron stamps, sieving, washing, and so forth, Agricola spoke of "separating."[15] In so doing, he did not differentiate between mechanical and chemical operations in metallurgy. Rather, he understood the sequence of metallurgical steps from the processing of ores via their melting to the separation of metals as a continuous succession of phases of a uniform procedure for producing metals. Even when Agricola used alchemical expressions such as "perfecting by fire,"[16] the context shows that the meaning of this expression is different from that in transmutational alchemy and contemporary chymical-pharmaceutical writings.[17] While the authors of the chymical-pharmaceutical writings of the seventeenth century understood the German word *läutern* (to perfect) as an intensification of qualities or a perfection, for Agricola it meant purification in the sense of separating those adhering corporeal parts of the ore that covered the metal and its metallic qualities. The use of a chemical means such as fire thus differed from mechanical means only to the extent that it was a finer instrument.

The term "separation" in writings on metallurgy thus referred to both mechanical and chemical procedures. This leaves open the question of how the obvious differences between them were explained: above all, the role of the substances added in the separation of metallic mixtures, and the role of the means of extraction in the extraction of gold and silver from their ores. Gold and silver were extracted, as we have seen, by adding lead or mercury, and for the separation of both metals from each other *aqua fortis* was used. On the extraction of gold with mercury, Agricola wrote that the gold concentrates and mercury (quicksilver) are mixed together until "the quicksilver can no longer be distinguished from the concentrates made by the wash-

---

15 Here we refer to the first contemporary German translation (Agricola [1557]), where the word *scheiden* is used to describe the preparation of ores by mechanical means in book VIII, the smelting of ores in book IX, and the separation of alloys in book X. In the Latin original (Agricola [1556]), Agricola usually uses the verb *separare*. H.C. and L.H. Hoover (Agricola [1950]) translate this word as "parting" when referring to the chemical operations and "separating" in other contexts. This thus neglects the broader meaning of the German word *scheiden,* which refers to both the mechanical and the chemical operations.

16 According to the contemporary understanding, all operations with fire were typically chemical ones.

17 "Although those who burn, roast and calcine the ore, take from it something which is mixed or combined with the metals; and those who crush it with stamps take away much; and those who wash, screen and sort it, take away still more; yet *they cannot remove all which conceals the metal from the eye* and renders it crude and unformed. Wherefore smelting is necessary, for by this means earths, solidified juices, and stones are separated from the metals so that they obtain their proper color and become pure, and may be of great use to mankind in many ways. *When the ore is smelted, those things which were mixed with the metal before it was melted are driven forth,* because the metal is perfected by fire in this manner." (Agricola [1950] p. 353; our emphasis). From the context, it becomes clear that Agricola sees the step of melting ores that follows their mechanical processing as a uninterrupted continuation of the mechanical preparation steps.

ing, nor the concentrates from the quicksilver. [...] and soon the quicksilver, which *has absorbed* all the gold, runs together into a separate place away from the rest of the concentrates made by washing."[18] In this statement, the production of gold amalgam is described as the mercury "absorbing" the gold. Other metallurgical writings talk of a process of "entering"—for example, of silver into lead—when describing the production of amalgams or alloys.

The separations of metals from each other in the wet way were described in very similar terms. In his book on ores and assaying of 1574, the German assay master Lazarus Ercker (1530–1594) described the separation of gold and silver from each other by means of aqua fortis or parting acid (nitric acid) as follows:

> When you wish to use parting acid in order to obtain the gold in auriferous silver, the silver must first be fire-refined in a test. [...] After the annealed plates have cooled, put them in a coated glass vessel or parting flask. Do not take more than five, or at the most six, marks of these thin, hammered, little plates of auriferous silver at a time, even though you have much silver, because there is danger of breakage and they occupy a lot of space in the flask. Pour onto them parting acid (after it has been purified by precipitation) to a depth of two good fingerbreadths above the silver. It will at once *start to work* of its own strength. Then place the parting flask with the silver in hot sand in a large clay dish (made of good material) standing on [one of the chambers of] a Heinz furnace so that the sand will be kept hot all the time. When the first parting acid has worked enough and no longer *attacks*, decant the *silver-laden parting acid* into another coated flask. [...] After the parting acid has separated all the silver from the gold and *has absorbed it*, pour the various lots of silver-laden parting acid together into a flask, as said before. Pour on the gold or gold calx that was left in the flask enough clean boiling-hot water to cover the gold amply, and put it back on the fire.[19]

In this recipe Ercker described the separation of silver from gold as an "attack" of the parting acid and the subsequent "absorption" of the silver by the acid. As a result, "silver-laden parting acid" was created, which was decanted from the remaining gold. The procedure was repeated three times in order to obtain pure gold. The possibility to recover the silver from the parting acid, which Ercker described next, was a significant condition for his understanding of the process. "When silver has been dissolved from gold by means of parting acid, the parting acid has absorbed it, and you wish to recover it again from the parting acid," he wrote, "it can be done in several ways."[20] The most common method used a copper vessel, as described above. When the silver-laden parting acid was poured into a copper dish, Ercker observed, the invisible silver will "at once be deposited visibly on the copper dish." The precipitated silver was then melted in the fire to "collect the silver more thoroughly and more compactly."[21] Among the methods of recovering silver from parting acid one yielded not only the

---

18 Agricola [1950] p. 243; our emphasis.
19 Ercker [1951], pp. 162-163, our emphasis. We quote from the English translation by A. G. Sisco and C. S. Smith. See also Ercker [1960] p. 163: "[...] worauf es bald aus eigener Kraft zu arbeiten beginnt. [...] Wenn dieses Scheidewasser hier genug gearbeitet hat, daß es nicht mehr angreift, so gieß das nunmehr silbrige Scheidewasser ab [...]. Wenn nun das Scheidewasser alles Silber aus dem Golde herausgelöst und in sich aufgenommen hat [...]." In his Probierbüchlein of 1595 the German master minter Modestin Fachs described this process in very similar terms; see Fachs [1678].
20 Ibid. p. 167.
21 Ibid.

precious silver but also "the parting acid in such condition that it can be used again for parting"; this method completely "freed" the acid from silver.[22] It is clear from this terminology that Ercker understood the dissolution of silver as a process in which the silver was preserved and merely moved into the acid. The acid, in turn, functioned like a mechanical instrument, which presumably divided the silver into invisibly small parts, and at the same time as a sort of receptacle. The result of the dissolution was a kind of mixture that consisted of the absorbing acid and the invisible silver. The recovery of both the dissolved substance and the acid—that is, the reversibility of the chemical dissolution—was crucial for this understanding.

An analysis of metallurgical writings shows that, long before the revival of the mechanical corpuscular theory in the seventeenth century and without any recourse to traditional theories of nature, metallurgical practice led to ideas that comprised aspects of the later concept of chemical compound. These ideas concur with this latter concept in assuming that the substances that enter a compound are preserved in it. There is also reason to assume that the reversibility of these chemical operations played a decisive role in the development of this understanding.[23] It seems plausible to furthermore assume that a craftsman's perspective on metallurgical operations might have been instrumental in the realization that these chemical operations were reversible. Artisanal production includes many reversible processes of mechanical disassembling and assembling. The structural analogy of these processes with the reversible chemical operations allowed such operations to be integrated into the established understanding of work processes in the arts. Metallurgy particularly lent itself to such an integration, since the reversibility of many of its operations could be realized without first reconstructing and mentally combining partial operations that were carried out separately in different processes.

Although a significant facet of the modern conceptual network of chemical compound and composition can be found in the metallurgical writings of the sixteenth and early seventeenth centuries, metallurgists and the authors of metallurgical writings did not further pursue the development of this concept, which marked the beginning of modern chemistry. Their interpretation of chemical processes remained subordinated to their technical descriptions of metallurgical operations. In the next section we discuss related efforts of iatrochemists and apothecary-chemists, who produced chemical remedies and elaborated chemical techniques of the production of salts for medical application.

---

22 Ibid. p. 169.
23 An indirect confirmation of this assumption can be found in a critical remark by the German alchemist Johann Rudolph Glauber, who, in contrast to the metallurgists' view, assumed a "transmutation" of the lead used in extractions of precious metals or in their separation. Glauber suspected that the fact that lead can be re-extracted from litharge was the source of the metallurgists' error: "[…] many refiners erre, supposing that corporeal Lead together with the imperfect metals that are mixed with it, goes into the cuple, not being yet turned into Litharge, because corporeal Lead is again melted from thence […]." (Glauber [1651] p. 240).

8.2 Reversible operations in pharmaceutical salt production

In the sixteenth and early seventeenth centuries, the distillation and extraction of plant and animal materials were the characteristic chemical operations of pharmacy. The focus of chemical-pharmaceutical practice was on the extraction of oils and juices and on the distillation of aqueous and oily substances from plants or vegetable materials such as herbs, blossoms, fruits, seeds, woods, resins, and balsams.[24] However, in the sixteenth century Paracelsus and his followers began to use more frequently minerals as a source of medicines in addition to vegetable and animal substances. What was at first only a sporadic production of new chemical preparations eventually induced a fundamental change in pharmacy.[25] During the seventeenth century, preparations of mineral acids and salts surfaced as an important sector of chemical-pharmaceutical practice. This new commercial chemical practice spurred a process of reflection that contributed to the formation of the modern concept of the chemical compound.

Traditional pharmacy had processed mainly vegetable and animal materials, mixtures which, in turn, yielded various fractions of mixtures of substances when distilled. By contrast, the new chemical practice dealt with numerous pure chemical substances, such as metals, earths (oxides or carbonates of light metals), common salt, alkalis, vitriols (iron and copper sulfates), and alum (aluminium potassium sulfate), which could either be used directly as medicines or processed further. The distillation of salt mixtures yielded a plethora of yet unknown salts as well as mineral acids.[26] The proliferation of the use of mineral acids such as nitric acid, already applied in metallurgy, sulfuric acid, and muriatic (hydrochloric) acid as solvents for metals, earths, fixed alkali, and the few pure chemical substances from the plant and animal kingdoms—the fixed alkalis (sodium and potassium carbonate) and volatile alkali (ammonium carbonate)—made possible the synthesis of a growing number of chemical compounds, especially salts. The following two tables provide an overview of salts synthesized from metals, alkalis, or earths and a mineral acid or acetic acid within the commercial production of medicines in the early and mid-seventeenth century.[27] The introduction of mineral acids as a new means of chemical-pharmaceutical preparation prompted a change not only in the general character of seventeenth-century pharmaceutical practice, but also in the way it was understood by the authors of chemical-pharmaceutical writings.

The production of artificial salts by dissolving a metal or other mineral pure substance in an acid was performed in various ways significant for the initial understanding of these operations. When metals dissolve, the acid functions as both reagent and solvent. The salt produced in this process has to be separated from the solvent, often

---

24 See also chapter 11, which discusses the continuation of this practice in the eighteenth century.
25 On the following, see Holmes [1989] p. 35ff.; Klein [1994a] p. 144ff., and [1994b] pp. 175ff. and 183ff.; Multhauf [1966]; and Schneider [1972].
26 These included the often-used salts sublimate (mercury(II) chloride) and calomel (mercury(I) chloride), which were used to treat venereal diseases, and butter of antimony (antimony(III) chloride).
27 These tables are taken from Klein [1994b] p. 176ff.

	Spirit of niter Nitric acid	Spirit of vitriol Sulfuric acid	Vinegar Acetic acid	Sublimate Mercury(II) chloride*
Mercury	Praecipitatio mercurii egregia (Croll) Vitriol of mercury (Valentinus) (Mercury(I) and Mercury(II) nitrate)	Turpethum minerale (Croll) (Mercury sulfate)		Sublimé doux (Beguin) Calomel (Mercury(I) chloride)
Silver	Calx of silver (Libavius) Sel ou Vitriol de Lune (Beguin) (Silver(I) nitrate)			
Copper or Copper oxide	Calx of copper (Libavius) (Copper(II) nitrate)		Calx of copper (Libavius) Sel ou Vitriol de Venus (Beguin) (Copper(II) acetate)	
Tin	Calx of tin (Libavius) Sel ou Vitriol de Jupiter (Beguin) (Tin(II) nitrate)		Calx of tin (Libavius) (Tin(II) acetate)	Spiritus fumans Libavii (Libavius) (Tin(IV) chloride)
Lead or Lead oxide	Calx of lead (Libavius) (Lead(II) nitrate)		Calx of lead (Libavius) Sel ou Vitriol de Saturne (Beguin) Sugar of lead (Lead(II) acetate)	
Antimony Antimony sulfide		Vitri antimonii correctio (Croll) (Antimony(III) sulfate)		Butter of antimony (Libavius, Beguin) (Antimony(III) chloride)
Iron		Sel ou Vitriol de Mars (Beguin) Vitriol of Iron (Valentinus) (Iron sulfate)		
Fixed alkali Potassium carbonate, sodium c.		Tartarus vitriolatus (Croll) (Potassium sulfate)		

* Sublimate was used instead of free hydrochloric acid at the beginning of the seventeenth century.

Figure 8.2: Syntheses of salts at the beginning of the seventeenth century.[28]

through distillation, in order to be identified as a new chemical compound. If the salt had a low boiling point, the result was a distillate or sublimate, just like vegetable and animal distillates.[29] In other cases, the synthesis of artificial salts bore more resemblance to the extraction of vegetable materials by means of solvents. The solvent,

28 The parenthesized names following the historical names of the substances indicate that the production of the given salt is described in the writings of this author.

	Spirit of salt		Spirit of niter		Spirit of vitriol
Fixed alkali	Sal diureticum, Sylv. Fiebersalz, Sal marinum regeneratum (Potassium chloride)	**Fixed alkali**	Artifical salpetre, Sal nitrium Potassium nitrate Cube-shaped niter (Sodium nitrate)	**Fixed alkali**	Tartarus vitriolatus Nitrium vitriolatum (Potassium sulfate) Salt of Glauber (Sodium sulfate)
Volatile alkali	Sal ammmoniac (Ammonium chloride)	**Volatile alkali**	Nitrium flammans (Ammonium nitrate)	**Volatile alkali**	Sal ammoniacum secretum Glauberi (Ammonium sulfate)
Alkaline earth (Calcium carbonate)	Phosphorus of Homberg (Calcium chloride)	**Alkaline earth** (Calcium carbonate)	Salpetre of lime (Calcium nitrate)	**Alkaline earth** (Calcium carbonate)	Gypsum arte factum (Calcium sulfate)
Tin	Oleum Jovis, Spiritus fumans Libavii (Tin(IV) chloride)	**Iron**	Crocus Martis (Iron oxide)	**Iron**	Vitriol of iron (Iron(II) sulfate)
Antimony (Antimony sulfide)	Butter of antimony, Oleum antimonii (Antimony(III) chloride)	**Copper**	Vitriol of copper (Copper nitrate)	**Copper**	Vitriol of copper (Copper(II) sulfate)
Copper (Copper oxide)	Oleum Veneris (Copper(II) chloride)	**Lead**	Vitriol of lead, Crystalli Saturni (Lead nitrate)	**Silver**	Vitriol of silver (Silver(I) sulfate)
Silver	(with NaCl) Calx of silver (Silver(I) chloride)	**Mercury**	Vitriol of mercury (Mercury nitrate)	**Mercury***	Turpethum minerale (Mercury(II) sulfate)
Mercury	(with NaCl) Sublimate (Mercury(II) chloride), Calomel (Merc.(I) chloride)	**Silver**	Crystalli Dianae, Crystalli Lunae, Höllenstein, Sel de lune (Silver nitrate)	**Zinc***	Vitriol of zinc (Zinc(II) sulfate)
Galmei* (Zinc carbonate)	Oleum lapidis calaminaris (Zinc(II) chloride)	**Tin***	Vitriol of tin, salt of tin, Crystalli Jovis (Tin nitrate)	**Antimony*** (Antimony sulfide)	Vitri antimonii correctio (Antimony) (III) sulfate
Iron*	Oleum Martis (Iron(II) chloride)				
Lead* (Lead oxide)	Oleum Saturni (Lead(II) chloride)				
Arsenic*	Oleum arsenici (Arsenic chloride)				

Figure 8.3: Syntheses of salts in the mid-seventeenth century.[30]

29 "Butter of antimony" (antimony(III) chloride), for example, which was used as a caustic agent to treat ulcers, was produced by Glauber in the middle of the seventeenth century by dissolving antimony oxide in hydrochloric acid and by distilling off the salt, which was an oily fluid. Or, "sublimate" (mercury(II) chloride) was a volatile substance obtained from the sublimation of mercury together with common salt and vitriol. "Spiritus fumans Libavii" (tin(IV) chloride), a smoky liquid, was also distilled from a mixture of tin with sublimate.

in this case an acid, was distilled off or evaporated, and the salts were obtained from the highly concentrated solution, usually during cooling.[31] Because of these analogies—combined with the fact that the chemical synthesis of salts was part of a uniform operational setup and method of handling substances applied to the three realms of nature—these chemical operations were initially placed in the same conceptual scheme as the analogous operations on mixtures of plant and animal substances. The "salts of metals," "vitriols," "olea of metals," and "sublimates" were understood as "essences" of metals, analogous to the vegetable "essences."

The new understanding of these operations did not proliferate until the acidic dissolutions, hitherto considered a preparatory step within chemical-pharmaceutical practice, shifted into the focus of practical-commercial interest over the course of the seventeenth century. It may suffice to illustrate the development of this new understanding using the historical actors' interpretations of the operation of acid dissolution with subsequent precipitation. For this operation, metals, sulfuric acid, common salt, and alkalis were used as means of precipitation. The prototype of this method was the precipitation of silver with copper in metallurgy. Most of the authors of texts on chemical-pharmaceutical matters who described precipitation with metals dealt in depth with the metallurgical separation of gold and silver, followed by the precipitation of silver. This may be one reason why the interpretation of the hidden processes underlying these operations differed fundamentally from their traditional understanding as the extraction of essences and enhancement of qualities and strikingly resembled the understanding of metallurgists. The acids were seen as instruments acting externally, which cut the dissolved metals into small unobservable pieces, and this image of acid as a mechanical instrument was complemented by its image as a receptacle that incorporates and holds the separated parts of the metal. However, the subsequent precipitation in acidic dissolutions posed explanatory problems. When a metal was dissolved in an acid, a transparent, fully homogenous solution resulted. If a further substance was added, a powdery metal calx precipitated from the solution. This led to questions such as why the pulverized metal pieces did not sink to the bottom without the addition of a further substance, the way in which the substance added triggered this process, and why only a few substances were suitable to induce precipitation.

The answers that authors such as Johann Rudolph Glauber (1604–1670) and Christopher Glaser (c. 1628–1672) gave to these questions in their pharmaceutical writings were quite similar.[32] Glauber explained the precipitation of precious metals by means of non-precious ones in the following way:

---

30 For purposes of comparison, the salts were listed in the same sequence as in Geoffroy's table. Salts with bases that Geoffroy did not mention are indicated with an asterisk.

31 The salts of nitric acid (metallic nitrates), sulfuric acid (metallic sulfates), and acetic acids (metallic acetates) usually formed crystals and were called "salts of metals" and "artificial vitriols" because of their resemblance to native vitriols and other natural salts. The residues resulting from the dissolution of metals in hydrochloric acid were watery or oily liquids (metal chlorides) called "olea of metals."

32 For life and work of Glaser and Glauber, see Gillispie [1970–1980] vol. V pp. 417f. and 419ff., respectively.

> [...] dissolve any metal in its appropriate Menstruum, [...] and then put into that solution
> another metal, such as the dissolvent doth sooner *seize on*, then upon that which it hath
> *assumed* and then you will finde, that the dissolvent doth *let fall the assumed metal or*
> mineral, and *fals upon the other,* which it doth sooner *seise on*, and dissolveth it as *being
> more friendly to it* [...].[33]

The pulverized metal pieces were thus not supposed to fall from the solution, since the solvent was supposed to absorb or "assume" them. Glaser wrote that the acid "fastens on" the metal,[34] and Le Febvre stated that the acid imbibes the metal and thus becomes satiated.[35] In the precipitation of one metal through the addition of another, the first metal was dropped because of the preference or predilection of the solvent for the second metal.

A variant of precipitation by means of metals was precipitation by means of alkalis. While in the precipitation of metal calces by metals the acid was seen as the active and elective substance, in the case of precipitation by alkalis, the alkalis themselves were also believed to play an active role. They should weaken or kill the power of the solvent that fastened the metal. Glaser, for example, wrote: "[...] that the *Corrosion* of the *Aqua-Regia* is destroyed by the Liquor of the Salt *Alkali* of *Tartar*, which, as other *Alkalies*, breaks the force of *Corrosive* Spirits, so that they are constrained to let fall to the bottom the body which they held in form of Liquor."[36] In both cases, dissolution was not seen as the mere mixing of the acid and the dissolved substance, but the mechanical function of the acid was complemented by the chemical relation between the dissolved substances.

The precipitation of metal calces by metals or alkalis was a special case of reversible operations in a number of respects. Fire, previously an omnipresent tool in chemical operations, played no role in precipitation. As, in this case, fire was no longer a cause of chemical transformations chemists could pay close attention to the relationships of substances.[37] In addition, metals were pure substances with similar observable properties, but clear differences in terms of their reactions and relationships with other substances. Comparisons of their reactions resulted in the creation of a specific order of precipitations of metals with metals or with alkalis from acidic solutions. Glaser gave the following hierarchy of precipitation in *aqua fortis*: silver by means of copper, copper by means of iron, iron by means of zinc, and zinc (zinc hydroxide) by means of fixed niter (potassium carbonate). Putting it in Glaser's words:

> The Silver dissolv'd in the *Aqua-fortis*, and poured into the Vessel of water, precipitates,
> and separates it self from its *Dissolvent*, by putting a plate of Copper into it; the Spirits of
> the *Aqua-fortis* immediately leaving the Silver to fasten on the Copper, which they dis-
> solve; and during the *Dissolution*, the Silver precipitates it self. [...] The Silver is found in
> the bottom. It must be wash'd, dry'd, and kept (if you please) in form of a *Calx*, or else

33 Glauber [1651] p. 120; our emphasis. "*Solvire welches Metall du wilt in seinem behörigen menstruo*
    [...] *und lege darnach in ein solche solution ein ander Metall / welches das solvens lieber angreifft alß
    dasjenige / welches dasselbe zu sich genommen hatte / so wirstu befinden / daß das solvens das
    angenommene Metall oder Mineral fallen lässet / und das ander / welches es lieber angreiffet / antasten
    und solviren wird.*" (Glauber [1646–1649] part II p. 96).
34 See Glaser [1677] p. 59.
35 See Le Febvre [1664] p. 161.
36 See Glaser [1677] p. 63.

reduc'd into an *Ingot* in a *Crucible*, with a little Salt of *Tartar*. But if into this second water, which is properly a *Solution* of Copper, you put a body more earthy and porous than Copper, as Iron is, the Copper precipitates, and the *Corrosive* Spirits of the *Aqua-fortis* fasten to the substance of Iron; which may likewise be precipitated by some Mineral more earthy and porous than Iron, as *Lapis Calaminaris* and *Zink*. Lastly, if you pour into this Liquor charg'd with these Substances some of the Liquor of fix'd *Nitre*, drop by drop, this latter will destroy the *acidity* of the *Aqua-fortis*, and precipitate those Minerals.[38]

Some of the relations between substances that Geoffroy later listed in his *Table des différents rapports* were thus noticed earlier.

An analysis of chemical-pharmaceutical writings shows that over the course of the seventeenth century acidic dissolutions shifted into the focus of practical-pharmaceutical interests, and that the interpretations of these operations began to deviate from traditional alchemical understandings. Similar to metallurgists' mechanical-instrumental understanding of the effect of acids, which implied the material preservation of ingredients, iatrochemists slowly began to understand the process of dissolution and the individual dissolution products as combinations of substances based on material relationships or affinities.

8.3 A chemistry of pure substances takes shape: Geoffroy's affinity table of 1718

Our outline of chemical operations using pure chemical substances in sixteenth- and seventeenth-century metallurgy and pharmaceutical salt production shows that the chemistry of pure substances was a domain of chemical practice no less than of theory. It existed in workshops and artisanal laboratories and never became an exclusively academic subject. As is characteristic of chemistry in general, the workshop and the laboratory were by no means different worlds separated from each other.[39] In eighteenth-century chemistry, materials, instruments, techniques, experiences, and conceptual knowledge flowed continuously back and forth from artisanal to academic sites. The men who inhabited these worlds ceaselessly crossed these boundaries as

---

37   The dissolution of gold in aqua regia and the precipitation of fulminating gold (gold(III) oxide with admixed ammoniac) with the alkaline oleum tartari (watery solution of potassium carbonate) had already been described by Oswald Croll, Basil Valentinus, and Jean Beguin. Further examples include precipitations from solutions of silver or copper in nitric acid, as described by Glauber. He wrote that "salt of tartar" could be used to precipitate a "green precipitate" (basic copper carbonate) from the dissolution of copper in nitric acid, which could be used as a dye instead of *verdigris*. According to Glauber, ammoniac could be used to precipitate a "black precipitate" (silver oxide) from a solution of silver in nitric acid. Nicolas Lemery described precipitations with a fixed or volatile alkali from a solution of zinc in nitric acid. The result was "Magistere de Jupiter" (Zinn(II) hydroxide), which could be used as a pomade. The "Magisterium des Bleis" (lead carbonate) had the same cosmetic use; its precipitation, by oleum tartari from the dissolution of lead in acetic acid, was described by Le Febvre and Glaser. Precipitations with sulfuric acid were described in particular detail in Glauber's *Furni novi philosophici*. He precipitated copper sulfate by dissolving copper in hydrochloric acid and subsequent addition of sulfuric acid. The distillation of sal ammoniac (ammonium chloride) with a fixed alkali or an earth (calcium carbonate) is an example of a commercial chemical-pharmaceutical operation that could be interpreted as the separation of a compound of a volatile alkali and nitric acid by means of a fixed alkali or an earth.
38   See Glaser [1677] p. 59f.
39   See, for instance, Klein [2005b].

well, and it is probably impossible to find a single academic chemist in this period who was entirely divorced from chemical technology. This kind of entanglement of knowing and doing did not change significantly with the gradual development of sites, mainly at academies and universities, where chemical investigations could be pursued without being subjected to immediate economic interests. The investigation of neutral salts at the Parisian Academy of Sciences, for instance, never lost touch with pharmaceutical and other practical concerns. The chemistry of pure substances was thus neither an activity confined to academic laboratories as opposed to work-shops nor one of academic chemists as opposed to practitioners.

Furthermore, the different contexts in which the chemistry of pure substances was embedded in the early modern period show that it maintained quite distinct enter-prises of theory and practice. Being part of both metallurgy and pharmacy, our term "chemistry of pure substances" denotes a common ground for a variety of different chemical practices rather than one distinctive field of chemistry with clearly estab-lished contours. As to academic chemistry, the study of metallurgical substances existed only as a facet of the much wider field of mineral chemistry, and the pharma-ceutical study of pure chemical substances was linked closely with that of plant and animal substances. In areas which look like a distinct domain of chemistry, as in the case of the investigation of neutral salts at the Parisian Academy, certain shifts in focus are responsible for this effect rather than the incipient formation of a well-defined discipline. Not surprisingly under such circumstances, there was not even a name used to denominate this occupation with pure chemical substances as a distinct chemical practice.

It is therefore obvious that the chemistry of pure substances can be defined only on the basis of its objects of inquiry. But it should be noted that in the eighteenth cen-tury it were the chemists themselves who distinguished these objects of inquiry from other ones in their practices of classification. We argue that their distinction corre-sponds exactly with the boundaries of objects of inquiry in the tableau of the *Méth-ode*. The question raised above was whether the authors of the *Méthode* were the first to see an inner bond among the many different activities with pure chemical sub-stances scattered through all domains of chemistry, or whether their distinction of the particular sphere of the chemistry of pure substances followed a tradition established earlier. Fortunately, there exists unmistakable evidence for such a tradition. The famous tables of chemical affinities testify unambiguously to the existence of this particular chemical practice. The first of these tables was the *Table des differents rap-ports* constructed by Etienne François Geoffroy (1672–1731),[40] published in 1718.[41] We can thus even determine when the distinction of operations with pure chemical substances first became manifest, namely approximately seventy years before the *Tableau* of 1787.

---

40 For life and work of E. F. Geoffroy, see Partington [1961–1970] vol. III p. 49ff.; and Gillispie [1970–1980] vol. V p. 352ff.
41 See Geoffroy [1718], and the English translation, Geoffroy [1996].

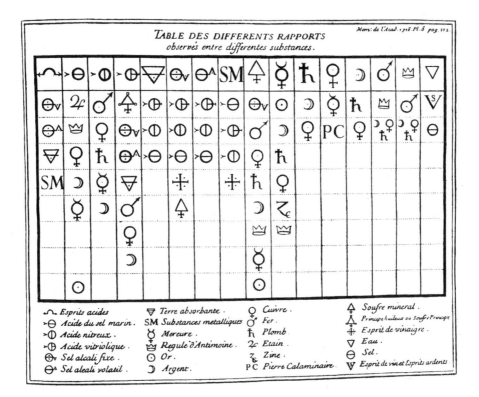

Figure 8.4: Geoffroy's affinity table from 1718.

Affinity tables first of all ordered certain chemical reactions, namely replacement reactions. They did so by representing the substances that combine with a reference substance in the order in which they replace each other. If, say, three substances are capable of combining with a certain reference substance, they were arranged in such a way that the first place was given to the substance that decomposed the compounds formed from the second and the third substance with the reference substance by means of combining itself with the latter; the second place went to the substance that combined with the reference substance and replaced the third; the third substance was placed last because it was not able to replace either of the preceding two substances. Take, for instance, the second column of Geoffroy's table (figure 8.4). At the head of the column, separated from the rest by a horizontal line, we find the symbol for muriatic acid (hydrochloric acid), the reference substance. Beneath we have the symbols of metals that combine with muriatic acid; from top to bottom, tin, regulus of antimony, copper, silver, mercury, and, after two empty fields, gold. This arrangement states that tin can replace all the other metals that combine with muriatic acid, regulus of antimony all except tin, copper all except tin and regulus of antimony, and so on, whereas gold is replaceable by all of them, and itself replaces none.[42] This mode of rendering replacement reactions by arrangement is characteristic of all affinity tables of the eighteenth century, regardless of whether columns of symbols or lists of names were used.[43] Even the corresponding tables of combining substances in the second book of Lavoisier's *Traité* are still arranged in this form.[44] As Geoffroy's table is a paradigm for all later affinity tables not only in this aspect, it makes sense to address the classificatory features of these tables by discussing the first table of 1718.

Confinement to pure chemical substances

The feature that must be emphasized first is the table's confinement to pure chemical substances. This confinement, of course implied in its focus on replacement reactions, presupposed that the stated replacement reactions were established empirically by chemical analyses and syntheses. Geoffroy's table meets this condition. It is in part a summary of investigations of analyses and resyntheses performed by the chemists at the Parisian Academy over several decades, supplemented by reversible metallurgical operations.[45] The same holds more or less true for the subsequent affinity tables that list an established core of replacement reactions, whereas individual additions often indicate cases still disputed. Substances pertaining to the undisputed core of replacement reactions were evidently pure chemical substances; the affinity tables themselves, which expanded considerably in the course of the eighteenth century, thus constitute a kind of inventory of such substances for historians.

The pure laboratory substance phlogiston

Among the substances pertaining to this undisputed core, we also find the famous phlogiston. Until Torbern Bergman's table from 1783, there is no instance of an affinity table without phlogiston.[46] In Geoffroy's table it is named *principe huileux ou soufre principe* and appears as the first substance in the column headed by sulfuric acid. Does the occurrence of phlogiston contradict our assumption that affinity tables deal solely with pure chemical substances? Did phlogiston not represent one of those old elements or principles which chemists regarded as constituents that could not be isolated in a pure form? Indeed, it shared one essential feature with these old principles, as it was thought to bestow certain properties upon the substances in which it was contained. But this, of course, is not sufficient to qualify it as a principle in the traditional sense; otherwise Lavoisier's oxygen must be categorized as such a princi-

42 For the empirical background of the arrangement in this column see Klein [1994a] p. 24.
43 There is at least one instance of a reverse arrangement, namely Gellert's table from 1751 (published again in Gellert [1776]) which could, therefore, be called a table of increasing solubility instead of decreasing affinity. See Duncan [1996] pp. 112, 123f.
44 See Duncan [1996] p. 168; see also Siegfried [1982] p. 29.
45 Geoffroy states this point explicitly—see Geoffroy [1718] p. 203 and Geoffroy [1996] p. 314. For Geoffroy's table as a whole and its immediate historical background, see Holmes [1996]; Klein [1994a] pp. 16–34, and [1995]. For objections of contemporary chemists regarding certain replacement orders of the table see Duncan [1996] p. 118f.
46 Furthermore, in the second half of the eighteenth century, beginning with Demachy's table of 1769, the matter of heat appeared in affinity tables as a pure laboratory substance.

ple as well. Rather, the decisive question is whether or not eighteenth-century chemists regarded it as a substance that could be separated and handled as a building block for the production of chemical compounds, and, if they did so, on what kind of evidence they based their assumption.

In the case of Geoffroy's table, the answer is rather clear. The combination of phlogiston with sulfuric acid, which is listed in the fourth column, was assumed to form sulfur, and this alleged synthesis of sulfur was thought to have been proven by an experiment Geoffroy performed and presented to the Parisian Academy in 1704.[47] Georg Ernst Stahl also rested his assumption of phlogiston on interpretations of experiments concerning the synthesis of sulfur, as well as the calcination of metals and the recovering of metals from their calces. These interpretations of experiments not only included decompositions and recompositions of compounds, but also implied that the perceptible substances involved in the experiments—sulfur, sulfuric acid, metals, and metal calces—were pure chemical substances. Similarly, all occurrences of phlogiston in the affinity tables of the eighteenth century must be read as interpretations of experiments that explain the resulting substances as elective recombinations of pure chemical substances of which phlogiston was just one—though an admittedly special one. In the words of Bergman, whose affinity table listed phlogiston as well, phlogiston was

> so subtle indeed, that, were it not for its combination with other substances, it would be imperceptible to our senses. It can, however, be made to migrate from one body into another, according to the laws of elective attraction [...].[48]

It is important to understand that the nature of such hypothetical substances was determined by the entire intellectual framework in which they fulfilled their role. Phlogiston was introduced and used by E. F. Geoffroy, G. E. Stahl, and others, not in the context of a philosophy of principles, but for the interpretation of local experiments dealing with the analysis and synthesis of pure chemical substances.[49] In that intellectual and experimental framework it was taken to be an ordinary substance itself. The same holds for the hypothetical substances in the *Tableau* of 1787. All of the postulated acidifiable bases represented in the first column of the tableau (see field I/b) were thought to be analogous to sulfur, carbon, and phosphorus.[50] Sulfur, carbon, and phosphorus, which formed acids with oxygen, could be obtained by decomposing the respective acids. Though hypothetical substances, as parts of the entire system of experiments and chemical order, the authors of the *Méthode* conceived of the acidifiable bases as pure chemical substances that could be handled in experimentation.[51]

---

47 See Klein [1994a] p. 25f.; and Kim [2003] p. 103.
48 Bergman [1783] p. 102f.
49 As is well known, phlogiston had a prehistory in the chemical theory of elements and principles which is beyond the scope of this chapter.
50 When referring to the *Tableau's* fields—field I/a, etc.—we refer to the diagrams of chapter 5, particularly figure 5.4.
51 Another instructive example for the determination of the nature of hypothetical substances by the framework in which they are conceptualized is given by Holmes with regard to certain unknown neutral salts, the existence of which was assumed by Macquer. See Holmes [1989] p. 54.

Rendering and classifying substances according to composition

Equally significant with regard to classification is another general feature of Geoffroy's table, namely its rendering of substances according to the principle of chemical composition. This conceptual and classificatory feature of the table is easily overlooked at first glance; for Geoffroy's table rendered all listed substances, those at the heads of the columns as well as those in the columns, equally and uniformly as building blocks of possible compounds without discriminating between simpler and more compound building blocks. In this latter respect, the table was relativistic. But this is not the whole story. For the table included not only the building blocks of chemical compounds, but also the entire list of compounds that can be formed from the reference substance and one of the substances listed in the subsequent column. Geoffroy's representation of replacement reactions and of the elective chemical affinities that direct replacement reactions presupposed that the beholder of the table already knew that the symbols contained in the column must be combined with the symbol for the reference substance. It did not explicitly render the compounds resulting from the listed replacement reactions. Rather, the recognition of the enlisted compounds presupposed some familiarity with these tables. The decisive point is that the symbol of each substance located beneath the reference substance of a column represented two different entities: the symbolized substance as a building block of the combination with the reference substance, on the one hand, and, on the other, the resulting chemical compound itself. Looking at Geoffroy's second column again, the tin symbol stood for this metal and at the same time for the *spiritus fumans Libavius* (tin(IV) chloride), which resulted from the combination of tin with muriatic acid; likewise, the symbol of the regulus of antimony stood for this metal and at the same time for *butter of antimony* (antimony(III) chloride), which resulted from this metal's combination with muriatic acid. To render chemical compounds in this way meant to define them exclusively by their composition, and grouping together and ordering these compounds in columns was a method of classification. Geoffroy's table thus classified chemical compounds exactly according to composition. (We will come back to this in the following chapter.) In all of these aspects, Geoffroy's table is representative of the subsequent affinity tables of the eighteenth century. The principle of ordering substances according to chemical composition is as generally a characteristic of these affinity tables as it is of the *Tableau* of 1787.

Wet way and dry way

There is another tacit presupposition in the table, the knowledge of which is essential for its adequate understanding. The listed replacement reactions belong to different chemical operations and thus take place under different conditions. Two main processes must be distinguished. First are the operations that include the dissolving of substances in acid solutions—what was called the "wet way." The reactions in the first eight columns pertain to the operations of this type. Second are the operations of combining and recombining substances under high temperatures, for example, by melting—what was called the "dry way." The reactions in columns nine through fifteen pertain to this second type of processes.[52] Geoffroy distinguished between these different processes through their arrangement in different parts of the table. But he did not further distinguish differences in chemical affinities in the wet and dry ways, depending on more fine-grained differences of temperature. The table's silence on these latter differences, which affected the conclusions that could be drawn from the listed replacement patterns for the interpretation of experiments,[53] permanently caused trouble during the eighteenth century and finally led Bergman to propose a major revision of the arrangement of the table in 1775.[54]

What we encounter with the wet and dry ways of replacement reactions represented in Geoffroy's table are the two main fields of chemical theory and practice with pure chemical substances discussed above: salt production in acid solutions, and metallurgical separations and combinations of metals. Thus, we can expect to meet again the two clusters of classes around the conception of the neutral salts and around that of metallic calces and alloys. Indeed, all the classes of these two clusters were present in Geoffroy's table, even constituting the bulk of its content and structuring its essential framework.[55] The most significant achievement of Geoffroy's table consisted precisely in this junction of the two clusters of classes representing the two main fields of the early modern chemistry of pure chemical substances. This junction was based on the notion of replacement reactions, which pervaded the two clusters of classes. Geoffroy realized that the combinations, separations, and recombinations of substances in the spheres of both chemical theory and practice can be regarded as such kinds of reactions, and hence apprehended in one coherent conceptual frame.

Thus, the affinity tables already displayed several of the features of essential significance for and characteristic of the later classification by the authors of the *Méthode*. First, the tables were constructed on the basis of replacement reactions and as such were confined to pure chemical substances. Second, since these tables represented chemical compounds along with their building blocks, the principle of ordering chemical substances according to composition must be taken as one of their essential features. Third, comparable with the role that the notion of oxides played for the classification of the *Méthode*, the notion of replacement reactions provided the groundwork for the affinity tables' unification of the cluster of classes around the conception of neutral salts with that around the conception of metals and metal alloys. It should, however, not be overlooked that this was achieved at the cost of excluding all reactions of pure chemical substances that did not exhibit replacement patterns.

---

52 The remaining column again represents dissolutions in water; that is, reactions in the wet way.

53 The affinity table's use in assisting in the interpretation of chemical operations performed in the laboratory was perhaps the main purpose that Geoffroy pursued in its construction. See the elaborate descriptions in Geoffroy [1718] p. 206ff., and [1996] p. 316ff.

54 On the irritations of eighteenth-century chemists due to the implicit rendering of these differences by affinity tables, see Duncan [1996] p. 121ff. The table in Bergman [1785] (the first edition of the Latin original was published in 1775), renders the replacement orders in the wet way and those in the dry way separately and juxtaposes the two respective replacement rows of the same reference substance in such a way that they form one vertical column, with the dry row beneath the wet row.

55 The metallic calces were systematically listed first in Bergman's table.

We complete our investigation of the historical context of the classification on which the authors of the *Méthode* based their taxonomic and nomenclatory reform by summing up the main results. By its confinement to a rather small selection of chemical substances, namely to pure substances, the mode of classification in the *Méthode* pointed to a particular domain of chemistry as its historical context—the early modern chemistry of pure chemical substances. The chemistry of pure substances, which comprised fields of chemical practice and theory, was performed in workshops as well as academic laboratories. It was devoted to a variety of issues around mining and metallurgy, medical chemistry and the manufacture of chemical medicines, pottery, porcelain making, and the preparation of salts. Embedded in a variety of practical and theoretical enterprises, the modern concepts of chemical compound and reaction emerged gradually, from the second half of the seventeenth century onwards and, with it, the understanding of the essential characteristics by which pure chemical substances distinguish themselves from the mixts of the Aristotelian and Paracelsian traditions, as well as from mechanical mixtures in the modern sense of the term. This understanding became manifest in the affinity tables of the eighteenth century, in the tradition of which the *Méthode*, as well as book II of Lavoisier's *Traité*, must be placed. The eighteenth-century affinity tables furthermore document that a chemistry of pure chemical substances was already emerging as a distinctive chemical practice at the beginning of the eighteenth century.

Classification of Pure Chemical Substances before 1787

In our analyses of the *Tableau*, particularly in chapters 6 and 7, we recognized and discussed several classificatory principles of a specific chemical character employed in the tableau's construction. The most important of them was doubtless that of classifying according to composition. Further principles were applied to subdivide the simple substances in classes such as metals, alkalis, and so on, and also to distribute the compounds into connected classes such as acids, salts, and the like. When discussing these additional classificatory principles, we presumed that many of these concrete classes were already well established long before 1787. A historical assessment of the *Tableau's* achievements requires a verification and further qualification of this assumption.

Since no comprehensive, explicit classification of pure chemical substances was ever constructed before the *Tableau*, an investigation of earlier ways of ordering these substances must look for other documents yielding information about such classificatory traditions. Therefore, in this chapter patterns and collective customs of classifying pure chemical substances prior to the *Tableau* will be traced by investigating documents that are either not explicit classifications of substances themselves, or are classifications of substances not confined to pure chemical substances. We will first study classificatory structures embedded in the affinity tables and then examine a mature and comprehensive classification of the mineral kingdom, namely Torbern Bergman's classification of 1779. In the discussion of Bergman's classification, we will take the opportunity to trace the history of the main classes of pure chemical substances.

9.1 Classification of substances in affinity tables

The affinity tables of the eighteenth century are of particular interest in the history of classifying pure chemical substances, for they not only testify to the unification of hitherto dispersed operations with such substances on the basis of the conception of replacement reactions, but also document conventions of ordering pure chemical substances in the eighteenth century.[1] Displaying stable structures in their arrangement, as well as the changing factors of implicit and explicit classificatory customs, these tables promise insights into the historical background of the *Méthode's* classificatory achievements. We will investigate the classificatory dimension of the affinity tables in two steps. First we will address details and general features of this dimension by discussing the paradigmatic table of Geoffroy. Second, in an excursus which can be skipped by those less interested in these details, we will further explore some remain-

---

1 Up to this point, the affinity tables of the eighteenth century attracted the attention of historians of science mainly because of the conception of chemical affinity embodied in them. They were rarely studied as documents giving testimony of chemical classification in this period. Among the many merits of Alistair Duncan's profound book on these tables is that it dedicates a section of its fourth chapter to the question of the classification of chemical substances in this age—see Duncan [1996] pp. 159–168. In the following we further extend and elaborate Duncan's approach.

ing questions, in particular the possible classificatory significance of certain arrangement patterns, by scanning such patterns in the subsequent affinity tables up to that of Bergman.

Geoffroy's table

Geoffroy's table—see figure 8.4 in the previous chapter—contains two cases of explicit classification of substances on a higher taxonomic rank. At the head of column one we find a symbol for *esprits acides*, that is, acids in general, and at the head of column eight, an abbreviation for *substances metalliques*. However, it is not easy to determine the precise circumference of these taxonomic units. As to the *Esprits acides* in columns one through four, acids are put in two different forms at the heads of the columns, once in generic form (column one)[2] and another time as single acid species (columns two through four). The replacement pattern shared by the salifiable bases—the alkalis, the earths, and the metals—is given in the generic column one. The other three columns additionally spell out the different replacement patterns displayed by the different substances of one of these bases, the metals, with respect to the three different mineral acids.[3]

From this case, an important general conclusion can be immediately drawn with respect to the table's classificatory implications: since the introduction of columns depended partly on the differences in replacement patterns among the single members of a class of substances, one must not necessarily take all columns as classificatory units. This holds *a fortiori* for later affinity tables where the columns are crowded with substances, the grouping together of which makes no classificatory sense.

The other case of an explicit classification on a higher taxonomic rank, the generic metal at the head of column eight, is less transparent. It seems advisable to take the *substances metalliques* not as a comprehensive class of all metals, but as a class of only those that form salts with acids.[4] As a consequence, the relation between the generic unit of *substances metalliques* and the metals on the right-hand side of the table (that is, to the metals in the "dry" processes) should be taken as not exactly determinable. This uncertainty should not be discussed away,[5] as it indicates a significant feature of the table: as a table not primarily designed for the classification of substances, but for rendering chemical affinities, it represented only predetermined building blocks (to use our expression from chapter 7) and ordered them exclusively

---

2   However, it remains an open question whether or not this generic acid also comprises the spirit of vinegar.

3   Accordingly, in columns two and three the alkalis and the earth are thought to precede the listed rows of metals. In column four the alkalis and the earth are listed anew because, in the case of the sulfuric acid, the *Soufre Principe* is thought to precede all salifiable bases.

4   Since not all of the eleven metals that occur in Geoffroy's table are listed in columns two through four, it remains an open question whether Geoffroy was claiming that all metals form salts with acids. Geoffroy listed no salt formed by zinc or *Pierre Calaminaire* with any of the acids. This absence is conspicuous, as a salt of these two metals (*Pierre Calaminaire*, a carbonate of zinc, was taken to be a semi-metal at the time) had been known since the mid-seventeenth century—see figure 8.3 in the previous chapter.

with respect to replacement reactions. In other words, affinity tables do not render comprehensive classificatory structures that would transcend such affinity patterns.

Apart from these two cases, only in the text accompanying Geoffroy's table can be found further explicit classifications of substances on a higher taxonomic rank, indicated by the use of plural forms such as middle salts (*sels moyens*), alkaline salts (*sels alkalis*), acid salts (*sels acides*), and absorbent earths (*terres absorbantes*). However, these terms are used so casually in Geoffroy's text that no certain conclusions can be drawn as to the significance of their occurrence or, inversely, of the absence of other such classificatory terms. Some of the classificatory distinctions involved in these terms will be discussed later. Now, we should take notice of two implications these terms have for an understanding of the classificatory meaning of certain symbols used in the table.

First, in addition to the two cases of classification on a higher taxonomic rank discussed so far, there are at least two more cases where symbols represent not a chemical species like a single kind of metal or a single kind of mineral acid, but higher taxonomic units comprising different species. The symbol for salts at the last place in the last column obviously represents such a higher taxonomic unit.[6] The other case is the symbol for the earth, which represents not a chemical species but a higher taxonomic unit, although the legend at the bottom of the table speaks of earth only in the singular form.

From the term *sels alkalis*, a further and more general conclusion regarding the classificatory meaning of the symbols can be drawn. Even if, as in the case of the alkalis, only species of a certain kind of substances are symbolized in the table, this does not imply that no higher taxonomic unit of such species was employed in its construction. Rather it may only mean that, again as in the case of the alkalis, the species of this unit display no common replacement pattern. The case of the alkalis further suggests that a higher taxonomic unit is indicated by graphic means, such as the similarity of the symbols for the species of the same unit, or by just arranging species' symbols as adjacent column headings—in the case of the alkalis, of columns six and seven.

This semiotic dimension of the arrangement of substances at the head of the columns deserves particular attention with regard to the table's classificatory distinctions and orders. The order of substances within the columns was ruled by the conception of replacement reaction. Of course, this does not hold for the order of the reference

---

5   There is, of course, no reason to assume that chemists regarded a metal—say, the copper used in salt formations—to be different from the copper used in metallurgical processes. Thus, the *substances metalliques* doubtlessly also represented all metals on the right hand side of the table, provided they formed salts with acids. Neither is there reason to doubt that the chemists of the age classified metals into one taxonomic unit in other contexts. Rather, the interesting point here is that one cannot be sure whether or not the table actually represented such a general class of metals.

6   Once again, however, it is not easy to determine precisely what kinds of substances are grouped together in this unit. As indicated by the terms middle salts, alkaline salts, and acid salts, the chemists of the age designated not only the neutral or middle salts with the term salts, but also the alkalis, which they called "alkaline salts," and further the acids, which they called "acid salts." (We will come back to this later.) Geoffroy gave no hint, neither in the table nor in the text, to what exactly the salt symbol referred.

substances at the column heads. True, the conception of replacement reactions also had some consequences for the reference substances as seen earlier: the single mineral acids would not have shown up at the head of columns two through four if the replacement pattern of the metals were the same for all acids. The same holds for the two alkalis, which could have been represented by one generic reference substance if the replacement pattern of the three mineral acids were the same in both cases.[7] But these consequences amount to no more than modifications of the principal arrangement of the substances at the heads of the columns, which was ruled by specific chemical conceptions as shown in the previous chapter—by the conception of the neutral salts and by that of metals and alloys. We argue that this means that the arrangement of the reference substances at the head of the columns is the chief means by which Geoffroy's table renders the clusters of classes around the conceptions of neutral salts and of metals and alloys.

Beginning with the right-hand side of the table and ignoring for the moment columns nine and sixteen, we have six columns (ten through fifteen) dedicated to the dry processes of composing and decomposing alloys. Geoffroy gave no hint as to whether or not the succession of these six reference substances had a particular meaning.[8] It is hard to imagine any classificatory significance of this sequence. The contingent arrangement of the group, however, does have such significance. It is certainly not by chance that no column with an entirely different reference substance was inserted between two columns with metals at their head. Thus, the *en suite* arrangement of the six reference metals of the dry way must be taken as the way in which the table explicitly renders the substances in columns ten through fifteen as the class of building blocks of a distinct kind of compounds, namely the class of alloys.

A similar rendering of connected classes by a specific arrangement of the substances at the head of columns can be observed on the left-hand side of the table. This time the classes in question are connected by the conception of neutral salts; that is, the classes of acids, earths, alkalis, and metals. As already discussed, different replacement patterns rather than classificatory considerations induced Geoffroy to post the three single mineral acids as reference substances in addition to the generic acid, and, similarly, the two different alkalis instead of one generic one. However, this does not disturb the arrangement of the reference substances of the first eight columns, which are ordered in such a way that the four classes of substances involved in the formation of neutral salts can easily be discerned as distinct units. The sequence of these units—first acids, then earth, alkalis, and, finally, metals—is strikingly reminiscent of the *Tableau* of 1787. There, in column I, the column of the simple substances, the class of the most simple substances was followed first by that of the acidifiable bases (field I/b) and then by the three traditional classes of the salifiable bases (fields I/c–e).[9] Since the acids are (relatively) simple substances in the frame-

---

7   There is furthermore the difference that the volatile alkali was thought to form salts only with the mineral acids, whereas the fixed alkali was thought to do so with two more substances as well.

8   The frequency with which mercury and sulfur are adjacent reference substances in eighteenth-century affinity tables suggests that it is no coincidence that the series of these six metals is led by mercury, which also enjoyed an exceptional position among the metals at the time.

work of the phlogistic chemical system and thus would be placed in the field occupied by Lavoisier's acidifiable bases, the arrangement of these fields is in striking conformity with that of the units of reference substances at the head of Geoffroy's columns one through eight.[10] In any case, the most interesting feature of this sequence consists in the *en suite* arrangement of the three classes of salifiable bases. Through this, the classificatory division of classes of salt-forming substances into acids and salifiable bases is also rendered in the tables' arrangement of the reference substances.

Our view, that the arrangement of the reference substances can be seen as the way in which Geoffroy's affinity table renders the clusters of classes around the conceptions of neutral salts and of metals and alloys, presupposes, of course, that independent evidence can be given that the classificatory distinctions involved were actually made at the beginning of the eighteenth century. However, before coming to such evidence, we will examine in an excursus whether the arrangements with classificatory significance in Geoffroy's table were really representative of eighteenth-century affinity tables.

Excursus: Classification by arrangement in affinity tables after 1718

With respect to several features of essential significance for classification, we already argued in the previous chapter that Geoffroy's table can claim to be representative of all affinity tables of the century. Firstly, all affinity tables were constructed on the basis of replacement reactions and were thus confined to pure chemical substances. Secondly, all affinity tables order chemical substances according to composition. In this excursus, we will investigate the issue of whether the way in which Geoffroy's table arranged the reference substances is also representative of the subsequent affinity tables. This question deserves attention because of the assumed classificatory significance of this arrangement pattern.

The question of whether Geoffroy's arrangements are representative of the subsequent affinity tables concerns, of course, only affinity tables with symbols in columns. Duncan listed fourteen such tables, which were published before the *Tableau* of 1787.[11] We can confine our examination to eight of these tables,[12] ignoring those that are more or less faithful variations of others,[13] or deviate so far from Geoffroy's arrangement that a comparison is impossible.[14]

---

9 When referring to the *Tableau's* fields—field I/a, etc.—we refer to the diagrams of chapter 5, particularly figure 5.4.

10 The circumstances for placing the metals before the other classes of salifiable bases in the *Tableau* of 1787 were addressed in chapter 5.

11 Duncan [1996] p. 112f. Concerned are his nos. 2, 4, 6, 7, 8, 9, 10, 12, 13, 14, 17, 17a, 17c, and 17d. Possibly no. 15 should be added if a copy could be found.

12 Grosse's table of 1730, Duncan's no. 2, republished in Demachy [1774] plate ii; C.E. Gellert's table of 1751, Duncan's no. 4, again published in Gellert [1776]; A. Rüdiger's table of 1756, Duncan's no. 6, republished in Demachy [1774] plate iv; J.P. Limbourg's table of 1758, Duncan's no. 7, republished in Demachy [1774] plate vi; H. M. Rouelle's table of 1763, Duncan's no. 9, published anonymously in vol. III of *Recueil des planches* of Diderot's *Encyclopédie*; Demachy's table of 1769, Duncan's no. 12, in Demachy [1774] last plate; B. G. Sage's table of 1773, Duncan's no. 14, in Sage [1773]; and T. Bergman's table of 1783, Duncan's no. 17a, in Bergman [1785].

Geoffroy 1718	4 AC	E	2 AL	SM	S	6 M	OS			16
Grosse 1730	5 AC	E	2 AL	S	ME	SM	5 M	3 OS		19
Gellert 1751	5 E	2 AL	5 AC	OS	S	OS	12 M	OS		28
Rüdiger 1756	4 M	S	AC	AR	OS	4 AC	AL	2 OS		15
Limbourg 1758	6 AC	2 AL	3 E	OS	S	2 OS	AR	6 OS	11 M	33
Rouelle 1763	4 AC	E	2 AL	2 SM	S	6 M	3 OS			19
Demachy 1769	6 AC	E	2 AL	BO	SM	3 OS	AR	S	OS	3 M · 20
Sage 1773	5 AC	S	AL	E	OS	SM	ME	2 OS		13
Bergman 1783	25 AC	5 E	3 AL	4 OS	S	5 OS	16 CX / 16 M			59

Legend: AC acid, AL alkali, AR arsenic, BO borax, CX metallic calces, E earth, M metal, ME mercury, OS other substances, S sulfur, SM metallic substances in general.

	Conception of neutral salts						
	Acids	Salifiable Bases					
Geoffroy 1718	ACIDS	EARTH	ALKALIS	SM	S	METALS	x
Grosse 1730	ACIDS	EARTH	ALKALIS	S x SM		METALS	x
Gellert 1751	EARTHS	ALKALIS	ACIDS	x	S x	METALS	x
Limbourg 1758	ACIDS	ALKALIS	EARTHS	x	S x	METALS (SM)	
Rouelle 1763	ACIDS	EARTH	ALKALIS	SM	S	METALS	x
Demachy 1769	ACIDS	EARTH	ALKALIS	x SM	x S x	METALS	
Sage 1773	ACIDS S	ALKALI	EARTH	x SM		METALS	x
Bergman 1783	ACIDS	EARTHS	ALKALIS	x	S x	CALCES / METALS	

Figure 9.1: Sequences of the reference substances in affinity tables: 1718–1783.

---

13 This holds for Duncan's nos. 10, 13, 17c, and 17d. Bergman's table in the first Latin edition of Bergman [1785], Duncan's no. 17, can also be ignored. To some extent, H. M. Rouelle's table, Duncan's no. 10, is only a variation of Geoffroy's. However, because it was published in Diderot's *Encyclopédie*, it should not be skipped.

14 This holds for Duncan's no. 8.

In our figure 9.1, the upper table shows the sequences of reference substances at the heads of the columns in the eight selected affinity tables, to which is added, at the top, the sequence in Geoffroy's table. In each line, the succession of the abbreviations for substances from the left to the right renders exactly the sequence of the heading reference substances and, thus, of the columns in the respective table. For the most part, abbreviations of class names such as acid, earth, etc. are listed, and the numbers record how many reference substances of the same kind follow without interruption. In this manner one can sum up the columns of a table and arrive at the number given in the last column of our table. With the exception of sulfur (S), substances which do not belong to the clusters of classes around the conceptions of neutral salts and alloys are indicated by OS, the abbreviation for "other substances." In the lines for Rüdiger's, Limbourg's, and Demachy's tables, arsenic and borax are listed separately because it is not entirely clear whether or not these authors recognized them as metals or as acids.[15] In the lines for Grosse's and Sage's tables, mercury is listed separately since it has the appearance that these authors wished to stress mercury's special position among the metals.

At first glance, the arrangements of the columns do not seem to follow any common rule. Upon closer inspection, however, one discovers a shared arrangement pattern in the six French tables. All of them begin with acids as reference substances and continue with the earth and alkalis—mostly the earth before the alkalis, but sometimes the other way around. Only Sage's table is an exception, for it inserts the column with sulfur as a reference substance between the acids and these two classes of salifiable bases. So far, even Bergman's table fits this pattern. In all the tables but that of H. M. Rouelle, who followed Geoffroy's arrangement more closely than anybody else, the third traditional class of salifiable bases, namely the class of metals (SM), does not follow immediately after these two classes. Rather, one or more columns with different reference substances—sulfur and mercury in the table by Grosse, borax in that by Demachy, and liver of sulfur in that by Sage—precede the column with the metallic substances. In Limbourg's table this SM column seems to be omitted altogether, but Limbourg, anticipating Bergman's later arrangement to some extent, listed the respective reactions with acids in the columns with the single metals as reference substances, which had usually been reserved for reactions in the dry way until that time. Finally, between the SM column and those with the single metals as reference substances, more columns with different reference substances are inserted, among which sulfur often occurs in accordance with the example of Geoffroy's table.

Bergman's table, notwithstanding its new mode of representing reactions in the wet way and dry way, obviously follows this French arrangement to a large extent. The same cannot be said of the tables by Gellert and Rüdiger. If the arrangement of the columns in the latter's table followed any rule at all, it is certainly hard to determine which one. Gellert's table gives evidence of a field of chemical theory and practice in which compounds of earths and metals form the core, whereas salt formation

---

15 For the uncertainties of eighteenth-century chemists with respect to the nature of arsenic and borax, see Kopp [1966] vol. IV p. 92f. and vol. III p. 340ff.

plays a rather marginal role. Therefore its arrangement must be interpreted against this new background, which cannot be done here.

The lower table in figure 9.1 tries to render the patterns of arrangement that we observed. The table by Rüdiger is left out because of the apparent lack of order already mentioned. All "other substances," this time arsenic, borax, and mercury included, are indicated by an x, regardless of whether a single substance or a couple of such substances is concerned. The exceptional position of Gellert's arrangement is obvious. But equally obvious is the high degree of conformity among all other tables. The most important deviation from Geoffroy's arrangement consists in Limbourg's and Bergman's new rendering of the wet way and dry way by integrating the SM column as reference substances in the columns with single metals. The place for the column with sulfur as a reference substance is another major issue of divergence.

With respect to classification, we can state first that all tables group the reference substances according to classes. This is obvious in the case of the classes pertaining to the fields of metallurgy and salt formation: the different acids, earths, and so on are arranged as neighboring reference substances. That this is not entirely trivial is proven by Rüdiger's table, which does not follow this rule consistently. With regard to the compounds listed implicitly by the tables (as discussed in the previous chapter) it seems unlikely that the tables represent a tacit classification of these substances as well. In the case of Geoffroy's table, the assumption of such a classification may seem sensible despite the fact that, as discussed earlier, the choice of reference substances depended not exclusively on classificatory considerations, but also on the patterns of replacement reactions displayed by the different substances. In later tables, however, we find compounds lumped together in one and the same column whose only commonality is that the reference substance is one of their components. In Bergman's table, for instance, the salts formed by one of the acids with the different salifiable bases are in company with the solutions of the acid in spirit of wine and in water. Such a grouping made no classificatory sense.

The arrangement of the cluster of classes around the conception of neutral salts deserves particular attention and is therefore indicated at the head of the lower table in figure 9.1. Aside from Gellert's table, we always find the same succession of these classes as in Geoffroy's table. The acids are at the front of the row, the earth/alkali columns follow next, and the SM column comes last (provided the SM class has not already been integrated into the metal columns, as in Limbourg's and Bergman's tables). However, apart from H. M. Rouelle's table, no table follows Geoffroy's example in arranging the SM column immediately after the columns for the two other classes of salifiable bases. Rather, columns with reference substances are inserted before the SM column, although the reference classes they contain do not pertain to the classes of substances around the conception of neutral salts.[16]

In summary, we can state that the classes of acids, earths, alkalis, and metals (as to the wet way) were arranged according to the conception of neutral salts. Moreover,

---

16 As mentioned above, a closer inspection could possibly prove that Grosse and Demachy still regarded arsenic, borax, and mercury as such substances.

the constant position of the acids in front seems to designate them as the substances that form salts with the substances of the three other classes. But it remains somewhat uncertain as to whether or not these affinity tables also rendered the alkalis, earths, and metals as classes that pertained to the higher taxonomic unit of salifiable bases. However suggestive these classificatory features of the affinity tables may be, most of them are rendered only by the tables' arrangement. They therefore remain questionable to the extent that no evidence was given to distinguish among the assumed classificatory units independent of the affinity tables. Thus, the explicit classificatory orders and distinctions of pure chemical substances prior to the *Tableau* from 1787 must be investigated next.

9.2 The classification of minerals before 1787

Although classificatory orders and distinctions of chemical substances can be found in almost all chemical textbooks of the eighteenth century, to our knowledge there is no single instance of a comprehensive and systematic classification prior to the *Tableau* of 1787 that focused on pure chemical substances. Rather, one finds either partial and isolated classifications of certain groups of such substances, for example salts, or classifications that tried to comprise all kinds of chemical substances, both pure ones and natural raw materials, many of which were mixtures from today's point of view. For a picture of explicit classificatory distinctions of pure chemical substances and the relations and arrangements of such taxa in the eighteenth century, one has to investigate which classificatory order was given to these substances in such broader classificatory attempts.[17]

Bergman's classification of the mineral kingdom

For our investigation of explicit classifications of chemical substances before the *Méthode*, an *Essay* by Torbern Bergman (1735–1784) of 1779 is a particularly suitable starting point.[18] The essay—whose 1783 English translation was entitled *An Essay on the General Usefulness of Chemistry, and its Application to the Various Occasions of Life*—was a survey of chemistry, its chief subject matters, and its various applications, addressed to a broad audience.[19] In its second part, the essay gave a description of the realm of natural bodies of the mineral kingdom which is nothing

---

17 This implied some obstacles that chemists had to overcome, for such all-comprising naturalistic classifications followed principles of classification that sometimes proved to be harmful for a classification of pure chemical substances. This holds for even the most general principle usually employed, namely the distinction and separation of substances according to their provenance; that is, to which of the three kingdoms of natural beings they belonged. True, the bulk of pure chemical substances of the age pertained to the mineral kingdom. But there were also very important substances from the two other kingdoms, in particular vegetable and animal acids and alkalis, which were separated from the related mineral substances (the mineral acids) by this most general principle. In what follows, we will ignore this separation.

18 On the life and works of Bergman, see Partington [1961–1970] vol. III p. 179ff.; and Gillispie [1970–1980] vol. II p. 4ff.

19 Bergman [1783].

other than the outline of a systematic classification of these substances and can, therefore, be summarized in a diagram like that in our figure 9.2.

On the highest taxonomic rank, Bergman distinguished six classes—the salts, the earths, the inflammable bodies, the metals, the waters, and the airs. For our purpose we can ignore the two last classes and focus on the four traditional classes of the mineral kingdom.[20] The distinction of these four classes is doubtless very old—Bergman himself ascribed it to the Arabian physician Avicenna (980–1037)[21]—and it may be surprising to see it still alive in the eighteenth century. Actually, at that time this overarching division of minerals was circulating in several variations. For example, Homberg classified the minerals, as discussed earlier, into metals, salts, stones, and earths, and did not mention the inflammable bodies; Boerhaave divided the minerals into the classes of metals, salts, sulfur, stones, earths, and semi-metals;[22] the German chemist Johann Christian Wiegleb distinguished five classes—the earthy bodies, the metals, the salts, the waters, and the inflammable bodies.[23]

Such divisions of the mineral kingdom into four or five main units are obviously not classifications according to composition. The criteria applied for these divisions were perceptible properties such as solubility in water, taste, lustre, malleability, combustibility, and so on.[24] We encountered these criteria back in chapter 7 in connection with the classes of simple substances in the tableau of the *Méthode*. Yet it is very important to notice the decisive difference between their employment there and here. In the *Méthode* these criteria classified a particular type of pure chemical substances, namely the simple substances. Their classification according to such criteria as solubility complemented the principle of classification according to composition employed for the second type of pure substances, the chemical compounds, which were clearly demarcated from the simple substances. This does not hold in the case of contemporary all-comprising classifications of the mineral kingdom like Bergman's classification of 1779, for the main classes of these classifications contain lower-ranking taxonomic units of simple substances as well as compounds, that is, natural raw materials such as stones, ores, and mixtures of salts and earths. They do not unify taxonomic units of substances that are analogous with respect to their chemical composition. Rather, both simple and more compound substances are grouped together into one and the same class on the basis of shared perceptible properties. And this seems to be absolutely appropriate for a classification of the mineral kingdom which ordered pure chemical substances alongside natural raw materials.

20 Bergman himself usually confined the classes of the mineral kingdom to the first four—see, for instance, Bergman [1782] p. 18ff. and Bergman [1791] p. 227. In eighteenth-century classifications of the mineral kingdom, "fossils" or "petrifactions" usually constituted an additional class.
21 See § 45 of Bergman [1791] p. 227.
22 See Boerhaave [1732] vol. I pp. 31–56.
23 See Wiegleb [1786] p. 4; the first edition of this *Handbuch* appeared in 1781. For an overview of classifications of the mineral kingdom in the mid-eighteenth century, see the table in Oldroyd [1974b] p. 508f.
24 For an explicit discussion of such criteria, see §§ 47–54 of Bergman [1791] pp. 228–233.

Class		Genus		Species	
Salts	simple	Acids	mineral	6 species	
			vegetal	several species	
			animal	2 species	
			all kingdoms	Aerial acid	
		Alkalis	fixed	Vegetable fixed alkali	
				Mineral fixed alkali	
			volatile	Ammoniac salt	
	compound	Neutral salts		several species	subdivided according to perfect / imperfect, and double / triple / quadruple
		Middle salts	earthy	several species	
			metallic	several species	
Earths	simple		absorbent	4 species	
			not absorbent	Siliceous earth	
	compound			several species	
Inflammable substances		Sulfur		several species	e.g. Sulfur, Phosphorus
		Oils	essential	several species	
			unguinous	several species	
		more genera ?		several species	
Metals		Proper metals	noble	3 species	as many species of metall calces
			ignoble	several species	
			neither nor	Mercury	
		Semi-metals		several species	

Figure 9.2: Torbern Bergman's classification of the mineral kingdom from 1779.

It was characteristic of the classificatory criteria employed in mineralogical classifications that they allowed primarily for physical properties and only secondarily for chemical ones. However, the disappearance of a distinct class of stones in Bergman's classification, one traditionally distinguished from the earths because of its hardness, is indicative of an increasing weight of chemical characteristics. Bergman grouped together stones with earths because they did not differ in solubility. The same holds for the subsumption of the semi-metals under the metals. The work of the Swedish mineralogist Axel Fredrik Cronstedt (1722–1765) marked a turning point in this respect.[25] Cronstedt aspired to a systematic mineralogy that distinguished and determined minerals according to their composition.[26] The Swedish tradition in mineralogy established by Cronstedt constitutes the historical background for why pure chemical substances, although a minority among the bodies of this kingdom, played such a dominant role in Bergman's classificatory attempts.[27] Traditionally in mineralogical classifications the main classes of minerals were neutral with respect to the distinction of pure chemical substances and natural mixtures. This also holds for the classification in Bergman's 1779 *Essay*. Yet almost all of the taxonomic units of species contained in his four classes were actually pure chemical substances, as can be seen from the fact that almost all of them can also be found in Bergman's affinity table.[28] However, the dominance of pure chemical substances did not change the principal mode of this classification, which remained one that distinguished the main classes of the mineral kingdom according to physical and chemical properties rather than chemical composition.

It is Bergman's focus on pure chemical substances that makes his classification so suitable a starting point for our investigation. Following its main features rather than going into every of its details, we can distinguish classificatory orders and distinctions whose history is exactly what we are looking for. In the following, we will, therefore, successively discuss these features and try to trace the tradition and historical development of each one. We will begin with the class of metals, come then to the class of inflammable bodies, discuss subsequently that of the earths, and, finally, deal with that of the salts.

Metals and calces

Within the class of metals, Bergman distinguished between proper metals and semi-metals. The former genus was subdivided into the noble metals with the species platinum, gold, and silver; the ignoble metals with several species; and a taxon of metals that are neither noble nor ignoble and contains only mercury. All metals and semi-metals could be subjected to calcination and each thus yielded its specific metallic

---

25 On the life and works of Cronstedt, see Partington [1961–1970] vol. III p. 173ff.; and Gillispie [1970–1980] vol. III p. 473ff.
26 On Cronstedt's classification of minerals, see Porter [1981] p. 558ff.; and Oldroyd [1974b].
27 Cronstedt's impact on Bergman becomes particularly clear in one of the last works by the latter, the *Meditationes de systemate fossilium naturali* of 1784 (English translation: Bergman [1791]). In this work, Bergman explicitly accepted the principle of determining the species and varieties of minerals according to their chemical composition. See § 21 of Bergman [1791] p. 215f.; and Porter [1981] p. 562.
28 The species of the compound earths were very likely not pure chemical substances, but mixtures, and constitute the most important exception in this respect. We will come back to this issue below in the context of classifications of earths. It is not clear whether or not Bergman also regarded the doubtful genera of the class of inflammable substances as such pure substances. There we find mixtures such as gunpowder, pulvis fulminans, and so on. See §§ 235ff. of Bergman [1783] p. 111ff.

calx.[29] These distinctions and their underlying criteria were not invented by Bergman. Rather, here we encounter the familiar cluster of taxonomic units around chemists' collective conception of metals.[30] These taxonomic distinctions emerged in the context of metallurgy, and chemists' incipient understanding of the substances as pure chemical substances can be traced back to the famous metallurgical treatises of the sixteenth century. A few remarks on metallic calces should be added since we did not encounter this kind of substance in affinity tables prior to that of Bergman.[31]

As is well known, the Lavoisierian metal oxides had been called "metallic calces" since the Middle Ages.[32] They were understood to be the result of the calcination process through which, as the name suggests, metals are transformed into a substance reminiscent of lime (the Latin is *calx*). As far back as antiquity, an analogy was seen between the processes of calcination and combustion. Both processes were regarded as disruptions in which the initial material was transformed into "ashes," as the metal oxides were called in antiquity. In contrast to the combustion of other materials, however, metals could be regained from their ashes, as metallurgical experience had proved. Therefore, after the emergence of the modern conceptual system of compound and reaction at the turn of the seventeenth century, the recovery of the metal was understood as a recombination of the calx with the component lost during calcination.[33] This latter component became known under the name phlogiston, which went back to G. E. Stahl's theory of calcination. According to Stahl's theory, which was unanimously shared by eighteenth-century chemists before Lavoisier, a metal was the compound of its calx with phlogiston, and, thus, a metallic calx was a simpler substance than a metal.

As calcination was not a reaction that could be understood as a replacement reaction, it was naturally absent from affinity tables. Only in Bergman's table of affinities are metal calces included, specifically in connection with the formation of salts. According to Bergman, who is credited with having been the first to state it clearly, acids formed salts not with metals, but rather with metallic calces.[34]

Inflammable bodies

The inflammable bodies of the mineral kingdom were traditionally classified into sulfur, liquid inflammable materials like petroleum, and tenacious or solid ones like mineral rosins (*resina terrae*), mineral pitch, coal, and so on.[35] Bergman's classifica-

---

29 See § 241 of Bergman [1783] p. 114. Whether these specific metallic calces could be reduced to one common calx or must be regarded as distinct calces was a disputed question up to the second half of the eighteenth century—see Kopp [1966] vol. III p. 142f.; and Cassebaum and Kauffman [1976] p. 450f. It was also an open question whether the metallic calces must be distributed into the class of metallic substances or into that of the earths. For many chemists regarded these calces as a particular earth—see, for instance, the article "chaux métallique" in Macquer [1766].

30 Only the unit of the alloys is left out of this classificatory outline. In the *Meditationes* of 1784, each metal has its alloys, ores, and calx—see § 145 of Bergman [1791] p. 277f.

31 Only Gellert's table of 1751 differentiated between metals and metallic calces in two cases.

32 See, for instance, Kopp [1966] vol. III p. 103ff.; Crosland [1962] p. 108.

33 Up to this point in time, the calcination of metals, regardless of whether by fire or by acids, was often understood as a mechanical pulverization.

34 See Kopp [1966] vol. III p. 103ff.; and Partington [1961–1970] vol. III p. 191.

tion in the *Essay* of 1779 by and large fits into these common classificatory divisions, except for its neglect of the tenacious and solid inflammable minerals.[36] More attention is due to the fact that Bergman questioned whether or not the usual genera of this class could be discerned distinctly in the traditional way. He wrote in the *Meditationes* of 1784, "In the strictness of language, all the genera of this class might be reduced to one, as the same principle of inflammability prevails in each of them."[37] Bergman obviously cast doubts on the possibility of clearly distinguishing genera of the inflammable mineral bodies, given the common perception that all of them were compounds of phlogiston in which the phlogiston prevails. His doubts thus appear inspired by classificatory considerations about the conception of composition.[38]

Probably the most striking feature of the class of inflammable bodies is the fact that sulfur appears in it twice—once as a genus and a second time as a single species. Bergman's generic sulfur and its species deserve special attention, especially in comparison to the class of acidifiable bases contained in the *Tableau* of 1787. Generic sulfur had a long history.[39] In ancient times, "sulfur" had been the generic name for all inflammable materials because it was regarded as their most important constituent principle. Around 1700, phlogiston became the modern successor of this generic sulfur, and it was therefore no coincidence that Geoffroy called it "sulfur principle." As a consequence of these theoretical transformations, sulfur lost its status as a principle and was instead taken to be a compound of an acid and phlogiston. But, and this is the interesting point, up to Bergman's day, "sulfur" was the name for the combination of sulfuric acid with phlogiston and, at the same time, for compounds of all kinds of acids that solidify when combined with phlogiston.[40] The old generic understanding of sulfur as a name for all inflammable bodies thus took on a new and precise meaning in the framework of the phlogistic chemical system. We argue that this generic sulfur was merely the phlogistic counterpart of the class of the acidifiable bases in the tableau of the *Méthode* (field I/b).

The history of phosphorus in the eighteenth century strongly corroborates the assumption that a phlogistic counterpart of this class was under way.[41] Bergman's determination of the phosphorus as "species of sulfur," striking to a contemporary reader, appears natural given this new generic understanding of sulfur.[42] According to G. E. Stahl's theory, phosphorus was a compound of phlogiston and an acid. How-

---

35 See, for instance, the articles "Brennbare Materien" and "Erdharze" in Gehler [1787–1796] vol. I p. 440 and vol. II p. 12, respectively. The classification of this group of minerals was affected by debates about a possibly vegetable origin of substances like amber and petroleum.

36 We can ignore this deviation, however, since Bergman, in other classificatory schemes, always included samples of these missing substances in the class of inflammable bodies. See §§ 141f. in Bergman [1782] p. 95f. and §§ 86ff. in Bergman [1791] p. 252f.

37 See § 89 of Bergman [1791] p. 253.

38 The phrase "principles that prevail," however, is strikingly reminiscent of the old theory of elements and principles, and the arguments Bergman used in the characterization of the genera (see §§ 84–87 of Bergman [1791] p. 252) remind us of what we could observe in Homberg's classification from 1702—see the excursus in chapter 6.

39 See Kopp [1966] vol. III p. 301ff.

40 See, for instance, Gehler [1787–1796] vol. I p. 440.

41 See Kopp [1966] vol. III p. 331f.

42 § 215 of Bergman [1783] p. 104.

ever, the question of exactly which acid formed phosphorus with phlogiston remained in dispute until the 1740s, when the German apothecary and chemist Andreas Sigismund Marggraf succeeded in identifying the phosphoric acid as a new peculiar acid rather than an impure mixture of one of the well-known mineral acids. Thus, phosphorus could be understood as a compound exactly analogous to sulfur. This analogy, and with it the phlogistic understanding of the generic sulfur, is explicitly expressed in the *Essay* of 1779, in which Bergman stated immediately after the paragraph on phosphorus: "The other acids no one as yet has been able to combine with phlogiston, in such a manner as for them to form solid sulfur."[43]

Only three years later Bergman published results on a further species of generic sulfur, namely "the aerial acid, saturated with phlogiston."[44] Aerial acid—renamed carbonic acid in the Lavoisierian system and thus recognized as a carbon oxide—saturated with phlogiston was the same thing as carbon. Thus carbon, the third substance in the class of acidifiable bases of the *Méthode*, reappears in the phlogistic system as the third species of generic sulfur. This third species of generic sulfur is particularly telling, since its interpretation as a compound of aerial acid with phlogiston included a phlogistic reinterpretation of Lavoisier's understanding of this acid as a carbon oxide. For it was Lavoisier who, back in the 1770s, within the developing framework of his anti-phlogistic chemical system, established the connection between carbonic acid—Bergman's aerial acid—and carbon. And this insight by Lavoisier was Bergman's only justification for classifying carbon as a species of generic sulfur.[45] The fact that Bergman's phlogistic reinterpretation of Lavoisier's discovery led immediately to the classification of carbon as a sulfur species confirms our argument that the genus of generic sulfur was nothing more than the phlogistic mirror image of the class of acidifiable bases in the anti-phlogistic classification of the *Méthode*.

Earths

Within the class of the earths, Bergman distinguished between simple and compound earths and subdivided the former into four species of absorbent earths and a fifth one, the siliceous earths. In his *Essay* of 1779 Bergman defined the compound earths as "those which consist of two or more simple ones combined."[46] Compound earths were commonly thought to represent a considerable part of the materials belonging to the mineral kingdom. In the *Meditationes* of 1784 Bergman distinguished exactly

---

43 § 216 of Bergman [1783] p. 105. Strikingly, phosphorus is not listed among the species of this generic sulfur in Bergman [1782] and [1791]. A possible reason could be that phosphorus was regarded as a substance of the animal rather than of the mineral kingdom.

44 See § 135 of Bergman [1782] p. 93: "phlogiston acido aereo satiatum;" also see § 159 of Bergman [1791] p. 283.

45 See Kopp [1966] vol. III p. 282–286.

46 See § 194 of Bergman [1783] p. 93. In this essay Bergman did not address the question of whether chemical compounds or mixtures are brought about by such combinations. In other writings, however, he distinguished between compound earths that are "intimately united" (*intime unitae*) and those that are mechanical mixtures (*mechanica miscela*). See § 83 of Bergman [1782] p. 57; see also § 79 of Bergman [1784] p. 249.

320 compound earths resulting from all possible mixtures of the five different earths and an exhaustive combinatory algorithm of their different proportions.[47]

The subdivision of the simple earths into five different species documents chemists' achievements in chemical analysis and identification of substances in the second half of the eighteenth century. Up to the 1750s, only two simple earths were undisputedly distinguished, siliceous and calcareous earth. The latter relates to the absorbent earth of Geoffroy's affinity table. Using the adjective "absorbent," chemists indicated that this earth could be dissolved in acids, which was impossible in the case of siliceous earth.[48] Prior to Bergman's table of 1783, therefore, siliceous earth did not occur in affinity tables, which listed earths in connection with reactions in the wet way. Geoffroy's use of the plural form "absorbent earths" indicates that, at the beginning of the century, the term denoted a group of dissolvable earths rather than a single earth.[49] This generic absorbent earth indiscriminately comprised all known dissolvable earths. Calcareous earth did not become a distinct, particular, absorbent earth until argillaceous earth (alumina), magnesia, and ponderous earth (baryta) could be distinguished and determined as three different particular dissolvable earths in the third quarter of the century.[50]

Chemists' successful discrimination of substances was embedded in a deep transformation of the conception of the matter earth, which played a pivotal role in ancient chemistry.[51] Earth, one of the Aristotelian elements, was also an ultimate principle in the seventeenth-century Paracelsian chemical philosophy of elements and principles, remnants of which were still alive in the early eighteenth century. In the works of Johann Joachim Becher (1635–1682), an influential German chymist,[52] earth even achieved the status of the most fundamental principle, from which the classical Paracelsian principles mercury, sulfur, and salt could be derived as modifications, namely as mercurial earth, as fatty or inflammable earth, and as vitrescible earth. In other strands of this chemistry of simple principles, earth, together with water, was a principle in addition to the Paracelsian principles. More important than such differences, however, was the ontological status of the earth principle in this philosophical framework. Like all other principles, it was conceptualized as a generating principle rather than a physical component part of perceptible substances (see chapter 2). This changed during the first half of the eighteenth century, when the ontological nature of the principle of earth was transformed into that of a primitive earth and understood as the essential component of all ordinary, perceptible earths, which owed their differences to further admixtures. Around the middle of the century, chemists no longer

---

47 See §§ 135–145 of Bergman [1791] pp. 270–277.
48 The dissolution of the siliceous earth in hydrofluoric acid was accomplished and reluctantly recognized in the last quarter of the eighteenth century—see Kopp [1966] vol. III p. 368f. Bergman mentioned this dissolution of siliceous earth, however, he explicitly maintained the name absorbent earths for the four other simple earths—see §§ 206–207 of Bergman [1783] p. 99f.
49 See Geoffroy [1718] p. 204.
50 Alumina by Marggraf 1754, magnesia by Black 1755, and baryta by Scheele 1774—see Kopp [1966] vol. IV pp. 61, 54, 43.
51 See Oldroyd [1974a], and [1975].
52 For life and works of Becher, see Partington [1961–1970] vol. II p. 637ff.; Gillispie [1970–1980] vol. I p. 548ff.; and Smith [1994].

took the fact that they were not able to isolate primitive earth as a proof of its distinctive ontological nature, but rather as a result of the still imperfect state of the chemical art.

At the turn of the seventeenth century, most chemists regarded siliceous earth as that kind of ordinary earth which came closest to the sought-for primitive earth. Convictions like these were still current in 1766 when Pierre Joseph Macquer published his famous *Dictionnaire de Chymie*.[53] But at that time, chemists animatedly discussed the question of whether different earth principles actually existed.[54] This discussion proved to be the final phase in the life of the earth principle. The background of this debate was the recognition of the distinct character and equal status of the particular earths distinguished so far, of which there were then four. Each of these substances resisted further decomposition, so none of them could claim superiority with regard to simpleness. Furthermore, there were no shared reasons to regard the characteristic properties of one of these earths as more primitive than those of the others. All that remained was thus to draw the conclusion that these different earths were undecomposable "primitive" substances. And it was not Lavoisier, but rather Bergman who drew this conclusion, in his *Sciagraphia regni mineralis* of 1782.[55] As he put it in 1784:

> But until proper experiments shall have fully developed the nature of such compositions
> [sc. the distinct earths] they must be, in respect to our knowledge of them, considered as
> primitive substances.[56]

With this recognition by Bergman, who gave voice to a much broader movement at the time, the earth principle faded away, leaving behind only ordinary earths—common chemical substances.

The deep conceptual change in the chemical notion of earth in the last third of the eighteenth century is exemplary for the slow transition in which the chemistry of ultimate principles gave way to the modern conceptual system of chemical compound and affinity.[57] The coexistence of notions that are in retrospect incompatible is characteristic of this phase of transition. Therefore, it is often difficult for historians to determine with which kind of entity a chemist was dealing, an ultimate generative principle or a bodily component of a substance. The historical actors' terms are not the arbiters in this case, as they used the relevant terms ambiguously (see chapter 2). This terminological difficulty particularly affects our understanding of classificatory attempts in early modern chemistry. As we showed above in chapter 5, discussions about the ultimate principles of classes of substances can easily be mistaken for classifications according to chemical composition. The case of the earths further shows

---

53 See Macquer [1766] vol. II p. 567.
54 Ibid. p. 563f.
55 See Bergman [1782] p. 57ff.
56 Bergman [1791] p. 245f.
57 The same conceptual change could be demonstrated with regard to the principles of acids, metals, and so on. The originally supposed principles of acids or metals, for instance, first gave way to a primordial acid and a primordial metal that gradually faded away in the course of the eighteenth century, leaving behind the ordinary substances. Traces of this can be found even after Bergman and Lavoisier—see Cassebaum and Kauffman [1976] p. 448ff. See also Siegfried and Dobbs [1968] p. 276f.

that not only ordinary chemical substances were subjected to classification, but the
ultimate principles as well. Both classifications must not be confused as concerning
the same kind of entities.

Salts

In the class of salts, Bergman distinguished four genera—the genus of acids, that of
alkalis, that of neutral salts, and that of middle salts (figure 9.2). Both the classification and the nomenclature of salts had been among Bergman's main concerns since
the 1760s.[58] He undertook several attempts to classify this group of substances,
which differ only slightly from one another.[59] It was the subdivisions of the neutral
salts and the middle salts which troubled Bergman in particular. He tried to classify
the substances of these taxonomic units on the basis of two different criteria—on the
one hand, according to saturation ("perfect/imperfect") and, on the other, according
to the order of chemical composition ("double/triple/quadruple"). As a consequence,
he arrived at a highly complex solution.[60]

With his distinction of four genera of salts, Bergman by and large followed an
established tradition (figure 9.3), except for his distinction between neutral and middle salts. Whereas in France all compound salts were called neutral salts since G. F.
Rouelle's *Mémoire* of 1744, regardless of whether they were formed by an acid with
an alkali or with an earth or metal, Bergman restricted the name neutral salts to salts
formed with alkalis. His subdivision of acids according to provenance and of alkalis
according to the state of aggregation under normal temperature conforms with a classificatory tradition that can be traced back at least to the seventeenth century.

Bergman's grouping of acids and alkalis into the class of salts, which may be
striking today, was quite natural given his distinction between the four classes of minerals, which followed an ancient tradition. All mineral bodies that were incombustible, soluble in water, and excited "a pungent taste on the tongue" were called salts.[61]
With regard to these characteristics, acids, alkalis, and neutral salts were indeed identical. In this framework, chemists regarded acids and alkalis not only as relatively
simple substances, but also as the proper salts, whereas they took the neutral salts to
be both chemical compounds and, as the names indicate, substances that must be

---

58 See Beretta [1993] p. 138ff.
59 Five years after the *Essay*, in his *Meditationes*, Bergman expressed some doubts as to whether neutral
    salts constitute a distinct genus or should be regarded as species of the acid, if not perfectly saturated,
    or of the alkali, if perfectly saturated—see §§ 67–68 of Bergman [1791] p. 241f. It also appears that he
    proposed to regard the middle salts, which he then named "analogous" salts, as species of the acid.
    Furthermore, he regarded double salts containing metals (metallic calces) or earths and an alkali as such
    analogous salts, and suggested classifying them as species of the alkali—see ibid. § 70 p. 243 and § 130
    p. 268. Taking together all paragraphs on salts (§§ 64–72 and 107–130), however, no coherent picture
    seems to emerge.
60 See the synoptic scheme in § 130 of Bergman [1791] p. 268. The problems involved in the combination
    of these two criteria are also obvious in his classification of 1779, where the division of the neutral salts
    according to complexity of composition is juxtaposed to that according to saturation with a "besides"—
    see § 186 of Bergman [1783] p. 89. For the notion of an order of chemical composition, which went
    back to G. E. Stahl, see chapter 6 and also part III, chapter 13.
61 Bergman [1783] p. 78.

	Simple Salts			Compounds of Acids with			
	Acids	Alkalis		Alkali		Earth	Metal
		volatile	fixed	volatile	fixed		
Homberg 1702	*Sels volatiles*		*Sel fixe*	*Sels ammoniacs*	*Sels moyens*		
	Sels acides	*Alcali v.*	*Alcali f.*				
N. Lemery 1713	*Sels acides*	*Alcalis*		*Sels salés*			
Geoffroy 1718	*Acides*	*Alcalis*		*Sels moyens*			
Boerhaave 1732	*Salia acida*	*Salia alkalia*		*Salia composita*		*?*	*?*
Rouelle 1744	*Acides*	*Alcalis*		*Sels neutres*			
Macquer 1766	*Acides*	*Alcalis*		*Sels neutres*			
Bergman 1779	*Acids*	*Alkalis*		*Neutral salts*		*Middle salts*	

Figure 9.3: Classification of salts, 1702–1779.

understood in relation to these simpler salts. The name "neutral salts" marked the compound salts as neither acid nor alkaline. Comprehending the compound salts in this way as neutral salts, the degree of saturation almost naturally became an issue and indeed was addressed long before Bergman's distinction between "perfect" and "imperfect" compound salts.[62] The name middle salts (*sels moyens*) originally highlighted a different feature, one which distinguished compound salts from the simple ones, as can be inferred from Homberg's divisions (figure 9.3).[63] Ignoring the established distinction between acids and alkalis, he distinguished the simple salts according to, in today's terms, their state of aggregation under normal temperature. He divided the salts into fixed and volatile ones and called those compound salts that were formed of both a fixed and a volatile salt *sels moyens*. Geoffroy dismissed this conception and gave the name *sels moyens* to those very compound salts which became known as neutral salts in the middle of the eighteenth century.[64] Bergman's reintroduction of the term middle salts did not pick up on Homberg's point. Rather, it seems he argued that it made no sense to apply the term neutral salts to salts whose components lacked antagonistic properties to be neutralized.[65]

---

62 G. F. Rouelle's distinction of neutral salts in Rouelle [1754] is particularly remarkable in this respect — see Kopp [1966] vol. III p. 70f. According to the ratio between the acid and the base of salts, Rouelle distinguished 1) "sel neutres avec excès ou surabondance d'acide," 2) "sels neutres parfaits ou salés," and 3) "sels neutres avec le moins d'acide" — Rouelle [1754] p. 574f. This classification can be taken as the starting point for the nineteenth-century distinction between "acid salts," "(neutral) salts," and "basic salts" — see Kopp ibid. p. 71. Particularly note that in the eighteenth century, the term "acid salts (*salia acida*, *sels acides*)" denominated acids and must not be mistaken for these "acid salts" of the nineteenth century.
63 See Homberg [1702] p. 40.

A brief look at Herman Boerhaave's classification of salts in his famous *Elementa Chemiae* (1832) shows that the French chemists' classification of salts was not immediately accepted by all influential chemists (see figure 9.3). Although Boerhaave occasionally used the term *salia neutra* in his textbook, he deliberately avoided calling the *salia composita* "neutral salts" in the paragraphs where he discussed the salts compounded of acids with alkalis.[66] Instead he mentioned that other chemists called these salts *hermaphroditi* or *alcalia vacua* or *acida implentia*.[67] Boerhaave's *Elementa Chemiae* further shows that chemists' classification of salts depended on context. Salts were naturally an important issue in the classification of the mineral kingdom, and in this natural historical context, Boerhaave subdivided the class of salts in the following way:

> Naturales hi [sc. sales fossiles] sunt, sal maris, sal Gemmae vel fossilis, sal de fontibus salinis, Nitrum, Borax, sal ammoniacus spontaneus, Alumen, Acidum vagum fodinarum.[68]

However, when he dealt with the use of salts as *menstrua*, he subdivided the group of salts in the following way:

> [...] distinguo Sales, adeoque & Menstrua salina in hasce commones Classes. 1. Alcalia fixa 2. Alcalia volatila 3. Acida vegetantia nativa 4. Acida vegetantia fermentantia 5. Acida vegetantia fermentata 6. in Acida vegetantia parata combustione 7. in Acida vegetantia parata destillatione 8. in Acida fossilia nativa 9. in Acida fossilia parata accensu 10. in Acida fossilia parata destillatione 11. in Salia sic dicta jam neutra, nativa, ut est Borax. Nitrum. Sal fossile, Gemmae, fontium, maris. Ammoniacus. 12. Alia quoque salia, quae ex his simplicibus composita sunt.[69]

Salifiable bases

Among the denominations of the compound salts, the name *sels salés* used by Nicolas Lemery deserves particular attention. This name expressed the view that the com-

---

64 See Geoffroy [1718] p. 207ff. Geoffroy gives no general definition of the notion of a *sel moyen*. Among the concrete *sels moyens* mentioned in this essay, one finds compounds of acids with earth and with metals, but not of acids with alkalis. That the latter are comprised under this notion can be concluded from the fact that chemists of the first half of the eighteenth century usually restricted the name sel moyen to precisely these compounds—see Rouelle [1744] p. 353.

65 But Bergman was also unhappy with the term "middle salts" for salts formed of earths or of metals with acids, as can be taken from the fact that he invented a new name for these substances in his classification of 1784.

66 See, for instance, Boerhaave [1732] vol. I pp. 763, 821 and 830. Apart from such rare exceptions, "neutral salts" in Boerhaave [1753] has no equivalent in the Latin original. This holds in particular for the definition of neutral salts (ibid. vol. I p. 555), which is a tacit addition of the translator.

67 See Boerhaave [1732] vol. I p. 788.

68 Boerhaave [1732] vol. I p. 43. "1. common salt, divided into sal gemmae or rock-salt, that of salt-springs, and sea salt. 2. Salt-petre. 3. Borax. 4. Sal-ammoniac. 5. Alum. and 6. the vague acids of mines." Boerhaave [1753] vol. I p. 105.

69 Boerhaave [1732] vol. I p. 763. "I distinguish salts, and saline menstruums, into the following classes; *viz.* (1) Fixed alkalis. (2) Volatile alkalis. (3) Native vegetable acids. (4) Fermenting vegetable acids. (5) Fermented vegetable acids. (6) Vegetable acids obtained upon burning. (7) Vegetable acids procured by distillation. (8) Native fossil acids. (9) Fossil acids procured by burning. (10) Fossil acids procured by distillation. (11) Neutral acid salts; as, borax, niter, pit-salt, sal-gem, sea-salt, and sal-ammoniac. (12) Other salts composed of these simple ones." Boerhaave [1753] vol. I p. 529.

pound salts were formed of two salts, an acid and an alkali, the particles of the latter being "filled" with the former.[70] Lemery's conspicuous application of this term to even salts formed by an acid and an earth or a metal was based on the fact that he took earths and metals to be alkalis.

> [...] on a appellé alkali tous les sels volatiles ou fixes [i.e. the traditional alkalis], & toutes les matieres terrestres qui fermentent avec les acides. [...] dans ces matieres ter-restres, dans les métaux, dans les coraux, dans les perles & generalement dans tous les corps qui fermentent avec les acides [...].[71]

At first glance, this broad understanding of alkalis seems to have the strange implication that Lemery regarded earth and metals as salts, too. However, this is not Lemery's point. The understanding of alkalis put forward here does not pertain to the old overarching classification of the mineral kingdom discussed so far, but to a seventeenth-century conception of chemical reactions that highlighted the antagonism of acids and alkalis.

As discussed in chapter 7 in connection with the classes of simple substances in the *Tableau* of 1787, from the seventeenth century on chemists defined the class of alkalis essentially in terms of their chemical behavior with acids. The effervescence of alkalis with acids was one important feature. An even more important one was the neutralizing effect of the former reaction. The resulting compound was neither acidic nor alkaline. However, if chemists regarded the class of alkalis as comprising all substances that neutralize acids—or "sweeten acids" as the English translator of Lemery put it—earths and metals had to be ordered into that class. This induced the chemists of the last decades of the seventeenth century to group together all substances forming neutral compounds with acids into one common taxonomic unit. Yet this taxonomic unit was nothing other than the precursor of Lavoisier's salifiable bases, the existence of which we could not certainly establish based solely on the affinity tables.

In the first half of the eighteenth century, beginning with Homberg, chemists occasionally named alkalis, earths, and metals the "bases" of salts.[72] Homberg's distinction between *alcalis salins*, *alcalis terreux*, and *alcalis metalliques* echoed and spelled out Lemery's *sels salés*.[73] In Macquer's *Dictionnaire* of 1766 the term bases (*bases des sels neutres*) was firmly established.[74]

Figure 9.4 shows the main steps by which the salifiable bases were classified in the eighteenth century. Against the background of Lemery's and Homberg's unification and classification of these substances, their discrimination and grouping in Geoffroy's table can doubtlessly be taken as a rendering of the taxonomic unit of the

---

70 "Quant à ce qu'on appelle sel salé, c'est un mélange d'acide & d'alcali, ou plûtôt un alcali soulé & rempli d'acide." Lemery [1713] p. 24.
71 Lemery [1713] p. 21. Taking the contemporary English translation as evidence, such a broad understanding of alkalis can be found back in the first edition of Lemery's discourse and is put forward there as an established and familiar notion. "Many stony matters, such as Coralls, Perles, Crabs-eyes, are called Alkalis, by reason of the effervescency that always happens, when you pour Acids upon them. Lastly, all things which do absorbe or sweeten Acids by ebullition are called Alkalis in Chemistry [...]." Lemery [1677] p. 9.
72 Homberg [1702] p. 44; see Kopp [1966] vol. III p. 69.
73 Homberg [1702] pp. 41, 44.
74 Macquer [1766] vol. I p. 195.

	Alkalis			Earths				Metals
Homberg 1702	*Alcalis salins* — *volatil*	*fixe*		*Alcalis terreux*				*Alcalis metalliques*
N. Lemery 1713	*Sels volatils*	*Sels fixes*		*Matières terrestres*				
Geoffroy 1718	*Sels alcalis volatiles*	*Sels alcalis fixes*		*Terres absorbantes*				*Substances metalliques*
Stahl 1723	*Flüchtiges urinos. Saltz*	*Fixes alcalisches Saltz*		*Erdische Wesen*				*Metallische Wesen*
Duhamel 1736		*Sel alcali du tatre*	*Base de sel marine*					
Macquer 1766	*Alcali volatil*	*Alcali fixe* — *végétal*	*minéral*	*Bases terreuses* — *argilleuses*	*calcaires*			*Substances metalliques*
Bergman 1779	*Volatile alkali*	*Fixt alkali* — *vegetable*	*mineral*	*Earths* — *argillac.*	*calcareous*	*ponderous*	*magnesia*	*Metallic calces*
Morveau 1782	*Ammoniac*	*Potasse*	*Soude*	*Alumine*	*Calce*	*Barote*	*Magnésie*	*Métaux*

Figure 9.4: Classification of the bases of neutral salts, 1702–1782.

salifiable bases.[75] As the subsequent instances demonstrate, the grouping together of salifiable bases became standard after 1718.[76] Our instances also show the way in which chemists further subdivided the salifiable bases. By distinguishing the base of sea salt as a particular fixed alkali (soda) that is distinct from the well-known fixed alkali derived from tartar (potash), Henri-Louis Duhamel du Monceau (1700–1782) launched the distinction between mineral and vegetable fixed alkalis.[77] Geoffroy's *terres absorbantes* were first subdivided into argillaceous earth (alumina) and calcareous earth; the latter was subsequently subdivided into three different earths. Finally, in Guyton de Morveau's *Mémoire* of 1782, we find the French names of the salifiable bases from which our contemporary names are derived.

We can sum up our investigations of the history of chemists' classification of minerals prior to Lavoisier by stating that this history confirms our analysis of the classif-

---

75 See Geoffroy [1718] p. 204f.
76 In conformity with the first column of Geoffroy's table, Stahl took the canonical succession of the salifiable bases as a succession of replacement. See Stahl [1765] chapter XXII.
77 See Duhamel du Monceau [1736]. It should be noted that Duhamel du Monceau himself did not use the terms "mineral fixed alkali" and "vegetable fixed alkali." For life and work of Duhamel du Monceau, see Gillispie [1970–1980] vol. IV p. 223ff.

icatory arrangements and distinctions indicated by the affinity tables. Our survey shows how deeply eighteenth-century chemical classifications of minerals were affected by the transition from the ancient philosophy of principles to the modern conceptual system of chemical compound and affinity. It shows in particular that the webs of taxonomic units that eventually shaped the fundamental network of the *Méthode's* classification—the web of classes around the conception of neutral salts and that around the conception of metals, their oxides, and their alloys—emerged slowly, and long before the Chemical Revolution.

A Revolutionary Table?

Having discussed eighteenth-century chemists' selection and grouping together of pure chemical substances, and particularly their collective ways of classifying them around the concepts of middle salts and of metals, alloys, and metal calces, we are now in a position to recognize the conditions under which the chemical order of the 1787 *Tableau* was constructed, to distinguish its own achievements from those on which it built, and to revalue how revolutionary the proposals of the "new chemists" actually were.

10.1 Reaping the rewards of a century

The coherent and systematic method for a new chemical nomenclature, to which Guyton de Morveau, Lavoisier, Berthollet, and Fourcroy aspired with their 1787 proposal, presupposed an equally coherent and systematic classification of pure chemical substances. Yet the authors of the *Méthode* could not simply resort to an established classification suitable for their purpose. First, their project was inextricably linked with Lavoisier's new chemistry. The nomenclatural proposal of the four authors was constructed in the framework of Lavoisier's anti-phlogistic chemical system and thus presupposed a classification in conformity with this new system. Second, as we have shown, there existed in the framework of the traditional phlogistic system many partial classifications of pure chemical substances, both explicit and implicit, as well as a few comprehensive classifications of the entire mineral kingdom. But there was no elaborated classificatory system devoted precisely to the realm of pure chemical substances. Therefore, even if the authors of the *Méthode* had modestly confined themselves to proposing only a nomenclature in accordance with the phlogistic chemical system, this goal, too, would have required a lot of classificatory work.

However, as our historical survey showed, by no means did they have to start from zero, but were able to build on the classificatory distinctions and distributions of both their predecessors and their contemporaries. And this was true not only for single groups of particular substances or subdivisions of certain taxonomic units. Rather, essential features of their classification of pure chemical substances had been developed in the course of the eighteenth century.

We argue that the *Tableau's* main classificatory structure was no novelty. First, the main structure, what we called the classification's backbone, consisted of two linked clusters of classes, one around the conception of neutral salts and the other around the conception of metals, their oxides, and their alloys—see figure 7.2 in chapter 7. As we showed in chapter 8, this linkage can be traced back to the beginning of the eighteenth century, as F. E. Geoffroy's table of affinities from 1718 testifies. By implication, these two clusters of classes themselves were far from being inventions of the authors of the *Méthode*. Instead, they predated even Geoffroy's table. The first one, composed of the classes of acids and acidifiable bases, the three classes of salifiable bases, and the class of the neutral salts, can be traced back to the last decade of the seventeenth century, as we were able to demonstrate. And the second one proved to be even older, going back to the sixteenth century, though the classificatory criteria employed in the sixteenth and seventeenth centuries differed from those used in the affinity tables and the 1787 *Tableau*.

Second, upon closer inspection, some of the *Tableau's* seemingly novel classes proved to be anti-phlogistic mirror images of classes already distinguished and constructed in the frame of the phlogistic chemical system. As shown in the previous chapter, the *Méthode's* class of acidifiable bases can be taken as such an anti-phlogistic mirror image of the sulfur genus in the traditional class of inflammable mineral bodies. The sulfur genus comprised sulfur, phosphorus, and carbon as its species — precisely the three substances of the *Tableau's* acidifiable bases. And phlogiston itself was, of course, the phlogistic counterpart of two apparently new and particularly important substances in the *Tableau's* class of the most simple substances, namely oxygen and calorique, the matter of heat. In classes like these we thus encounter taxonomic units that were new and old at the same time. They were new in that they presupposed the new anti-phlogistic understanding of the substances concerned, and they were old in a classificatory respect since they united exactly the same substances as their phlogistic counterpart. The awareness of this double-faced nature of certain classificatory features of the *Tableau* is essential for its historical assessment. Even the two webs of classes that form the backbone of the whole classificatory design comprise classes, and even pairs of classes, of such a double-faced nature. The relationship of these classes to their phlogistic mirror images epitomizes the core of the Chemical Revolution, as we will see later.

The classification of the *Méthode* could also build upon the affinity tables, as well as on classifications in mineralogy like those in the Swedish tradition, as shown in our analysis of Torbern Bergman's attempts. The chemical affinity tables established the tradition of selecting and arranging pure chemical substances, and this tradition must be taken as the most important historical background of the *Méthode's* tacit selection of such substances. It is particularly such tacit ways of distinguishing and ordering pure chemical substances that the *Tableau* of 1787 inherited from these tables. The rendering of the three traditional classes of salifiable bases as a taxonomic unit by arranging them *en suite* (field I/c–e), for instance, displays familiarity with the means of representation developed by and characteristic of the affinity tables.[1]

However, the classification of the *Méthode* was not only indebted to but also distinguished from the former affinity tables and other classificatory efforts of the eighteenth century. The first distinguishing feature that must be emphasized is its comprehensiveness. The *Tableau* was the first relatively comprehensive and explicit

---

1   In highlighting the inheritance the *Tableau* could build upon, one should not ignore altogether those aspects of this legacy that appear in hindsight to be ballast rather than an advantage. We name just two such aspects: first, as the appendix at the bottom of the table shows, the authors of the *Tableau* were as unable as their predecessors to deal adequately with the vegetable and animal substances they regarded as pure chemical substances. Second, as the inclusion of alloys shows, the notion of stoichiometric composition was as absent from the *Tableau* as it was from the affinity tables and other chemical classifications of the eighteenth century.

classification of the realm of pure chemical substances. The authors of the *Méthode* had no precursors who were aware of and actually realized the possibility that a coherent classificatory system could be constructed out of the extant tradition of the chemistry of pure substances. Their classificatory scheme was, however, not brought about by merely combining the various classificatory pieces of their predecessors. Rather than a mere merger, the *Tableau* was a transformation of the extant attempts to classify pure chemical substances into a systematic and consistent whole. This transformation rested essentially on the consequent application of the principle of classifying according to composition. Its consistent application for the construction of a classification of pure chemical substances was without precedent.

A feature of the *Tableau* that had been always stressed as of highest significance—namely, its introduction of a class of simple substances—was particularly indicative of this consistent application of the principle of classification according to composition. This class foreshadowed Lavoisier's famous definition of a chemical element in the "*Discours préliminaire*" of his *Traité* from 1789. Though pragmatic and, thus, in a way relativistic itself, this conception of simple chemical substances brought to an end the traditional view that regarded all substances entering into a chemical composition equally as building blocks, regardless of their own state of (relative) simpleness or (relative) compoundedness.

The *Méthode* also transgressed the limits of previous classificatory attempts by extending the range of pure chemical substances to be classified. This is obvious with respect to the different classes of gaseous substances. Here the classification took into account a relatively young but then paramount field of chemical investigations that was awaiting incorporation into the system of pure chemical substances. This must be regarded as another major achievement of the *Tableau* independent of the fact that, as discussed earlier in chapter 7, the insertion of these classes of gaseous substances into the classificatory design has more the appearance of a mechanical juxtaposition than of a true integration. The *Tableau* further included a few new taxonomic units, which, however, complemented rather than altered the traditional classificatory pattern. This is the case for the class of simple compounds of sulfur, carbon, and phosphorus with metals or alkalis (field VI/b) and for the class of compounds of metal oxides with various substances (field IV/c).[2]

Less obvious than these extensions may be the *Méthode*'s transgression of a fundamental limitation of the classifications of the affinity tables. The main purpose of these tables was the rendering of replacement reactions. They were thus confined to those pure chemical substances actually involved in or resulting from such kind of reactions. A particularly drastic result of this confinement was the fact that metal calces did not appear among the substances involved in reactions in the dry way, since calcination by combustion was not regarded as replacement reaction. The two new classes of certain compounds of metal oxides and compounds of sulfur, carbon, and phosphorus (field IV/c and VI/b) can be taken as another case in point, one which

2   When referring to the *Tableau's* fields—field I/a, etc.—we refer to the diagrams in chapter 5, particularly figure 5.4.

reveals another side of this limitation. In affinity tables, these two classes would have been an impossibility. The tables' rendering of replacement orders implied an ordering according to common reference substances. This naturally excluded the formation of classes of compounds that were analogous with respect to chemical composition but had no common component, for the reference substance was always a common compound.[3] Thus, by abandoning the replacement order as a principle, the classification of pure chemical substances gained not just an extension of its scope, but also new possibilities of grouping together.

As shown above, the notion of replacement reaction enabled Geoffroy to link the most important operations of two major fields of chemical theory and practice within one framework, namely that of metallurgy and that of salt production. The notion of replacement thus constituted the basis on which the cluster of classes around the conception of metals and alloys and the one around the conception of neutral salts could be brought together in the affinity tables. Classification of the *Méthode* could therefore not abandon the notion of the replacement reaction as an organizing principle without substituting for its linking function. The authors of the *Méthode* linked the two clusters of classes in a twofold way, as discussed in chapter 7. First, they united these clusters' classes of simple substances in column I. Second, they brought together their analogous classes of compounds, the acids and the metal oxides, in column III. Both classes of compounds were regarded as oxides of simple substances in the framework of the anti-phlogistic chemical system. The linkage of the two most important clusters of classes in this classification—that is, the backbone of this comprehensive and consistently elaborated system—thus seems to depend on the new chemical system.

10.2 Classification and Chemical Revolution

We arrive here at the point where the connection between classification and Lavoisier's Chemical Revolution can be clarified. In the eyes of Lavoisier and his collaborators, the analogy between the acids and the former metal calces or oxides depended on their understanding of these substances as oxides of simple substances. However, acids and metal calces were also analogous substances for adherents of the phlogistic chemical system, albeit for different reasons. In their framework, acids and metal calces were regarded as relatively simple substances that were analogous because they formed compounds with phlogiston in an analogous manner, the acids forming the species of the generic sulfur and the metal calces forming the metals. Accordingly, the two classes of the resulting substances, that of sulfur, carbon, and phosphorus and that of the metals, were regarded as analogous classes of compounds because of their analogous formation.

---

3   All of the compounds of metal oxides listed in field IV/c are by implication compounds of oxygen. However, in the phlogistic chemical system, metal oxides were regarded as (relatively) simple substances and not as oxides. Moreover, even with respect to the anti-phlogistic *Tableau*, one can argue that field IV/c represents the taxonomic unit of compounds of different metal calces rather than that of compounds that have oxygen as their common component.

Chemists' understanding of metals as metal calces saturated with phlogiston and of sulfur (later also of phosphorus and carbon) as a compound of an acid and phlogiston constituted the very core of the phlogistic chemical system. This understanding was both the result and the centerpiece of G. E. Stahl's great integration of the hitherto separated domains of chemical theory and practice by means of the phlogistic conception of combustion. The assumed analogy of the classes of metals and of generic sulfur, as well as that of the classes of acids and metal calces, was an integral part of Stahl's phlogistic system. Its power and attraction rested essentially on the linkage of the different webs of classes via these analogies.

We argue that the Chemical Revolution changed the understanding of these analogies, but did not touch the analogies themselves. This is the crucial point in our view: the analogy between acids and metal calces or metal oxides, which allowed the linkage of the two main clusters of classes in the *Méthode*'s classification, was not a fruit of the new chemical system but an inheritance from the phlogistic system. We once again come across the double-faced character of a feature of the mode of classification represented in the tableau, which is of essential significance. The Lavoisierian theoretical conception of the analogy presupposed the new chemical system, but the analogy itself and its linking function, essential for classification, belonged to the phlogistic system.

The continuities and discontinuities between the Lavoisierian and phlogistic chemical systems become visible when we confront the *Tableau*'s taxonomic core structure with its phlogistic mirror image. For this purpose, we represent both taxonomic structures in form of tables (figure 10.1). The comparison of the two tables shows almost at a glance how the Lavoisierian taxonomy relates to its phlogistic mirror image: a simple shift transforms the phlogistic mirror image into its anti-phlogistic counterpart. One has only to exchange the places of the two pairs of analogous classes from column one to three and vice versa.

What does this strikingly simple switch from the old taxonomic system to the new "revolutionary" one mean? According to Thomas Kuhn, comparing taxonomic structures yields the yardstick for judgments about scientific revolutions.[4] He proposed to regard two theories as commensurable if

> their taxonomic trees can be fit together in one of the following two ways: (1) every kind in the one can be directly translated into a kind in the other, which means that the whole of one tree is topologically equivalent to some portion of the other; or (2) one tree can be grafted directly onto a limb of the other without otherwise disturbing the latter's existing structure. In the first case one scheme is subsumed by the other. In the second, a new scheme is formed out of the previous two, but one that preserves intact all of the earlier relations among kinds. If neither case holds then you are in the [...] realm of incommensurability.[5]

The question is then, is the possibility of generating a phlogistic mirror classification by just shifting two couples of classes of the *Tableau* indicative of the commensurability of the two chemical systems or, on the contrary, of their incommensurability?

---

4   See, for instance, Buchwald and Smith [1997] p. 371ff.; and Buchwald [1992].

1 Anti-phlogistic

	I Simple substances	II	III I + Oxygen	IV	V III + I	VI
a						
b	S/C/P-class		Acids		Salts	
c	Metals		Metal calces			Alloys
d	Earths					
e	Alkalis					

2. Phlogistic

	I Simple substances	II	III I + Phlogiston	IV	V I + I	VI
a						
b	Acids		S/C/P-class		Salts	
c	Metal calces		Metals			Alloys
d	Earths					
e	Alkalis					

Figure 10.1: Shift of the pairs of analogous classes with linking function.

We argue that this possibility constitutes commensurability rather than incommensurability: the kinds of the one system are directly translatable into the kinds of the other. This holds not only for phlogiston, which was translatable into the oxygen and

---

5   Buchwald [1992] p. 42f. The required transformation of the taxonomic web of the *Tableau* into a taxonomic tree would imply some problems, which, however, do not affect our argument. First, the result of this transformation deviated not only from the historical actors' way of classifying, which is certainly legitimate for analytical purposes, but also entailed a loss of information: no longer would the analogy of the classes of salifiable bases and metals be rendered, nor the relation of this pair of classes to that of the acids and the metal oxides. Second, there seems to be a tension between encaptic classification in the form of a taxonomic tree and classification according to composition. In the latter way of classifying, in the most simple case of binary compounds, each compound or combination belongs to at least two different possible taxa: that of all combinations of the one component and that of all combinations of the other. And these possible taxa are not nested, but do overlap. Thus, in contrast to encaptic classifications, overlap is not *a limine* excluded by formal means, but must be avoided *in concreto*, for example, by leaving one of two overlapping possible classes empty—see, in the *Tableau* for instance, the possible classes in fields VI/d–e which would overlap with the class in field VI/b if not left empty.

calorique of the Lavoisierian system, but for all taxonomic units that constituted the classification's core structure. All implications of the shift described above can and could be spelled out in detail by us as well as by the historical actors themselves.

A similar argument could thus be derived from an analysis of the many debates between the adherents of the phlogistic and the anti-phlogistic systems. One example is the attempt of contemporary chemists to preserve, by similar simple transformations, proven arrangements of substances from becoming dependent on a decision for or against one of the rival chemical systems. The revision to which William Nicholson subjected Bergman's affinity table in 1790 is a good case in point. He skipped Bergman's column with vital air (oxygen) as a reference substance with the remark: "[...] in the anti-phlogistic theory, the column entitled Phlogiston being taken in the reverse order, will express the elective attraction of Vital Air."[6] There can be no doubt that the historical actors fully understood what their opponents were talking about. They did not live in different worlds.

We can summarize the theoretical change manifest in the two classificatory structures as follows: Lavoisier's new chemical system entailed a radical new understanding of two pairs of analogous classes of substances, which is expressed by the symmetric exchange of their places within the classificatory web. Acids and metal calces lost their former status as simple substances, and salifiable bases and metals, formerly compounds, gained this status. This deep revision of fundamental chemical assumptions, engendered by the new theory of combustion, may earn the designation "revolution," if other criteria for scientific revolutions are chosen than Kuhn's. However, this deep revision did not affect the principal classificatory structure in which the pure chemical substances had been ordered since the beginning of the eighteenth century. This classificatory structure had a more basic status in the early modern chemistry of pure chemical substances than did the fundamental assumptions revised by Lavoisier. In other words, underlying both the phlogistic and the anti-phlogistic chemical systems was a shared conceptual structure that remained untouched by Lavoisier's "revolution." Thus, in the classification of pure chemical substances that developed in the early modern period, we encounter an ontological deep structure of chemical thinking that was remarkably stable. With respect to this ontological deep structure, no revolution took place at the end of the eighteenth century.

10.3 The classification's transitoriness

The confrontation of the *Tableau*'s classificatory structure with its phlogistic mirror brings to light another feature of the *Méthode*'s classification that might otherwise escape one's attention; for the comparison of the two classifications shows that the classificatory structure derived from the conception of neutral salts need not necessarily fit with a classification according to composition. In the phlogistic scheme (the

---

6   Duncan [1996] p. 166. "This is a neat illustration of the way in which affinity tables can represent the observed facts without commitment to any particular theory of composition," was Alistair Duncan's telling comment.

lower table in figure 10.1), we encounter a perfect harmony of the two classificatory principles. All the different classes of components of neutral salts, the acids and the three classes of salifiable bases, are united in one and the same column, the column for simple substances.[7] In the anti-phlogistic scheme, on the other hand, it cannot be overlooked that the two principles are at variance. The classification according to composition not only leads to a displacement of the acids into column III, but has also the consequence that the classes of salifiable bases were no longer adjacent in one and the same column, as only two of them, the earths and the alkalis, were still regarded as classes of simple substances.

One may perhaps object that this latent conflict between the two classificatory principles employed in the construction of the *Tableau* is only an artifact of our analytical reconstruction. There is, however, clear evidence that this conflict troubled the authors of the *Méthode*. As mentioned earlier in chapter 5, at the point in time when they composed the classification, it was already known that one of the supposedly simple substances, ammoniac, could be decomposed. The authors of the *Méthode* did not question this.[8] Why then did they not draw the consequence from this fact? Why did they nevertheless keep this substance in the column for simple substances? We don't know the reasons for this inconsistency. But we know the consequence of removing ammoniac from column one. It would make it obvious that the old class of alkalis, a cornerstone of the cluster of classes around the conception of neutral salts, no longer had a place in the proposed classification.

With this in mind, we arrive at the transitory character of the *Méthode*'s classification. This classification portrays the state of art in the chemistry of pure chemical substances at a certain point in time, and each change in this state must have repercussions for it. Its expiry date depended on the pace of further developments in this domain of chemical investigations. This was as obvious to the historical actors as it is to us.[9] But presumably nobody expected back in 1787 that major revisions would prove necessary only two years later. In his *Traité* of 1789, Lavoisier proposed classificatory distinctions and distributions of pure chemical substances that amounted to a dismissal of the *Méthode*'s compromise between the two conflicting classificatory principles, classification according to composition and the construction and arrangement of classes in conformity with the conception of neutral salts. For a demonstration of the transitory character of the *Méthode*'s classification, it is not necessary to scrutinize Lavoisier's classificatory attempt.[10] It suffices to mention here that this attempt actually manifests what was said about the latent tensions between the classificatory structure derived from the conception of neutral salts and the principle of classification according to composition.

---

7   Taking into account, however, that, as discussed in chapter 9, all adherents of the phlogistic chemical system before Bergman assumed that metals rather than metal calces form salts with acids, such a perfect harmony can be claimed only for Bergman's classification.
8   See Guyton de Morveau et al. [1787] p. 82.
9   See, for instance, Lavoisier [1789] p. 194, Lavoisier [1965] p. 177.
10   See Perrin [1973]; and Siegfried [1982] on Lavoisier's revision of the order of simple substances; see also Roberts [1991].

In his *Traité*, Lavoisier published no all-comprising classificatory tableau, but a series of 33 single tables representing taxonomic units of pure chemical substances, the rank of which is not always clear.[11] Whatever Lavoisier's reasons for this form of representation might have been, it was in any case very suited to keeping these tensions from becoming evident. But it could not really hide them, as two cases may demonstrate. The first one concerns the acidifiable bases. These substances were now subdivided into six simple ones, three of them still postulated substances, and twenty compounded ones which were, with the exception of one, aqua regia, postulated substances as well.[12] In other words, the principle of classification according to composition enforced the disintegration of the former class of acidifiable bases and the distribution of the pertinent substances into different taxonomic units. The second case concerns the alkalis and displays an open inconsistency. The three alkalis were removed from the list of simple substances, but no new taxonomic unit was allotted to them. Ammoniac can be found in the unit of *Combinaisons Binaires de l'Azote avec les Substances Simples* as a compound of niter with hydrogen.[13] The two fixed alkalis, on the other hand, were not distributed into any taxonomic unit of compounds since their composition was still unknown.[14] But, in tables that represent salts as being composed of a certain acid with the traditional salifiable bases, all three alkalis still figured as such bases.[15] True, the listing of the alkalis as bases together with the earths and the metal calces is formally neutral with respect to the question of whether they are compounds or simple substances. Yet the latter was suggestive not only for readers but obviously for the author himself; for, and this is the open inconsistency, the three alkalis also occur explicitly as simple substances in three tables of binary compounds.[16]

It is of interest to note that Lavoisier did not draw nomenclatural consequences from his classificatory revisions. The volatile alkali was still called ammoniac despite the rule that compound substances should be given compound names. The nomenclature proposed by the authors of the *Méthode* was thus apparently more lasting than some of the classificatory distinctions and arrangements of its *Tableau*. It may therefore seem appropriate to finish this chapter with a last glance at this nomenclature.

10.4 Classification and nomenclature

In our description of the formal structure of the *Tableau's* classification in chapter 5, we drew attention to an unique feature of its graphic organization, namely its arrangement of the taxonomic units in the form of a matrix. The genera, along with the species, are related vertically to other taxonomic units that are of the same order.

---

11 The second book of the *Traité* consists of nothing but these tables with commentaries.
12 See Lavoisier [1789] pp. 192 and 196, Lavoisier [1965] pp. 175 and 179.
13 See Lavoisier [1789] pp. 212 and 216, and Lavoisier [1965] pp. 194 and 198.
14 For Lavoisier's and others' conjectures about the composition of the fixed alkalis, see Siegfried [1982] p. 41f.; and Perrin [1973].
15 See the series of tables beginning with that in Lavoisier [1789] p. 232, and Lavoisier [1965] p. 212, respectively.
16 See Lavoisier [1789] pp. 220, 222, and 226, and Lavoisier [1965] pp. 202, 204, and 207.

Horizontally, the species are related to their compounds if they are simple substances, or, if they are compounds, to one of their components and at the same time to other compounds of this component. It is this horizontal linkage of taxonomic units that is of immediate significance for the nomenclatural purposes of the table, for the determination of the names for substances takes place along this horizontal coordinate. Apart from a few exceptions, each horizontal line of a simple substance and its compounds is rendered as a linguistic line of derivations developed from the simple substance's name according to certain rules. Take, for instance, line 7 in figure 5.2: *Soufre—Acide sulfurique/sulfureux—Gaz acide sulfureux—Sulfate/Sulfite de potasse—Sulfur de fer.*

The application of such linguistic rules presupposed (as was also remarked above) classes of substances that were already distributed into the five principal types of compounds (column II–VI). This distribution and, *a fortiori*, the establishment of these five orders of compounds, were a matter of neither linguistic laws or derivations nor formal classificatory rules. Rather, they were classificatory distinctions, distributions, and arrangements resting on chemical assumptions and convictions held by the authors of the *Méthode*. As was shown in chapter 9, the core of these assumptions and convictions developed in the course of the eighteenth century. It was in particular the traditional clusters of classes around the conception of neutral salts and that around the conception of metals, metal calces, and alloys that provided the structure within which systematically coordinated names could be elaborated.

Knowledge of this historical background of the classification's core structure is essential for an adequate conception of the relationship between the classificatory and nomenclatural achievements of the *Méthode*. After what we showed of this background in chapters 8 and 9, it can no longer be an open question as to whether the *Tableau's* classification was a fruit of the new methodical nomenclature or,[17] on the contrary, one of its prerequisites, or whether a coevolution of nomenclature and classification took place. Being rooted in the ontological deep structure of eighteenth-century chemistry of pure chemical substances, which can be traced back to the decades around 1700, the core classificatory structure of the table cannot be regarded as resulting from or emerging together with the new nomenclature. The classificatory framework presupposed by the new nomenclature did not rest on language, but on the experimental practice and reflections of more than one hundred years.[18]

The fact that the nomenclature presupposed the classification and not the other way around is also of immediate importance for the question of what role Lavoisier's Chemical Revolution played in the nomenclatural reform of the *Méthode*. This transformation shaped the proposed nomenclature to precisely the extent to which it modified and complemented the traditional classificatory structure. The Chemical Revolution was not the foundation of the *Méthode*'s classification, even though it shaped certain essential features of it. Nor was the new chemical system the decisive prerequisite condition for the new nomenclature, although it unmistakably left its

---

17   So contended, for instance, by Beretta [1993] p. 204f.; and Riskin [1998] p. 214f.
18   See, for instance, Holmes [1995].

mark on it. This is corroborated particularly by the classificatory feature with doubt-lessly the greatest significance for this nomenclature, namely the distinction and determination of chemical substances according to their composition. This way of identifying and ordering chemical substances was not a fruit of the Chemical Revolu-tion but, as shown in previous chapters, of the preceding chemistry of pure substances which can be traced back to the days of Geoffroy. Considering what was said in chap-ter 4 about the nomenclature's role in the propagation of the anti-phlogistic chemical system, one might arrive at the conclusion that the acceptance of Lavoisier's Chemi-cal Revolution owed more to the new nomenclature than the nomenclature did to his overthrow of the phlogistic system.

There is further evidence that the main feature of the new nomenclature was inde-pendent of the anti-phlogistic chemical system. It was provided by earlier attempts to reform the chemical nomenclature, which eventually led to the proposal of 1787. Per-haps the most instructive of such attempts was published by Guyton de Morveau in a *Mémoire* of 1782, at a point in time when he was a convinced adherent of the phlogis-tic chemical system.[19] The tableau added to this *Mémoire* (figure 10.2) concerned an essential part of the later nomenclature, the denominations of the classes around the conception of neutral salts. It was precisely the chaotic variety of the traditional names of the substances in these classes that made reform so urgent.

In a very abbreviated and dense form, the table listed the principles used to name substances of these classes and gave examples of such names. The principles appeared in the three columns on the left-hand side and the examples were given in the two further columns. The table displayed the principles of denomination along the established classes of substances involved in the formation of neutral salts. The first column listed the names of the acids; the next, the main component of the names of the resulting salts; and the third, the names of the different salifiable bases from which the other component of the salt name must be derived. By means of these col-umns it was thus ruled that 1) acids must be given a binomial name,[20] the second component of which marks the species; 2) salts, too, must be given a binomial name, the first component of which marks the genus and is derived from this specifying component of the acid name, whereas the second one marks the species and is derived from the name of the salifiable base that entered into the combination; 3) the salifiable bases must be given simple names. Looking at our figure 5.2 in chapter 5, it immediately becomes clear that the nomenclatural principle of Guyton de Morveau's table is essentially the same as that of the *Méthode*. Apart from the fact that Guyton de Morveau's names did not allow for different degrees of saturation, the main differ-ence between the two nomenclatures is the grammatical form of the specifying com-ponents in the names of acids and salts. Whereas in the *Méthode* the specifying component of an acid name was always an adjective, and that of a salt name always a noun in the genitive case, in both cases those forms were used interchangeably by

19 See Guyton de Morveau [1782]. See also Smeaton [1954]; and Crosland [1962] p. 153ff.
20 It should be noted, however, that, in contrast to the *Méthode*, the binomial acid name in Guyton de Morveau's proposal of 1782 does not indicate the anti-phlogistic assumption that acids are compounds.

TABLEAU DE NOMENCLATURE CHYMIQUE,

Contenant les principales dénominations analogiques, & des exemples de formation des noms composés.

RÉGNES.	ACIDES.	Les Sels formés de ces Acides prenant les noms génériques de	BASES ou substances qui s'uniffent aux Acides.	EXEMPLES pour la claffe des Vitriols.	EXEMPLES pris de diverfes claffes.
Des trois Régnes.	Méphitique ou Air fixe.	Méphites.	Phlogistique.	—	Soufre méphitique ou Plombagine.
	Vitriolique.	Vitriols.	Alumine ou Terre de l'argille.	Soufre vitriolique ou foufre commun.	Nitre alumineux.
Minéral.	Nitreux.	Nitres.	Calce ou Terre calcaire.	Vitriol alumineux ou Alun.	Muriate calcaire.
	Muriatique ou du fel marin.	Muriates.	Magnéfie.	Vitriol calcaire ou Sélénite.	Acéte de magnéfie.
	Régulin.	Régules.	Barote ou Terre du Spath pefant.	Vitriol magnéfien ou Sel d'epfom.	Tartre barotique.
	Arfenical.	Arfeniates.	Potaffe ou Alkali fixe végétal.	Vitriol barotique ou Spath pefant.	Arfeniate de potaffe.
	Boracin ou fel fédatif.	Borax.	Soude ou Alkali fixe minéral.	Vitriol de potaffe ou Tartre vitriolé.	Borax de Soude ou Borax commun.
	Fluorique ou du fpath fluor.	Fluors.	Ammoniac ou Alkali volatil.	Vitriol de Soude ou Sel de Glauber.	Fluor ammoniacal.
			Or.	Vitriol ammoniacal.	Régule d'or.
Végétal.	Acéteux ou Vinaigre.	Acetes.	Argent.	Vitriol d'or.	Oxalite d'argent.
	Tartareux ou du Tartre.	Tartres.	Platine.	Vitriol d'argent.	Saccharate de platine.
	Oxalin ou de l'Ofeille.	Oxalites.	Mercure.	Vitriol de platine.	Citrate de mercure.
	Saccharin ou du Sucre.	Saccharates.	Cuivre.	Vitriol de mercure.	Lignite de cuivre.
	Citronien ou du Citron.	Citrates.	Plomb.	Vitriol de cuivre ou Vitriol de Chypre.	Phofphate de plomb.
	Ligneux ou du Bois.	Lignites.	Etain.	Vitriol de plomb.	Formiate d'étain.
			Fer.	Vitriol d'étain.	Sébate martial.
Animal.	Phofphorique.	Phofphates.	Antimoine (au lieu de Régule d')	Vitriol de fer ou Couperofe verte.	Muriate antimonial ou Beurre d'antimoine.
	Formicin ou des Fourmis.	Formiates.	Bifmuth.	Vitriol antimonial.	Galacte de bifmuth.
	Sébacé ou du Suif.	Sébates.	Zinc.	Vitriol de bifmuth.	Borat de zinc.
	Galactique ou du Lait.	Galactes.	Arfenic.	Vitriol de zinc ou Couperofe blanche.	Muriate d'arfenic.
			Cobalt.	Vitriol d'arfenic.	Saccharate de cobalt.
			Nickel.	Vitriol de cobalt.	Formiate de Nickel.
			Manganèfe.	Vitriol de Nickel.	Oxalite de manganèfe.
			Efprit de-vin.	Ether vitriolique.	Ether lignique ou Ether de Goertling, &c. &c. &c.

N. B. Lorfque les acides particuliers déjà entrevus dans la molybdène, l'étain, &c. feront plus connus, on en formera les noms d'acide molybdique & molybdite, d'acide ftannique & ftannite, &c. Il en fera de même des nouvelles bafes. Le nouveau demi-métal trouvé par M. Bergmann dans les fois cafians, pourra être nommé hydrecille, caché dans le fer.

Les dix-huit acides, les vingt-quatre bafes & les produits de leur union, forment ainfi quatre cents foixante-quatorze dénominations claires & méthodiques, indépendamment des fépars ou compofés à trois parties, dont les noms viennent encore dans ce fyftème, comme fépar de foude, fépar ammoniacal, pyrite d'argent, &c. &c.

Figure 10.2: Guyton de Morveau's *Tableau de Nomenclature Chymique* from 1782.

Guyton de Morveau in 1782. Needless to say, such differences have nothing to do with adherence to either of the two rival chemical systems.

Classification was a necessary condition for the nomenclature of the *Méthode*. However, it was not a sufficient condition. It is well known that Guyton de Morveau and the other authors of the *Méthode* borrowed ideas from non-chemical sources to establish suitable principles for an unambiguous and consistent chemical nomenclature. The most important of these principles states that simple substances should be given a simple name[21] and compounds a name composed out of simple names in such a way that it fits the general genus-species scheme and at the same time indicates of which components the compound in question consists.[22] Above all, the *Logique* of Etienne Bonnot Condillac (1715–1780) and the botanical nomenclature of Linnaeus provided ideas and paradigms for the development of such rules and are therefore discussed as such a source in almost every work on the *Méthode*.[23] The significance of Linnaeus' botanical nomenclature is mainly its binomial naming, which apparently served as a model for the authors of the *Méthode*.[24] From Condillac's ideas Lavoisier took the arithmetical concept of an ideal scientific language that would, in principle, use names in the same way calculation uses numbers. Lavoisier called this language an "analytic method."[25] An indispensable prerequisite condition of such an analytic method was a nomenclature that ensured that names represented the concepts of the objects in question in such a way that conclusions could be drawn by merely following the constituents of the names. Much work has been done on the impact that this assumption of Condillac's ideas had on the theoretical conceptions of Lavoisier and his allies,[26] as well as on the moral and political dimensions of the new anti-phlogistic chemical system.[27] We prefer to close this chapter by calling attention to a rather elementary problem linked to this assumption.

In most fields of knowledge Condillac's concept of an ideal language, reminiscent of Leibniz's idea of a "universal characteristic," would prove realizable only in symbolic language and not in ordinary language. It is not by accident that the *Méthode* contains, as an appendix, a specific symbolic system for rendering chemical substances, which was proposed by Jean-Henry Hassenfratz (1755–1827) and Pierre Auguste Adet (1763–1834).[28] This system had no success. However, the move toward a symbolic language of chemistry itself seems very foresighted considering

---

21 See de Guyton de Morveau [1782] p. 373f. ("Corollaire 1°").
22 Guyton de Morveau et al. [1787] p. 19f.
23 On the importance of Condillac's *Logique* (1780) for the authors of the *Méthode*, see Guyton de Morveau et al. [1787] p. 7; see also Albury [1972].
24 Torbern Bergman's influence on the French chemists, in particular on Guyton de Morveau, is of interest in this respect, since Bergman was a pupil of Linnaeus. True, Bergman's own nomenclatural proposals are entirely different from those of Guyton de Morveau. But the common feature of the two nomenclatural attempts consists in the employment of binomial naming for rendering genus and species — see Crosland [1962] p. 144ff.
25 For the meaning of Lavoisier's use of "analytic method" in this context, see Guyton de Morveau et al. [1787] p. 7ff.
26 See, for instance, Beretta [1993] chapter 4, and [2000]; and Roberts [1992].
27 See, for instance, Anderson [1984]; and Golinski [1992].
28 On Hassenfratz and Adet, see Bensaude-Vincent [1994] p. 36; and Gillispie [1970–1980] vol. VI p. 164f. and vol. I p. 64f., respectively.

that the first version of our chemical formulae was introduced into chemistry less than thirty years after the *Méthode*. Therefore it seems natural to wonder why the authors of the *Méthode* believed that, in just the case of chemistry, such an "analytic method" could be accomplished on the basis of ordinary language. Which understandings of chemistry encouraged them to try the urgent reform of chemical nomenclature in the spirit of Condillac? Among the many possible reasons, two very elementary ones should not be overlooked.

As discussed throughout part II, over the course of the eighteenth century the knowledge of the chemical composition of a substance was increasingly thought to constitute the center of its understanding, which had previously been occupied by knowledge about its properties. This change was based on the conceptual system of compound and affinity that emerged at the turn of the seventeenth century. This new understanding of chemical substances resonated with the feature of Condillac's ideal language that attracted Lavoisier above all, namely its additive structure. Consisting essentially of a systematic construction of complex names by combining simple names, the analytic method taken from Condillac appeared as if it were invented for chemistry on the basis of the conceptual system of compound and affinity. However, it should not be overlooked that this method owed a great deal of its suitability for chemistry to the fact that the quantitative aspect of a substance's composition still played a secondary role. The much-praised introduction of different suffixes as indications of different degrees of saturation is, in hindsight, indicative of the unavoidable difficulties this method had to face when quantitative considerations came to the fore, especially in organic chemistry.[29]

Besides the additive structure of the conception of chemical compound, we wish to highlight another peculiar feature of eighteenth-century chemistry that was preadapted to Condillac's ideal language and, at the same time, to Linnaeus's binomial nomenclature: the concept of a binary composition of chemical compounds, which can be traced to the days of Homberg, Lemery, and E. F. Geoffroy.[30] This concept had a restrictive effect on chemical nomenclature. The authors of the *Méthode* could confidently undertake the reformation of the chemical nomenclature in conformance with Condillac's ideas because they could be assured in advance that only neat binomial names would be designed, and not the monstrous names we find today on medication instruction sheets.

---

29 See Guyton de Morveau et al. [1787] p. 40f.; Crosland [1962] p. 181; and Beretta [1993] p. 210. We will come back to this in chapter 14.

30 Possibly the most impressive manifestation of this assumption can be found in the tables of book II of Lavoisier's *Traité*. Appearing less strange in the framework of the phlogistic chemical system, the assumption that all compounds are of a binary composition is rather striking in the anti-phlogistic framework, in which salts are thought to be composed out of three simple substances. For an explicit reflection of this problem, see Guyton de Morveau et al. [1787] p. 21f. In the framework of the phlogistic chemical system, the same problem arises when metals are taken as salifiable bases rather than metal calces.

Part III
A Different World: Plant Materials

Introduction to Part III

In part II, we studied an ontology of substances deeply embedded in chemical experimentation—cycles of analyses and syntheses and replacement reactions—and a distinctive conceptual network constituted around 1700, which was composed of the concepts of chemical compound, composition, affinity, analysis, and synthesis. The substances belonging to this ontological regime were either identified and classified according to chemical composition, in the case of chemical compounds, or according to essential properties if they were simpler components of chemical compounds. In the latter case—which applies to the simple substances of Lavoisier as well as to the substance building blocks rendered in the chemical affinity tables published from 1718 onward—chemists selected and highlighted those properties of substances that they considered of outstanding importance in the substances' roles as components of chemical compounds.

This ontology of substances and mode of classification was restricted to a specific area of chemistry, however, which we have designated the "chemistry of pure substances," where "pure" means substances that were traceable in series of chemical operations. The majority of materials studied in eighteenth-century chemical laboratories and lecture halls—the raw minerals and the materials extracted from plants and animals—did not fit into that ontology. In this third part, we study an exemplary part of this distinctive chemical terrain: plant materials.[1] As in the previous part, we will ask questions concerning the provenance of plant materials, their changing identities, and the ways chemists sorted out and ordered plant materials from the eighteenth until the early nineteenth centuries. From our attempts to answer questions like these, we will draw conclusions about chemists' ontology of plant materials.

In the eighteenth and early nineteenth centuries chemists experimented with, labeled, and classified plant materials in a broad variety of ways. Plant materials were of interest to them as medicines, dyestuffs, food, and other commodities; as "simplicia" and natural "juices" of plants; as the simplest principles of plants defined by the ancient philosophy of ultimate principles, or, from the middle of the eighteenth century onward, as their more compound, proximate principles; and, around 1800, as organic substances embodying a force of life. Eighteenth-century chemists classified plant materials in a broad variety of ways, depending on their theories, their experimental experience, and the focus of their interests in different practical and intellectual contexts. But from a broader comparative historical perspective, eighteenth-century chemists' multifarious ways of ordering plant materials display common features. Plant materials were always identified, labeled, and grouped together by referring to their natural origin, the mode of their extraction, their perceptible physical or chemical properties, and their practical uses. In contrast to the classification of "pure chemical substances" according to chemical composition, eighteenth-century chem-

---

1   It should be noted that the following account largely applies to animal materials, too. We concentrate here on plant substances for pragmatic reasons, but also because that area was much more elaborated than that of animal chemistry.

ists classified plant materials according to provenance and perceptible features. As we will see in the following chapters, this broadly shared naturalistic and artisanal mode of classifying plant materials remained untouched by the Chemical Revolution, and eventually was changed in the 1830s.

Apart from these agreements, our analyses of chemists' ways of classifying plant materials over more than a century also uncover ontological shifts and accelerations of ontological shifts. They bring to the foreground a trajectory punctuated in the 1750s, the 1790s, and the 1830s. In contexts of conceptual inquiry, early eighteenth-century plant materials, which were commodities of the apothecary trade and other arts and crafts, were elevated epistemically as compound components or proximate principles of plants in the 1750s, reduced to organic compounds in the 1790s, and replaced by carbon compounds in the 1830s. The third and last transformation of the epistemic constitution of materials and the mode of their classification was accompanied by a deep transformation in the materiality of organic compounds. After roughly 1830, the term "organic compound" no longer referred to the same type of substances as it had in the century before. Whereas in large areas of classical botany and zoology it appears reasonable to assume that the objects to be classified, namely single plants and animals, remain the same independent of changes in people's understanding and practice, this assumption does not hold in the history of chemistry, mineralogy, and most other experimental sciences.[2] As we will see in the history of organic chemistry, the material objects to be classified changed as well.

Chapter 11 deals with the dimensions of plant materials and the practices and contexts of their classification that hold for the entire eighteenth century and even the early nineteenth century—the fact that most plant materials were commodities, and the chemists' selection of provenance and perceptible features as the most significant criteria of their classification. It opens with an overview of the range, provenance, and artisanal application of plant materials in the eighteenth century. This is followed by a discussion of chemists' ordering of materials according to the three natural kingdoms and an analysis of their mode of individuating and identifying plant materials. A last section illuminates in more detail the pharmaceutical and artisanal contexts of eighteenth-century chemists' circumscription and classification of plant materials.

The subsequent chapters 12 through 16 examine changes in the classification and ontology of plant materials. In so doing, they focus on the way chemists ordered plant materials in contexts of conceptual inquiries, and on the changing meaning and identity of plant materials in these specific contexts.

Chapter 12 discusses early eighteenth-century chemists' interest in the most simple elements or principles of plants, and in plant analysis that aimed to separate the simple principles. It further illuminates early eighteenth-century chemists' demarcation of the class of ultimate principles of plants, as well as other kinds of chemical

---

2   Given the role played by horticulture and breeding as a practical context of classification in eighteenth- and nineteenth-century botany and zoology, one may claim that even in these areas constitutions of single objects to be classified were not entirely independent of practice and understanding.

preparations made in the laboratory, from the pharmaceutical *simplicia* and botanical vegetable juices.

Chapter 13 begins with an analysis of the first major change in eighteenth-century chemists' ontology and classification of plant materials; namely, the increase in significance of the more compound components or proximate principles of plants in the 1750s, which became manifest when chemists grouped together the proximate principles of plants. This new mode of classification required the abolition of the former distinction between the pharmaceutical *simplicia*, or botanical juices of plants, and chemically extracted plant materials. Chemists' dismissal of this traditional distinction implied a redefinition and epistemic elevation of vegetable commodities and pharmaceutical *simplicia* as compound chemical principles or components of plants. The chapter also discusses the empirical and theoretical preconditions for that revaluation. The increase in attention to the proximate principles of plants around the middle of the eighteenth century was spurred by a new understanding of plants as organized living beings. This change in the significance of plants, which is discussed in the next section of this chapter, was embedded in ongoing transitions in natural history. It led to the rethinking of the traditional distinction between the three natural kingdoms and the grouping together of plants and animals as "organized" or "organic bodies," along with their separation from minerals as "unorganized" or "inorganic bodies." These developments culminated in a second ontological shift and change in the mode of classification in the 1790s. It was then that chemists began to dismiss mineral substances from the proximate principles of plants and to group together plant and animal substances as "organic substances."

The analysis of the classification of plant materials in the *Méthode de nomenclature chimique*, presented at the beginning of chapter 14, creates an important link between parts II and III of this book. This analysis first shows the ambition of Lavoisier and his collaborators to extend classification according to composition to plant substances, and, second, their failure in terms of collective acceptance. The next section of the chapter continues this analysis of their failure inasmuch as it points out the theoretical limits of the Lavoisierian analytical program. As Lavoisier's theory of the elemental composition of organic substances did not include the assumption of stoichiometric organic compounds, it was unsuitable as a working tool for the envisioned analytical mode of identification and classification of organic compounds.

Hence, it is not surprising that chemists continued to identify and classify plant substances in the traditional naturalistic and artisanal way well into the nineteenth century. Chapter 15 discusses the continuity of ambiguities in chemists' practice of classifying plant materials before and after the Chemical Revolution. It further deals with the gradual attempts of some French chemists to implement the Lavoisierian approach in their practice of classifying organic substances, and the criticism of these attempts by the Swedish chemist Jöns Jacob Berzelius and the French chemist Michel Eugène Chevreul.

Chapter 16, finally, discusses the first identification and classification of organic substances according to composition and constitution in 1828 by the two French chemists Jean-Baptiste Dumas and Polydore Boullay. It further discusses the deep ontological changes that came with this new mode of identification and classification.

11

Diverse Orders of Plant Materials

11.1 Commodities from the vegetable kingdom

The plants and most plant materials studied by eighteenth-century chemists belonged to the world of commodities, production, and trade. Chemists experimented with entire plants and their roots, leaves, flowers, fruits, seeds, woods, barks, and other plant organs used as herbal drugs in the apothecary trade. They studied experimentally the balsams, gums, resins, gum-resins, wax, honey, manna, sugar, camphor, and other vegetable raw materials, which were imported by merchants from foreign countries, sold by grocers, druggists, and apothecaries, and described in the official pharmacopoeias as part of *materia medica*. Whereas the official pharmacopoeias usually listed only the Latin names of the materials, pharmaceutical lectures and apothecary textbooks presented more detailed histories of *materia medica*. The pharmaceutical lectures of the French chemist and apothecary Etienne François Geoffroy (1672–1731), for example, presented a detailed history of a total of 44 different kinds of vegetable raw materials, or "juices," sold and processed in the apothecary trade.[1] Among these were opium, sugar, manna, tartar, indigo, starch, sago, various kinds of balsams (*balsamum Judaicum, Syriacum, Hieruchuntinum, Constantinopolitanum*; balsam of Capivi; balsam of Peru; balsam of Tolu; turpentine, etc.), *oleum cacao*, palm oil, camphor, various kinds of gum (*copal officin, gummi ammoniacum, gummi guajaci*, gum arabic, cherry-tree gum, ivy-tree gum, etc.), various kinds of resins (*benzoinum, elemi officin, labdanum*, mastic, scammony, *styrax calamita offic.*, etc.), and various gum-resins (*euphorbium officin, galbanum offic., gummi gutta, opoponax offic., sagapenum offic., sandaracha offic.*, etc.).

In addition to such vegetable raw materials belonging to the class of pharmaceutical *simplicia* (see below), chemists also experimentally investigated vegetable substances whose extractions from plants and vegetable raw materials occurred in chemical and pharmaceutical laboratories, or sometimes in local workshops and distilleries. This latter group of extracted plant materials included the fatty oils, obtained by the expression or decoction of a plant, the aromatic or essential oils procured by distillation, the distilled waters of plants, other distilled products such as spirit of wine and distilled vinegar, the native salts of plants obtained by evaporating the juices of fresh or fermented plants, alkaline salts left after the combustion of plants, infusions and decoctions obtained by boiling herbal drugs in water, tinctures made by digesting plants in spirit of wine or alcohol, vegetable extracts fabricated by evaporating decoctions and tinctures, essences, quintessences, and aromatic spirits made by distilling tinctures. Almost all of these extracted plant materials were applied as chemical remedies and mentioned in the eighteenth-century pharmacopoeias and handbooks of commodities. Only a few were not ubiquitous in the contemporary apothecary trade—such as the *spiritus rector* highlighted by Herman Boerhaave (1668–1738), several of the empyreumatic oils procured in the distillation of plants,

---

1   See Geoffroy [1736] pp. 346–366.

several of the newly discovered vegetable acids, several principles isolated in the second half of the eighteenth century (the coloring principle, the bitter principle, the narcotic principle, and the acrid principle)—and even fewer (such as the vegetable earth) were not applied pharmaceutically at all.

In the first half of the eighteenth century in particular, chemists studied further the various kinds of plant materials resulting from the mixing and chemical processing of different herbal drugs with each other or with other kinds of materials. To these composite plant materials belonged, for example, the composite artificial balsams, composite elixirs, composite aromatic spirits, and composite spirits. Furthermore, chemists also studied plant substances obtained by the chemical processing of a single extracted plant material, such as pure alcohol and the so-called "ethers" made from spirit of wine or alcohol and an acid. The number of these simple processed substances increased in the course of the eighteenth century, but it remained very small compared to the vegetable raw materials, the single extracted plant materials, and the composite plant materials.[2] Both the composite plant materials and the chemically processed simple substances were used as commodities, too, especially in the apothecary trade.

Although the vast majority of vegetable materials studied in eighteenth-century chemical experiments were remedies, chemists also experimented with other vegetable commodities, especially food and dyestuffs. Vegetable extracts, sugar, honey, flour, starch, sago, French liqueur and eau de vie, beer, wine, and other kinds of alcoholic beverages were both objects of scientific inquiry and commodities, explored and refined by eighteenth-century chemists or even commercially produced by chemist-manufacturers. Back in the early eighteenth century, the famous Dutch chemist Herman Boerhaave had already described chemical processes for the manufacture of syrups, jellies, and other kinds of nutritious vegetable extracts. As these preparations could be preserved over a long period, he recommended them for long travels, remarking that "perhaps nothing would more conduce to the health of the British and Danish sailors, than a due provision of this kind."[3] The most prominent and successful example of the chemical exploration of food in the eighteenth century is connected with the substitution of beet sugar for the more expensive imported cane sugar. It was initiated in the 1740s through experiments on plants by the German chemist and apothecary Andreas Sigismund Marggraf (1709–1782).[4] Among the German followers of Marggraf, his pupil Franz Carl Achard (1753–1821) was the most successful.[5] In the 1790s he received a salary and an estate from the Prussian King Friedrich Wilhelm II to establish a beet sugar manufactory.

Natural dyestuffs, many of which were of vegetable origin, and the processes of dyeing and calico printing occupied a prominent place in eighteenth-century chemical experimentation as well as in chemical teaching.[6] Chemists did not only study

---

2   In the early nineteenth century, several of these processed plant materials were identified as pure carbon compounds.
3   See Boerhaave [1753] vol. II p. 18.
4   For life and work of Marggraf, see Gillispie [1970–1980] vol. IX p. 104ff.
5   For life and work of Achard, see ibid. vol. I p. 44f.

these issues at remote academic sites and for the acquisition of knowledge about natural objects. They were also actively involved in the technology of dyeing and the commercial production of dyes. In France, for example, the chemist Jean Hellot (1685–1766) held the position of Inspector General of Dyeing as early as the 1740s, and in the second half of the century Pierre Joseph Macquer and Claude Louis Berthollet became inspectors of dyeing at the manufacture of the *Gobelins*, where they performed quality control experiments in the manufactory's own laboratory. After the middle of the eighteenth century similar links were established in Germany.[7]

In the eighteenth century chemists were studying plant materials in the context of different practices and interests, and the diversity of these practices and interests conditioned their different modes of classifying plant materials. Although we will shed some light on eighteenth-century chemists' classifications of plant materials in contexts of pharmaceutical and other artisanal practices, our main concern is classification in the context of conceptual inquiry. It is in this latter context that major changes of classification occurred in the 1750s, the 1790s, and the 1830s, which allow conclusions to be drawn about shifts in chemists' ontology of plant materials.

With respect to methodology, a study of chemists' classification of plant materials covering the entire eighteenth and the early nineteenth centuries entails a couple of problems that should be addressed briefly. Whereas chemists explicitly reflected on issues of classification and often presented the classes of plant materials in the form of comprehensive lists in the second half of the eighteenth centuries and the period afterward, this was not the case before 1750. In the first half of the eighteenth century chemists did not tackle questions of classification of plant materials in any explicit and systematic way. Nor did they introduce a systematic terminology, such as "class," "order," "genus," and "species," to demarcate the different taxonomic ranks of plant materials. But early eighteenth-century chemists did identify and classify plant materials. Classification was expressed in their terminology and in the organization of lectures, chemical textbooks (especially in their practical or experimental parts), and chemical-pharmaceutical manuals. As most chemical and chemical-pharmaceutical books were divided according to the three natural kingdoms, historical information on chemists' ways of classifying plant materials can be obtained by analyzing the parts on plant chemistry.[8]

11.2 Natural historical modes of identification and classification

Well into the nineteenth century, chemists divided substances into three large groups according to their origin from the three natural kingdoms. The experimental parts of chemical textbooks especially were organized in this natural historical mode of clas-

---

6   See Nieto-Galan [2001].
7   See Hufbauer [1982].
8   On eighteenth-century plant chemistry see also Holmes [1971], and [1989]; and Löw [1977]. Holmes and Löw focused on experimentation and the "analysis of plants" in connection with the extraction of plant materials from plants. Löw also addressed issues of classification, but without presenting a systematic historical analysis of the subject. For plant analysis in the late eighteenth century, see also Simon [2002], and [2005]; and Tomic [2003].

sification.[9] In the chemistry of plants, chemists followed the naturalistic order not only on the highest taxonomic rank of substances, but held to the naturalistic approach all the way down to the identification of kinds or species of plant materials. Below the taxonomic rank of mineral, plant, and animal substances, chemists identified and grouped together plant substances by combining at least three different taxonomic criteria: natural origin, mode of extraction, and observable physical and chemical properties. In addition to these pervasive criteria, the practical application of plant materials played an important role in specific contexts of application, as can be seen in the identification and demarcation of various kinds of vegetable remedies and vegetable dyestuffs.

"Natural origin" meant the origin of a plant material from the vegetable kingdom and a particular species of plants. In the case of plant materials obtained from trade, the country of provenance also informed their identification. Balsam of Peru, balsam of Copaiba, balsam of Tolu, oil of cacao, palm oil, gum arabic, gum-Senegal, cherry-tree gum, ivy-tree gum and cane sugar are names for vegetable commodities that reveal the importance of the materials' origin for their identification.[10] In connection with this mode of identification, the question of whether the specificity of a plant material separated from a particular species of plant corresponded with botanical classification—that is, whether the entire genus or even class of plants always contained the same kind of material—was raised repeatedly. Botanists like Linnaeus believed in the possibility of such generalization, presupposing that a natural system of plants could be established at all.[11] The chemist Guillaume François Rouelle (1703–1770) stated that "all plants of the same family always have the same virtues and properties. They differ only in minor ways; thus chemistry converges with the generic characters assigned by botanists."[12] Many other eighteenth-century chemists, however, were more skeptical about such generalization, or did not extend their inquiries to issues of botanical classification.

The grouping together of varieties of plant substances such as balsams, resins, gums, and so on relied on the perceptible properties they shared. All kinds of balsams, for example, had a strong aromatic smell, a liquid or syrupy consistency, colors varying from light yellow to brown, and dissolved not at all or only a little in water. Whereas physical properties served for the identification and demarcation of plant

---

9   See chapter 1.
10   See the examples in Geoffroy [1736]; and Neumann [1759].
11   See Müller-Wille [1999] p. 143f.
12   Rouelle [n.d.] p. 108. In the following we quote from Guillaume François Rouelle, *Le Cours de chimie de Rouelle* (Paris: Bibliothèque Interuniversitaire de Médicine, MS 5021, n.d.). For an outline of Rouelle's private chemical lectures in his house in the rue Jacob, see Rouelle [1759]. The chief sources of our knowledge on Rouelle's chemical teaching from 1742–1768 are manuscript copies preserved in European and American libraries and private collections. The fact that the dozens of different manuscript copies were written by different pupils of Rouelle, and often underwent later editing, raises the question as to whether Rouelle's original organization of the lectures and his classification of materials has been preserved. The systematic comparisons that hitherto have been made refer to manuscript copies of notes taken by Denis Diderot in 1754–1757 (the "Diderot manuscript tradition"). The experimental part of the Diderot manuscripts is divided according to the three natural kingdoms, and accordingly presents "processes" on plants, animals, and minerals. On the manuscripts belonging to the Diderot tradition see Jacques [1985] pp. 43–53; and Rappaport [1960] pp. 68–101.

materials both in chemical laboratories and in the mundane world of merchants and artisans, over the course of the eighteenth century chemists and apothecaries more strongly emphasized the significance of the chemical properties of plant substances. Thus, for example, G. F. Rouelle's distinction between resino-extractive and extracto-resinous substances relied on differences of solubility in water and alcohol.[13] With respect to laboratory products, chemists also used as taxonomic criteria the modes of chemical extraction and preparation as expressed in names such as "distilled waters," "distilled oils," "expressed oils," "decoctions," "infusions," and "extracts." Eighteenth-century chemists and apothecaries agreed in their acceptance of all three taxonomic criteria—natural origin, the mode of extraction, and physical and chemical properties—and hence shared the principal modes of identification and naming.

Individuation

Throughout the eighteenth century and in the early nineteenth century the nomenclature of plant materials and the grouping together of their varieties under a generic name followed natural historical and artisanal points of view. Chemists identified and classified plant materials according to their natural origin, their perceptible properties, their method of extraction, and their various practical uses. But what counted as one single plant material? How did eighteenth-century chemists decide whether an oil they had extracted from rosemary, for example, was a single substance or a mixture of different kinds of substances?

In everyday life, as well as in the domains of botany and zoology, the question of what the single or individual entities are that we group together as a kind or species of things and label "chair," "table," "rose," or "sparrow," is almost ridiculous.[14] We know what an individual chair or an individual sparrow is, and we often care so little about such individuals that we even do not designate them "individuals" in ordinary language. But this is different with materials. The boundary of one material vis-à-vis another does not necessarily stand up to sensory perception, and may vary from case to case depending on what we do with materials and what kind of goals we connect with their manipulation. How did eighteenth-century chemists decide in their experiments whether a sample of a material was a pure single substance or a mixture of several different substances? How did they circumscribe experimentally the boundary of a single plant substance? Why were natural balsams, for example, considered pure

---

13 See Rouelle [n.d.] p. 279f. On G. F. Rouelle's experiments on the extraction of plant substances by means of solvents, see also Holmes [1989] pp. 80–82; and Löw [1977] pp. 63–65.

14 Although "individuation" is a philosophical term, most philosophers have not paid attention to the problem of individuation addressed here; that is, how experimental scientists determine the boundaries of a single object. This problem is notorious in the experimental sciences, which often study invisible objects accessible only via experimentally produced traces and require the interpretative work of ascribing these traces to one or more objects. Philosophers have studied individuals and individuation in quite different contexts. Dupré discussed the problem of establishing rules for assigning individuals to a species (see Dupré [1993] pp. 27 and 44–53). Kuno Lorenz considered individuation as a problem of reference that can be treated by means of formal semantics (see Mittelstrass [1980–1996] vol. II pp. 227–229). Munitz and others have focused on the problem of identity of individual entities (see Munitz [1971]). For historians of science, a more useful philosophical discussion of problems of individuation may be that offered by Strawson [1979].

plant substances in the eighteenth century, whereas in the nineteenth century chemists no longer considered them as chemical individuals, but rather as mixtures of various resins, volatile oils, and other substances? Questions like these concerning the individuation of chemical substances may at first glance appear anachronistic, as they were not tackled systematically by eighteenth-century chemists, including Antoine-Laurent Lavoisier (1743–1794) and his collaborators. But in their practices eighteenth-century chemists did have to make decisions concerning this question. In the domain of chemistry, which we have circumscribed as the chemistry of pure substances, the individuation of substances was embedded in cycles of experiments; here individual or single substances were traceable, recurrent entities in reversible decompositions and recompositions and replacement reactions. By contrast, in plant chemistry chemists individuated materials by combining a number of different criteria pragmatically and tacitly.

First of all, eighteenth-century chemists perceived the uniform consistency, color, smell, and taste of a plant material as a strong indication that it was a single thing. The fact that a balsam, for example, was considered by merchants to be one single kind of commodity and was applied pharmaceutically as one medicine also contributed to that conviction. In the case of solid materials, uniformity of crystals was another criterion for a single pure substance. Furthermore, when a material was obtained wholesale from a plant without disintegrating in the extraction process, this was also a reason to think of it as one single entity. This was also the case when materials were extracted from a plant in the laboratory without altering their homogeneous appearance during the process of separation. Subsequent tests of the chemical properties of the material using solvents could further prove its uniformity. In the course of the eighteenth century it was in particular the testing of solubility, first with water and spirit of wine, and then with a greater number of solvents later in the century, that confirmed chemists' assumptions about the purity of a sample, or that convinced them that an apparently homogeneous sample of a material was actually a mixture of different substances.

From a comparative point of view, in particular from the perspective of the developments in the later organic chemistry of the 1820s and 1830s—as expressed in the work of chemists such as Jöns Jacob Berzelius (1779–1848), Michel Eugène Chevreul (1786–1889), Jean-Baptiste Dumas (1800–1884), Polydore Boullay (1806–1835), Auguste Laurent (1807–1853), and Justus Liebig (1803–1873)[15]—questions concerning the individuation of chemical substances are eminently historical questions. Such comparison reveals that not only did chemists' modes of classification change over history, but so did the single or individual entities they sorted out and grouped together. The questions of what counted as one single substance and what methods were used to define its boundaries, both experimentally and epistemically, are among the key questions of a historical ontology and must be asked if we want to

---

15 For life and work of Berzelius, see Gillispie [1970–1980] vol. II p. 90ff.; of Chevreul, ibid. vol. III p. 240ff.; of Dumas, ibid. vol. IV p. 242ff.; of Laurent, ibid. vol. VIII p. 54ff.; of Liebig, ibid. vol. VIII p. 329ff.; and Brock [1997]; of Boullay, see Partington [1961–1970] vol. IV p. 345.

understand historical transformations of chemistry on a broader cultural scale.[16] Compared to the methods of individuation of organic substances that emerged in the transition period of organic chemistry between the late 1820s and the early 1840s, eighteenth-century chemists' methods of individuating plant substances shared many features with the way naturalists and craftsmen distinguished between a single material and a mixture of several materials. The main criteria were a homogeneous versus heterogeneous appearance and the maneuverability of a material as one coherent entity versus its disintegration by manipulation. This naturalistic and artisanal mode of individuation of plant materials also differed profoundly from the experimental, analytical mode of individuation in the contemporary chemistry of pure substances mapped by chemical tables.

11.3 Pharmaceutical and artisanal modes of classifying plant materials

In addition to the division of materials according to the three natural kingdoms, eighteenth-century chemists and apothecaries divided materials into remedies delivered by nature and remedies prepared by human art, the former being called "simples" (*simplicia*) and the latter "composite remedies" (*composita*) and chemical "preparations" (*praeparata*). "Simple Medicines," the French chemist and apothecary E. F. Geoffroy defined,

> are those which are form'd spontaneously, or by the Assistance of Nature alone; and those are called Compound, which are owing to the Art and Industry of Men, and to the Mixture of various Simples put together.[17]

Simplicia were natural materials and commodities of the apothecary trade applied either directly as medicines or as ingredients for the making of *composita* and chemical preparations. The class of vegetable simples comprised entire plants applicable as drugs; their organs, such as roots, leaves, flowers, fruits, and barks; and vegetable materials, most of which were imported by merchants from the colonies. The latter group included raw materials such as resins, gums, wax, balsams, camphor, fatty oils, honey, manna, and sugar. As most of these names were generic, the number of vegetable materials was quite impressive. Among the balsams, for example, E. F. Geoffroy distinguished *balsamum Judaicum, Syriacum, Hieruchuntinum,* and *Constantinopolitanum*; balsam of Capivi; balsam of Peru; balsam of Tolu; turpentine, and so on. All of these vegetable juices either flowed spontaneously from the barks and other organs of plants or were obtained by simple mechanical means, such as incision of the bark. Apothecaries and chemists grouped these vegetable commodities together with natural herbal drugs as *simplicia*, since, like the drugs, they were raw materials of pharmaceutical art given, or almost given, by nature. "Simple medica-

---

16 Several historians of chemistry have asked this question before, but in a different way, as a question about chemical purity (see, for example Brock [1993] pp. 173–176). As long as chemical purity is not seen exclusively as a matter of experimental techniques, but also as a concept and an epistemological and ontological issue, this manner of phrasing the question does not differ in content from the question asked in this paragraph.
17 See Geoffroy [1736] p. 2.

ments are those which one uses as nature provides them," the chemist-pharmacist Antoine Baumé (1728–1804) asserted, "or at least which have undergone only slight modifications."[18]

By contrast, composite medicines and chemical preparations were true products of art procured in the pharmaceutical laboratory. *Composita*, in the original Galenic meaning, were made by mixing together different kinds of *simplicia*. In the late seventeenth and the eighteenth centuries, when chemically procured remedies had become widely accepted, the class of *composita* often included chemical remedies, meaning remedies made by manipulations defined as chemical operations, such as distillations and decoctions.[19] Or sometimes the chemical preparations were ordered into a distinct class which supplemented the class of composite Galenic medicines. The tripartite division into *simplicia*, Galenic *composita*, and chemical *praeparata* is found in most of the eighteenth-century pharmacopoeias, the official apothecary books of the time.[20] Hence, *composita* and *praeparata* comprised all kinds of medicines produced in the apothecary's shop, ranging from powders, plasters, ointments, and other Galenic medicines to distilled waters, distilled oils, tinctures, essential salts, and other materials separated chemically from a single plant, to mixed or composite preparations concocted from several different kinds of plants and other ingredients (see figure 11.1).

With respect to the chemical remedies, eighteenth-century pharmaceutical books highlighted the fact that these materials were procured by means of chemical operations, but they did not introduce further chemical subdivisions. Rather, in eighteenth-century pharmaceutical manuals, official pharmacopoeias, and other kinds of apothecary books, all kinds of products of chemical operations in the laboratory were, as a rule, designated "chemical remedies," without distinguishing systematically between products of chemical separation and products of chemical mixing. Chemical remedies comprised simple plant substances like vegetable salts and oils extracted from one plant, mixtures of different kinds of herbal drugs ("composite chemical preparations"), and chemically altered (for example, fermented) plant substances. What all of these materials shared was that they were, in Geoffroy's words, specific products of "art and industry," manufactured in the apothecary's laboratory.

In the eighteenth century the artisanal world of the apothecary trade was an enduring practical and institutional context for chemists' experimental studies and classification of plant materials. But chemists also presented other practically oriented classifications depending on their occupations and interests and the audience

---

18 See Baumé [1762] p. 7. For life and work of Baumé, see Gillispie [1970–1980] vol. I p. 527.
19 See Baumé [1762].
20 See, for example, the divisions in the following pharmacopoeias: [1732], [1741], [1748], and [1781]; Boyer [1758]. For a discussion of classificatory divisions in pharmacopoeias, see Cowen [2001]; Earles [1985]; Kremers and Urdang [1961]; Schmauderer [1969]; and Schneider [1972]. In some cases, the term "preparations" referred to both Galenic *composita* and chemical remedies. For example, the German chemist Georg Ernst Stahl divided the catalogue of medicines presented in his *Fundamenta pharmaciae chymicae* (1728) into *simplicia* and *praeparata*, the latter comprising Galenic medicines such as powders, *confectiones*, plasters, and ointments, and chemical remedies, such as distilled oils, decoctions, elixirs, and tinctures; see Stahl [1728].

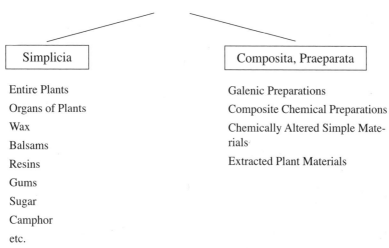

Figure 11.1: Eighteenth-century pharmaceutical classification of plant materials.

for chemical teaching. In his *Chemical Lectures*, the English chemist Peter Shaw (1694–1764), for example, grouped together grain, seeds, pulses, nuts, mast, roots, juice of grapes, yeast, and wine-lees, since all of these materials were used "for the Service of Brewing and Distilling."[21] He also presented a class of "wines and spirits," including "various Preparations for the commodious carrying on of Trade, and the Business of ordinary Life."[22] Furthermore, he presented a class of colors that cut across the divide among the three natural kingdoms, despite the fact that Shaw otherwise recognized this divide as a fundamental mode of classifying substances. In the case of colors, however, his goal was "to improve those several Arts that depend upon the use of Colours, Dyes, and Stains,"[23] and hence he selected classificatory criteria suited to this goal. When chemists became more intensively engaged with the technology of colors and dyeing in the course of the eighteenth century, they suggested many subdivisions of colors.[24] And they further elaborated additional classifications of materials in various practical contexts, or what they then designated "applied chemistry."

Eighteenth-century chemists were engaged in many different artisanal and commercial activities, which contributed to their experimental know-how and connoisseurship of materials as well as to their individuation, identification, and classification. The requirements of chemical teaching also informed chemists' order of plant materials. The pharmaceutical criteria of classification were foregrounded

21 See Shaw [1734] p. 185. For life and work of Shaw, see Gillispie [1970–1980] vol. XII p. 365f.
22 Shaw [1734] p. 200.
23 Ibid. p. 165.
24 See Nieto-Galan [2001] pp. 1–22.

when the audience consisted mainly of physicians and apothecaries, but the teaching of state officials, manufacturers, and interested craftsmen required attention shifts that came with the different forms of ordering. Chemists' various occupations and interests necessitated an evaluative gaze that selected and highlighted specific properties of materials, and lead to a broad range of modes of classification. In the case of remedies, it was the observed or alleged medical virtues of the materials, in the case of dyes their coloring effects, and in the case of food their nutritive properties. But whatever the more specific classification of materials was in contexts of practical or applied chemistry, it was always based on observable properties of materials or on virtues ascribed to them.

Pharmaceutical distinctions in chemical teaching prior to 1750

The pharmaceutical distinction between *simplicia, composita*, and chemical preparations had a strong impact on eighteenth-century chemists' classification of plant materials, in particular in the first half of the century. In the following we present two examples that show how this distinction entered chemical lectures and textbooks prior to 1750. The first is a quite obvious case of such classification, and the second is one that reveals its pharmaceutical impact only at a second glance. The two examples of chemists' distinction between simple medicines, on the one hand, and composite medicines and chemical preparations, on the other, are remarkable not only as yet another expression of the interconnection of pharmacy with eighteenth-century plant chemistry, but also with respect to a different mode of classification that emerged after 1750: the selection and grouping together of the more compound components or proximate principles of plants. For this new classification of plant materials, which can be found only in the second half of the eighteenth century, required the abolition of the former separation of simple plant materials (*simplicia*) from chemically extracted plant materials.

The first example comes from the quite influential chemical lectures of the German chemist and apothecary Caspar Neumann (1683–1737), held in the 1720s and 1730s in Berlin.[25] The organization of the two German editions of Neumann's chemical lectures demonstrate in a most manifest way early eighteenth-century chemists' pharmaceutical points of view regarding the classification of plant materials.[26] The edition by Zimmermann[27] and the Züllichau edition,[28] whose organization is almost

---

25 For life and work of Neumann, see Gillispie [1970–1980] vol. X p. 25f.

26 Holmes and Löw have argued that Neumann was one of those eighteenth-century chemists who established an analysis of plants based on extractions by means of solvents, and who was already grouping together proximate components of plants in the 1720s and 1730s; see Holmes [1989] pp. 75–77, and Löw [1977] pp. 57–60. However, Neumann did not edit his lectures himself. There are three different editions of Neumann's lectures, two in German and one in English. Whereas the organization of the two German editions follows pharmaceutical criteria, the English edition by William Lewis is organized according to groups of proximate principles of plants. However, this English edition is explicitly said to be "abridged and methodized." Holmes consulted only that "methodized" English edition, whereas Löw quoted from secondary sources. See Neumann [1740], [1756], and [1759].

27 Neumann [1740].

28 Neumann [1756]. We refer here to the shortened Züllichau edition of 1756.

identical, are both divided into five parts. Parts one and two describe chemical operations and their products under the headings "liquid remedies" (Züllichau) or "liquid preparations" (Zimmermann), and "solid" (or "dry") "remedies" or "preparations," respectively. Under the label "liquid remedies" or "preparations" are grouped together distilled waters, combustible and saline spirits, acids, alkalis, resinous and other kinds of tinctures, saline solutions, expressed oils, distilled oils, artificial balsams, and simple and mixed extracts. To the "solid remedies," described in part two, belong the alkaline salts, acid and middle salts, and a number of mineral preparations. Parts three to five describe materials divided according to the vegetable, animal, and mineral kingdoms. The part headed "vegetable kingdom" enlists, apart from spices, barks, coffee, tobacco, and many other herbal drugs, a number of natural plant materials—such as *Balsamum Peruvianum, Camphora*, twenty kinds of gum (*Gummi ammonio, Gummi Gutta, Gummi Mastichis, Gummi Scamonii, Gummi Tragacanthe*, and so on), *Manna Calabrina, Melle* (honey), *Cera* (wax), *Opio* (opium), *Sacharum* (sugar), *Suber*, and *Terebenthina*—and several other commodities manufactured on a large scale in workshops. To the latter belong *Cerevisia* (beer), *Cinis clavelettus* (potash), *Sapo* (soap), *Spiritus vini* (spirit of wine), and *Vinum* (wine).[29] It is quite obvious that the basic classificatory scheme underlying this division into five parts was the pharmaceutical divide between chemical remedies and *simplicia*. Whereas parts one and two grouped together chemical remedies procured in the chemists' and apothecaries' laboratories, and divided them according to their consistency, parts three to five assembled *simplicia* stemming from the three natural kingdoms along with several commodities made in workshops. These three latter parts actually were catalogues of alphabetically ordered commodities and *simplicia* of the same kind as the lists of *simplicia* contained in the contemporary pharmacopoeias. Like the lists of pharmacopoeias, both editions presented the Latin names of the remedies in alphabetical order.

Less obvious than in Neumann's lectures is the pharmaceutical impact on Pierre Joseph Macquer's plant chemistry.[30] In his *Elemens de Chymie-Pratique*, which was a very influential chemical textbook in the mid-eighteenth century, Macquer described 52 processes performed with plants, eighteen of which dealt with making composite chemical preparations such as medical soaps and mixtures made from fats and metals and used as plasters. But Macquer separated the products of chemical extraction from the products of chemical mixing, as well as from fermented, that is, chemically altered, plant materials by ordering their descriptions into different chapters. Moreover, he drew a curious second demarcation that is difficult to understand from our modern perspective, or even from the perspective of chemists' way of classifying plant materials a decade later. Macquer placed natural balsams, resins, camphor, bitumen,[31] wax, honey, sugar, manna, and gums in a separate chapter headed "particular analyses of several substances which belong to the vegetable kingdom."[32]

---

29 It should be noted that this inclusion of manufactured commodities in the class of pharmaceutical simples was not common at the time. It was based on the fact that apothecaries often purchased these materials.

30 See Macquer [1751].

In so doing, he set apart these vegetable materials, which belonged to the class of pharmaceutical simples, from chemically separated materials such as distilled oils, extracts, acid spirits, and alkaline salts. A few years later, in the *Plan* for a chemical course published together with Antoine Baumé, this division was reiterated.[33] Having first explained the order of the chemical analysis of plants — which at low temperatures yielded the more compound, proximate principles of plants, such as essential salts and oils, and at high temperatures and dry distillation, simpler parts such as empyreumatic oils and charcoal — Baumé and Macquer then added a lecture with the heading "on the analysis of several particular substances from the vegetable kingdom," which was concerned with such commodities as natural balsams, resins, wax, gums, and gum-resins. From the perspective of the later classification of the proximate principles of plants, Macquer's and Baumé's separation of natural balsams, resins, wax, gums, and gum-resins, on the one hand, from distilled oils, extracts, acid spirits, and other laboratory products, on the other, makes no sense. However, it does make sense when we see this ordering in the light of the contemporary pharmaceutical divide between simples and chemical remedies. The former group of natural plant substances belonged to the pharmaceutical simples, whereas the latter group of laboratory products belonged to the chemical remedies.

From chemists' and apothecaries' traditional perspective, *simplicia* and chemically procured vegetable materials were quite different remedies. In France in particular, chemical remedies had a peculiar social value. In the seventeenth century, the medical professors of the Parisian Faculty of Medicine had fiercely rejected the use of chemical remedies. Struggling for the acknowledgment of chemical remedies, authors of seventeenth-century chemical textbooks such as Etienne de Clave (fl. 1640), Nicaise Le Febvre (c. 1615–1669), and Christopher Glaser (c. 1628–1672 or 1678), had proclaimed the superiority of chemically prepared remedies over *simplicia* and Galenic *composita*. One consequence of these social controversies was that the French chemical textbooks strongly focused on chemical pharmacy.[34] Another consequence was that, against the background of the old claim for the superiority of chemical remedies, the grouping together of chemical remedies with herbal *simplicia* still met with peculiar resistance in early eighteenth-century France. This attitude may have been further corroborated from a naturalistic point of view. As we will show in the next chapter, natural plant materials classified as *simplicia* coincided with the vegetable juices of botanists. Whereas the latter botanical materials were given by nature, chemically procured plant substances were products of art, intervention, and of nature forced.

---

31 Contrary to chemists' common opinion, Macquer argued that bitumen was not a mineral but a plant substance.
32 Ibid. vol. II pp. 188–231.
33 See Macquer and Baumé [1757].
34 See de Clave [1646] p. 9; Le Febvre [1664] p. 22; and Glaser [1676] p. 6f. In the tradition of French chemical textbooks, N. Lemery was the first author to be less emphatic concerning the medical virtues of chemical remedies. "I would not expect Chimerical effects from Chymical Remedies," he proclaimed (Lemery [1677] preface).

Ultimate Principles of Plants: Plant Analysis prior to 1750

In the previous chapter we discussed aspects of chemists' ontology of plant materials and of their classification of these materials that are for the most part characteristic of the entire eighteenth century and the early nineteenth century. Throughout this period, chemists experimented with quotidian plant materials, most of which were commodities applied as remedies, food, and dyestuffs. And they identified and classified these materials according to natural origin, mode of extraction, perceptible properties, and practical uses, the latter criterion depending on contexts of application. But in addition to this first level of chemical inquiries into plant materials, which was concerned primarily with the perceptible properties and applications of plant materials, there was a second level of philosophical or conceptually driven inquiries that studied the imperceptible dimension of plant materials.[1] This conceptual agenda of eighteenth-century plant chemistry, which included the chemical analysis of plants and plant materials, informed additional modes of classification of plant materials. These conceptually driven modes of classification of plant materials are at the center of the subsequent chapters of this third part. Whereas the overall natural historical outlook of chemists' classification of plant materials did not change in the eighteenth century, and classifications according to the practical artisanal applications of plant materials continued to play an important role well into the nineteenth century and beyond, chemists' ways of ordering plant materials from a conceptual perspective changed quite dramatically.

12.1 Separation of the ultimate chemical principles

In the late seventeenth century and the first decades of the eighteenth century, chemists' analyses of plants pursued the goal of separating the most simple elements or principles of plants. This philosophical objective of early plant analysis is shown most clearly in a project of the French Academy of Sciences, which began in 1670 and was disbanded as a failure shortly after 1700.[2] In the project, the chemical analysis of plants was part of a larger natural and experimental history of plants, which included plants' anatomy and physiology as well as the discovery and explanation of their medical and other useful virtues. Based on the chemical philosophy of ultimate principles and the assumption that the medical virtues of plants were effects of the ultimate principles of plants, the chemist-apothecary who was in charge of the chemical experiments, Claude Bourdelin (1621–1699), aimed to separate the simplest chemical principles of plants.[3] In so doing, he relied exclusively on the established method for separating the ultimate principles—that is, dry distillation. As a result of

---

1   See also chapter 2, in which we relate these inquiries to "experimental history," "technological improvement," and "experimental philosophy."
2   On this project see Dodart [1731]; Holmes [1971], [1989], and [2004]; and Stroup [1990].
3   For life and work of Bourdelin, see Gillispie [1970–1980] vol. II p. 353.

this analytical method, all species of plants yielded more or less the same analytical products. This could be seen as an empirical confirmation of the philosophy of ulti-mate principles. But the result also demonstrated that in this way the medical virtues of many plant substances were destroyed. Therefore physicians, naturalists, and phi-losophers involved in the project, such as Claude Perrault (1613–1688) and Denis Dodart (1634–1707), who was also the leader of the project, observed Bourdelin's experiments with skepticism.[4] They thought it possible that the sap, extracts, tinc-tures, and other more compound materials extracted from plants were the true carri-ers of their properties and medical virtues.[5] Bourdelin was little impressed by their objections. In the laboratory of the Academy he performed dozens of plant analyses by means of dry distillation each year, attempting to separate their ultimate principles and to determine their different quantities, until his death in 1699.

Seventeenth- and early eighteenth-century chemists believed that all natural bod-ies—plants, animals, and minerals—consisted of more or less the same kinds of sim-ple ultimate principles. In the seventeenth century, French chemists such as Etienne de Clave, Nicaise Le Febvre, Christopher Glaser, and Nicolas Lemery proposed five elemental constituents or principles of natural bodies, namely water or *phlegma*, an acid spirit or mercury, an inflammable sulfur or oil, a fixed salt, and an earth.[6] By dry distillation most plants yielded water, a volatile acid, and a volatile oil, as well as a fixed alkaline salt and a fixed earth, both remaining in the retort. The majority of late seventeenth- and early eighteenth-century chemists considered these five materials to be manifestations of the ultimate chemical principles; that is, almost pure principles. Hence, plants played an important role for the corroboration of the French chemical philosophy of five principles: they were taken as representatives of all natural bod-ies.[7]

As with any other theoretical concept, the concept of ultimate chemical principles was defined not only by its reference to chemical operations and to substances that can be observed in the laboratory, but also by its relation to other concepts. In the sev-enteenth and early eighteenth centuries, the concept of ultimate principles was loaded with meaning going back to alchemical philosophy. As we have shown in chapter 2, in seventeenth-century alchemy and chymistry "principles" were not quotidian per-ceptible substances. Rather, they were defined as "semina" invested with a set of qualities that generated perceptible substances in co-action with the "elemental matrix." In most seventeenth-century chymical philosophies—especially those not

4 For life and work of Dodart and Perrault, see ibid. vol. IV p. 135f. and vol. X p. 519ff, respectively.
5 See Dodart [1731] pp. 153–161. In his considerations on the goal of the chemical analysis of plants, Dodart repeatedly stated that the goal was not the separation of simplest or first principles but rather the separation of "principes prochains." "The effects of plants often depend on the union of principles," he added (ibid. p. 157f.). He also pointed out that the more compound parts that carry the virtues of plants may be altered by fire, and he proposed to also use extractions by solvents as a means of chemical analysis. On the controversy about methods of chemical analysis among the participants of the Academy's project, see Stroup [1990] pp. 89–102.
6 On the French chemical textbook tradition and the theory of five chemical principles presented in them, see Debus [1991]; Kim [2001], and [2003] pp. 17–62; Klein [1994a] pp. 36–56, [1998b]; and Metzger [1923]. See also chapters 2 and 6.
7 This latter assumption—plants as representatives of all natural bodies—must be highlighted with respect to the developments after 1750 (see following).

combined with corpuscular theories—principles were not physical components of bodies, but constituting agents or generators of natural bodies. Although by the end of the seventeenth century most chemists considered the ultimate chemical principles to be physical parts of natural bodies, they still did not define them as ordinary substances. This becomes obvious, for example, in their growing skepticism that chemical analysis was indeed capable of separating the ultimate chemical principles. In contrast to ordinary materials observed in nature and studied in laboratories, the ultimate simple principles were hidden constituents of perceptible materials. As they were further defined as the ultimate simple causes of the properties and medical virtues of perceptible substances, their number had to be small, ranging from three to five, and sometimes nine (in Boerhaave's case).

If we compare the ontology of ultimate chemical principles in the first half of the eighteenth century with the meaning and reference of apothecaries' *simplicia*, the divide stands out clearly. On the one side of the divide are hidden entities and on the other observable, mundane things; objects bullied out of nature and things immediately given by nature; causes of things, few in number, and a huge number of particular things; indecomposable simple elements, and very compound substances; objects known only by the chemical philosopher and things familiar to artisans. Against this ontological background it becomes fully understandable why early eighteenth-century chemists carefully separated the class of ultimate principles of plants from all other kinds of plant materials, and in particular from resins, gums, balsams, wax, and other materials that segregated more or less spontaneously from plants and were well-known commodities of the trade. This separation was further conditioned by naturalists' concept of "juices of plants," whose reference coincided with that of *simplicia*. In the next section we study by way of example the convergence of the pharmaceutical concept of *simplicia* with naturalists' concept of juices, as well as the separation of chemical principles from both *simplicia* and juices of plants in the early eighteenth century.

12.2 Simplicia, vegetable juices, and the ultimate principles of plants

Early eighteenth-century chemists' division between raw plant materials and substances extracted and prepared from plants in the chemical laboratory was further stabilized by the concept of the "juices of plants." The concept of vegetable juices, which was current in eighteenth-century botany, had largely the same reference as the pharmaceutical concept of *simplicia*. Eighteenth-century botanists conceived of plants as hydraulic bodies containing different kinds of juices in their various organs. Back in the seventeenth century, Marcello Malpighi (1628–1694) and Nehemiah Grew (1641–1712) had claimed that the vessels and tubes of plants were replete with juices such as gums, resins, mucilages, oil, watery sap, and vegetable milk.[8] This

---

8   See Malpighi [1901]; and Grew, *An Idea of a Philosophical History of Plants* (1682), pp. 11–24, in Grew [1965]. For life and work of Grew and Malpighi, see Gillispie [1970–1980] vol. V p. 534ff and vol. IX p. 62ff., respectively.

conception was disseminated in the European chemical community, in particular by the famous Dutch chemist and physician Herman Boerhaave.[9] In the theoretical part of his *Elementa Chemiae*, Boerhaave included a chapter on the "history of plants" that implemented the naturalists' view. "A vegetable is an hydraulic body, (containing various vessels, replete with different juices)," we read right at the beginning of the chapter.[10] The honey collected from flowers by bees; the wax, manna, and balsams deposited on the surface of leaves; the sap flowing from fresh stems in spring; and the gums, balsams, and resins separating spontaneously from barks in summer were now grouped together as different forms of juices of plants. Being transferred into a botanical context, the pharmaceutical class of *simplicia* was renamed as an object of a "history"—that is, a description of plants in the *historia* tradition.[11]

But Boerhaave's history of plants was not merely a descriptive natural history. It also included considerations about the chemical differences in the native juices of plants and their successive transformation during plants' maturation and seasonal changes. According to Boerhaave, plant juices were "gradually changed and further elaborated" chemically in the plant.[12] An example was the native oil of the bark:

> becoming gradually inspissated by the sun's heat, it appears in the form and thickness of a balsam, and changes its name accordingly; by a still longer continuance, and a more intense heat, it grows yet thicker, and becomes a kind of semi-rosin; and by a further increase, or continuation of the same causes, the oil at length acquires both the nature and name of rosin. [...] The resin it self being further concocted, and consequently hardened, is called colophony.[13]

In the practical part of his *Elementa Chemiae*, Boerhaave also discussed the resemblance between materials known as *simplicia* in the apothecary trade, or as natural juices in botany, and chemical preparations made in the laboratory. For example, in a long foreword to his recipe for the extraction of the native oil of almonds in the laboratory, he connected instructions for a proper collection of herbal drugs with musings about the relationship between the oils expressed in the chemical laboratory and the wax, gums, resins, and balsams naturally flowing from plants and sold as commodities in the trade. Expressed oils may become thick by long standing, he asserted, or "may even become solid as we see in wax."[14] He further extended this reasoning from wax to gum, resin (or rosin), and balsam: "Lastly, we see that old trees are oppressed with their own oil, and thence suffocated, thro' the abundance of fat, as the pine, the fir, etc.; where this oil appears in the form of a gum, but in others under that of rosin, oil, or balsam."[15] According to these statements, the difference between fatty oils expressed in the chemists' laboratory and the natural wax, gums, resins, and balsams was only a difference in form or consistency. That is, they were almost the same kind of materials.[16]

---

9   On Boerhaave's plant chemistry, see also Klein [2003b].
10 See Boerhaave [1753] vol. I p. 134.
11 Ibid. p. 146. For the *historia* tradition, see chapter 2.
12 Ibid. p. 140.
13 Ibid. p. 145.
14 Ibid. vol. II p. 57.
15 Ibid. p. 59.

Yet despite his occasional comparison of the native juices of plants with materials obtained in the laboratory, and also despite his chemical explanation for the formation of different native juices of plants, Boerhaave did not really abolish the divide between *simplicia* and native juices of plants on the one hand, and the chemically extracted plant materials on the other. He never summarized his comparisons, and he never presented a list of plant materials comprising both the natural juices of plants and chemically separated materials. On the contrary, Boerhaave sorted out and clearly set apart from the natural juices of plants a particular group of chemically separated plant materials. These were the substances defined in the tradition of chemical philosophy as the ultimate "chemical principles" of natural bodies. In the theoretical part of his *Elementa Chemiae* he presented a short list of these simple principles contained in plants:

> A *spiritus rector*, or presiding spirit, a *sovereign oil*, the true seat of this spirit, an *acid salt*, a *neutral salt*, an *alcaline salt*, either fix'd or volatile, an *oil mixed with salt* after the manner of a soap, and a saponaceous juice hence arising, an *oil firmly adhering to the earth*, so as scarce to be separable therefrom; and lastly *earth* itself, the genuine firm basis of all the rest: these are the principles, or matters, which a well-conducted chemistry has hitherto produced from plants.[17]

It becomes clear from the synoptic style of this enumeration of substances that Boerhaave aspired to an explicit grouping together of principles. He listed a total of eight different "principles" or "matters" of plants: a *spiritus rector*, a sovereign oil, an acid salt, a neutral salt, an alkaline salt, an oil mixed with salt according to the manner of a soap, an oil firmly adhering to the earth, and earth. These eight principles were also mentioned in the practical part of his textbook, where he added elemental water to his list.[18] Boerhaave explicitly derived his concept of a "presiding spirit" of plants from the alchemical concept of "spiritus rector," which he linked with the seventeenth-century French chemical tradition of simple principles. But whereas the French chemists assumed only five ultimate principles—phlegma, spirit, oil, salt, and earth—Boerhaave enlarged the class of simple elements, distinguishing between three different kinds of elemental salt and two kinds of elemental oil.

Despite the fact that his class of chemical principles overlapped with his class of natural juices—oils and the *spiritus rector* appear in both—Boerhaave demarcated the simple chemical principles from the natural juices of plants. His list of eight principles is presented at the end of a chapter on the history of plants without even alluding to the possibility that the juices of plants might be unified with this list. On the contrary, Boerhaave even emphasized explicitly that his history of plants was finished before he went on to present his list of chemical principles.[19]

---

16 Boerhaave drew similar conclusions from a few additional experiments. See ibid. vol. III pp. 74 and 80.
17 Ibid. vol. II p. 147; our emphasis.
18 Ibid. vol. III p. 56.
19 "Thus much we thought proper to explain concerning the *history of plants*, *before* proceeding to shew in what manner they are treated by chemistry: and this may suffice" (ibid. vol. II p. 146; our emphasis).

12.3 Meanings of "plant analysis"

Our interpretation of Boerhaave's plant chemistry—especially our argument that Boerhaave set apart the more compound natural juices of plants and their simple principles—is somewhat at odds with previous interpretations proposed by F. L. Holmes and R. Löw.[20] We briefly discuss Holmes' and Löw's understanding as it is also of central importance for our argument concerning chemists' collective ontology of plant materials. Both Holmes and Löw unequivocally portrayed Boerhaave's plant chemistry, presented in his *Elementa Chemiae*, as "analysis of plants." They further characterized his plant chemistry as a landmark in a continuous development that slowly replaced the philosophical goal of isolating the ultimate principles of plants with the more pragmatic analytical program of separating their compound components or proximate principles. In so doing, they highlighted those experiments of Boerhaave's that described extractions by means of solvents.

Holmes's focus on Boerhaave's extractions of plant materials by means of solvents is part of his depiction of plant analysis as a long tradition that began in 1670 with the Paris Academy of Sciences' project of plant analysis, was renewed around 1700 by the work of the French chemist and apothecary Simon Boulduc (1652–1729), and eventually led to the successful extraction of pure plant substances such as morphine and strychnine in the early nineteenth century.[21] Holmes argued convincingly that Boulduc began to reinterpret familiar pharmaceutical methods of extracting plant substances as plant "analysis."[22] Far less convincing, however, is his further assertion that Boulduc had established a "new analytical program,"[23] which "broke away" from the standard order of analysis of plants that focused on ultimate principles, and instead highlighted the compound or proximate principles of plants.[24] We neither interpret Boulduc's experiments as revolutionary—his experiments stood in a pharmaceutical tradition and mostly pursued pharmaceutical goals—nor agree with Holmes's view that Boulduc established a new program of plant analysis. Instead, we argue that something like a program for the analysis of the more compound components of plants existed only from roughly 1750 onward.

The problem of interpretation arises in part from the fact that eighteenth-century chemists' "analysis" of plants—that is, experiments performed for the acquisition of knowledge about the composition of plants—was intertwined with the goal of separating and preparing pharmaceutically useful materials. Eighteenth-century chemists' experiments pursued multiple goals, which we labeled in chapter 2 as experimental history, technological improvement, and experimental philosophy. In many cases they did so even in one and the same experiment. Given chemists' philosophical agenda, it is very unlikely that they separated tinctures, extracts, resinous parts, and

20 See Holmes [1989]; and Löw [1977].
21 For biographical information on Boulduc, see Partington [1961–1970] vol. III p. 44.
22 Holmes [1971], and [1989] p. 68ff.
23 See Holmes [1989] pp. 68 and 73. Boulduc designated several of his experiments as "analysis," but he did not use the term analytical "program."
24 Holmes [1989] p. 70.

other more compound materials from plants exclusively for pharmaceutical goals, without ever drawing any conclusions concerning the imperceptible composition of plants. But inversely, it would be equally mistaken to interpret each separation of substances from plants as "analysis," for the primary goal of many extractions of plant substances in the eighteenth century was pharmaceutical.

Figure 12.1: Plant-chemical preparations in a seventeenth-century laboratory. From Barlet [1653] p. 332.

Here it is particularly instructive to pay close attention to the historical actors' terminology. Boerhaave and other early eighteenth-century chemists only rarely used the term "analysis" in their presentations of plant-chemical operations, especially before the middle of the eighteenth century. It is also telling that in the practical parts of early eighteenth-century chemical textbooks—which were also a significant point of reference for Holmes and Löw—chemists often substituted the terms "operation" and "process" for "experiment." Whereas the former terms highlighted the practical, productive goals of laboratory activities, "experiment" was an epistemological term used primarily in the context of "experimental philosophy"—that is, when the acquisition of natural knowledge and knowledge about imperceptible entities was in the foreground. Herman Boerhaave, in particular, systematically used the term "experiment" in the theoretical part of his book when he described experiments on fire, air, and other subjects of natural philosophy.[25] But he always used the terms "process" and "operation" in the practical part of his textbook. Similarly, the practical part of his textbook only rarely used the term "analysis."[26] Even Macquer used the term "analysis" in his *Elemens de Chymie-Pratique* only when he emphasized the acquisition or confirmation of knowledge in the context of the theory of composition.[27] But when his attention switched to the material products of chemical experiments and their artisanal and commercial application, he designated the very same operation "extraction," "separation," "expression," "drawing out," "making," "obtaining," "procuring," or, in the cases of metals and ores, "assaying."[28]

We conclude from this terminology, as well as from Boerhaave's and other chemists' explicit statements about the practical uses of substances separated from plants, that the related experiments were not unambiguous "analyses of plants" (meaning experiments performed exclusively for the acquisition of knowledge about the composition of plants). Moreover, among the eighty-eight experiments described in the practical part of Boerhaave's *Elementa Chemiae*, sixteen were not concerned with separations of substances at all, but with the production of composite chemical remedies.[29] Another part of the operations was presented as alterations rather than as separations of natural components. And among the separations, many aimed at the separation of simple principles by means of the old technique of dry distillation and even that of calcination.

Our argument that the compound components of plants became a significant collective objective of European chemists only after approximately 1750 is evinced, in

---

25 See Boerhaave [1753] vol. I pp. 205–593.
26 He used the term "analysis" especially at the beginning of the practical part of his textbook. For this see also Klein [2003b].
27 For example, in the introduction to his *Elemens de Chymie-Pratique,* Macquer used the term "analysis," emphasizing that the operations presented in this practical part also "confirm the fundamental truths laid down in the Theory" (Macquer [1751] vol. II p. ii).
28 For example, the heading of the first paragraph on vegetables reads "Of the Substances obtained (*retirer*) from Vegetables by Expression only." It describes "processes" of "expressing" the juice of a plant and of "drawing" (*tirer*) the essential oils of fruits. The next "process" describes how "to make" extracts of plants, and how to "extract" (*retirer*) from seeds the matter of emulsion; see Macquer [1751] vol. II pp. 1–21.
29 See Klein [2003b] p. 554.

particular, by the fact that only from that time on did chemists group together the compound components of plants. As we will explain in more detail in the next chapter, we take chemists' collective efforts to identify and group together the compound components or proximate principles as an empirical manifestation of their revaluation of plant analysis and an increase in the significance of compound components of plants. Similar or other empirical indications for the existence of a new program of plant analysis, individual or collective, are lacking in the first part of the eighteenth century, and Holmes did not present arguments for his claim to the contrary. Rather, he conceded explicitly that Boulduc's work "did not immediately induce other chemists, even within the Academy, to give up the older approach."[30] Likewise, with respect to Boerhaave, Holmes also conceded that he had not abolished the method of dry distillation of plants. But he explained this away as follows: "A mixture of conservatism with a sensible belief that the potential for improving the customary distillation methods for plant analysis was not yet exhausted may explain the slowness of early eighteenth century chemists to adopt Boulduc's approach."[31]

As becomes almost clear from this statement, a new analytical program of plant chemistry did not yet exist in the 1730s. We agree with Holmes's view that chemists' techniques of separating substances from plants developed gradually, but we wish to add that their attitudes and the objectives of plant analysis shifted more quickly in the years around 1750. This punctuated evolution becomes visible, in particular, in chemists' grouping together of the more compound proximate principles of plants from about 1750 on. Only from this date on did chemists group together substances like natural balsams, gums, resins, wax, and camphor with plant substances separated in chemical laboratories. Well-known commodities of the apothecary trade imported from foreign countries were assembled with the gummy, mucous, resinous parts of plants more recently extracted in the laboratory, and with the essential oils, the expressed oils, the distilled waters, and the essential salts that were long known chemical remedies. All of these materials, which were worlds apart from a traditional pharmaceutical point of view, were then unified under one single label: the "immediate" or "proximate principles" of plants. The next chapter examines this accelerated ontological shift along with chemists' way of classifying the proximate principles of plants.

---

30 Holmes [1989] p. 73.
31 Ibid. p. 75.

The Epistemic Elevation of Vegetable Commodities

13.1 Chemists' grouping together of proximate principles of plants after 1750

In accordance with the significant role played by the pharmaceutical use of plant materials and by pharmaceutical teaching, early eighteenth-century chemists had adopted a pharmaceutical mode of classifying plant materials. The organization of their chemical-pharmaceutical manuals and the plant-chemical parts of their chemical textbooks manifested their pharmaceutical points of view. Early eighteenth-century chemists divided plant materials into natural raw materials—such as resins, gums, balsams, and sugar, which belonged to the pharmaceutical class of *simplicia*— and chemically extracted plant materials and other chemical preparations, which were part of the class of chemical preparations and *composita*. We take this pharmaceutical mode of classification as an empirical indicator of the great significance chemists attributed to the pharmaceutical value of plant materials.

In the second half of the eighteenth century, however, chemists abolished the former pharmaceutical divide between vegetable raw materials and chemical preparations, and instead grouped several *simplicia* together with certain kinds of single plant materials chemically extracted from plants. Despite their difference in provenance, the chemists then understood these materials as compound components or proximate principles of plants (see figure 13.1). This change in the chemists' mode of classification does not imply, however, that their pharmaceutical interests in plant materials had diminished or vanished.[1] Rather, we argue in the following that the abolishment of the former pharmaceutical distinction between vegetable raw materials and chemical preparations hinged on a new constellation of factors: accumulated experience in the chemical laboratory with solvent extractions, a new conceptualization of plants as organized beings, and the proliferation and acceptance of the Stahlian theory of a graduated order of composition.

What counted as the most significant chemical components or principles of plants changed considerably around 1750. Whereas by 1700 chemists were almost exclusively concerned with the ultimate simple elements or principles of plants—apart from a few individual exceptions, such as Simon Boulduc—, after 1750 the majority of chemists became more interested in the compound components or proximate principles of plants. The increase in collective attention the compound components of plants received from 1750 on did not entirely replace the search for the ultimate principles of plants. Instead, chemists began to establish an order of analysis by distinguishing between two kinds of plant-chemical analysis: first, the analysis of entire plants and the organized parts of plants, which aimed at separating the more compound or proximate principles of plants; and, second, the further analysis of the proximate principles into their ultimate components or simple principles. Whereas the lat-

---

1   Simon [2005]; and Kim [2003] p. 212 have asserted that chemists' pharmaceutical interests declined in the second half of the eighteenth century. At first glance, the change of chemists' mode of classifying plant materials seems to evince that claim, but the following analysis will show that the development was more complex.

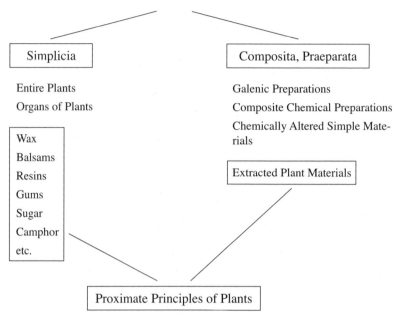

Figure 13.1: Chemists' grouping together of proximate principles of plants.

ter kind of elemental analysis employed the technique of dry distillation, the former required techniques that avoided high temperatures, such as distillation in the wet way, mechanical expression, and, most importantly, extraction by means of solvents.[2]

The first lists of proximate principles of plants were published almost simultaneously in the 1750s in France, Germany, and England. The form of condensed lists, as well as the more frequent use of taxonomic terminology, reveals a transition to more systematic ways of classification. In 1755, Gabriel François Venel (1723–1775) presented a list of twelve proximate or "immediate materials" (*matériaux immédiats*) of plants in an *Essai sur l'Analyse des Végétaux*.[3] A year before, the German chemist Johann Friedrich Cartheuser (1704–1777) had published a *Streitschrift* on the principles of plants in which he had put together a similar list,[4] followed by a list

---

2   F. L. Holmes presented all later eighteenth-century plant analysis as the analysis of proximate principles of plants by means of solvents. Therefore we wish to emphasize that almost all eighteenth-century chemists we have studied actually also further analyzed the substances extracted from plants by dry distillation. In so doing, many of them even tried to study the different proportions of the distillation products. Lavoisier's elemental analysis of plant and animal compounds continued to some extent an old tradition that was never given up when chemists became more interested in the proximate principles of plants. But Lavoisier also changed both the technique and the conceptual resources for analysis of the elemental composition of plant and animal substances. See also our explanations in chapter 14.

3   Venel [1755]. For life and work of Venel, see Gillispie [1970–1980] vol. XIII p. 602ff.

4   For biographical information on Cartheuser, see Kopp [1966] vol. II p. 116.

by Rudolf August Vogel (1724–1774) in 1755.[5] In his translation of Caspar Neumann's lectures from 1759,[6] William Lewis took up this new approach, even though he demarcated the compound components of plants less clearly from their more simple principles than his French and German colleagues (see figures 13.1–5).[7]

All of these chemists discussed the question of the appropriate method for the chemical analysis of plants. All were well aware that the dry distillation of plants and their combustion would destroy their more compound proximate principles. Instead they recommended mechanical means, such as expression and grinding, wet distillation (in which the temperature did not exceed the boiling point of water added to the retort), and extraction by means of solvents. As F. L. Holmes and R. Löw have shown, the use of solvents such as water and spirit of wine for the analysis of plants proliferated in the middle of the eighteenth century.[8] In France, G. F. Rouelle emphasized in his chemical lectures at the *Jardin Royale des Plantes* that mild solvents, unlike strong fire, would extract the components of plants without altering them.[9] In almost emphatic rhetoric, G. F. Venel, a pupil of Rouelle, praised the use of solvents as the most promising route for the acquisition of knowledge about the true components of plants.[10] He clearly distinguished between "analysis by combination" (*l'analyse par combinaison*) using solvents,[11] and analytic distillation by fire (*distillation à feu violent ou analytic*),[12] also designated "ultimate analysis" (*analyse ultérieure*), which aimed at the separation of the simple elements of plant substances and plants.[13] Based on G. E. Stahl's theory of an order of chemical composition,[14] Venel claimed that analysis by means of solvents took into account "the different orders of composition" (*les différens ordres de combinaison*).[15] It separated from a plant the unaltered immediate materials (*matériaux immédiats*)[16] — that is, its com-

---

5   For biographical information on Vogel, see Partington [1961–1970] vol. III p. 608.

6   See Cartheuser [1754]; Vogel [1755]; and Neumann [1759]. Wiegleb called Cartheuser's booklet a pamphlet, or *Streitschrift*. See Wiegleb's translation of Vogel (Vogel [1775] p. 24).

7   Lewis emphasized in the title of his translation of Neumann's chemical lectures that it was "methodized." Hence, we ascribe the organization of the book to Lewis, which is also confirmed by its striking difference in organization from the German editions. In the list of contents, which quite certainly is Lewis's, the following native principles of plants were listed under "Chemical history of the vegetable kingdom": gums, resins, essential oils, camphors, expressed oils, and essential salts. In the text, when describing the techniques of extracting these native juices of plants, extraction by solvents is highlighted (Neumann [1759] p. 266). But at the beginning of the chapter on the native principles of vegetables, phlegma or water and an "earthy substance" are added (ibid. p. 264). The text further spoke of "a saline matter" instead of essential salts. It is quite obvious that phlegma, earth, and saline matter were the traditional simple principles. We interpret this inconsistency as in accordance with the fact that Lewis made compromises between his own views, which are represented clearly in his list of contents and organization of the book, and his obligations as translator of Neumann. On Neumann's theory of vegetable principles, see also Exner [1938] p. 48.

8   See Holmes [1971], and [1989]; and Löw [1977].

9   Rouelle [n.d.] p. 109.

10 Venel [1755]. It should be noted that Venel, too, used mechanical means of separation and distillation at low temperatures.

11 Ibid. p. 319.

12 Ibid.

13 Ibid. p. 320.

14 Venel was a great admirer of Stahl; see, for example, the article "chymie" in Diderot's *Encyclopédie* (Venel [1753]). On Stahl's theory, see later in this chapter.

15 Venel [1755] p. 320.

16 Ibid.

pound parts—rather than its most simple constituents.[17] Having highlighted the technique of analysis by means of solvents, he wrote:

> These immediate materials of the last combination, which considered chemically constitute a vegetable, are the distilled waters, the essential oils, the balsams, the gum-resins, the resins, the extracts, the intermediate bodies between the last two [bodies] called by M. Rouelle, who observed them first, the resino-extractive and extracto-resinous [materials], the oils per decoction, several species of mucilage, the essential salts, the coloring parts, and so on.[18]

Although Venel may not have considered his list of immediate materials of plants to be exhaustive (which is suggested by the "and so on"), he grouped together substances separated from plants in the laboratory—namely, distilled waters, essential oils, extracts, oils by decoction, essential salts, resino-extractive and extracto-resinous materials, and the coloring parts of plants—with several *simplicia* or natural juices of plants—namely, natural balsams, gum-resins, resins, and mucilages (or gums). Dismissing the traditional pharmaceutical distinction between *simplicia* flowing naturally from plants or obtained per incision and chemically separated substances, Venel considered all of these substances to be natural compound components of plants.[19]

Natural balsams, gum-resins, resins, and mucilages as well as distilled waters, essential oils, extracts, oils by decoction, and essential salts extracted from plants by techniques other than dry distillation were not of any outstanding *conceptual* significance to eighteenth-century chemists prior to 1750. They were commodities chemists studied in the same way as other vegetable, animal, and mineral drugs. In the eyes of early eighteenth-century chemists these materials were first of all *materia medica* and subjects of pharmaceutical teaching. In their laboratories they also subjected these natural and chemical remedies to chemical analysis by dry distillation. But they did not ascribe a distinctive epistemic value and conceptual significance to these materials. After 1750, however, these vegetable commodities became elevated epistemically, capturing chemists' attention as proximate principles of plants. As we can see by comparing the tables in figures 13.2–6, not all individual chemists grouped together exactly the same kinds of substances, and the classifications changed over time. For example, ten years after his first classification of "immediate materials" of plants, Venel published a revised list that no longer contained the resino-extractive and extracto-resinous materials distinguished by G. F. Rouelle, but included eight new substances instead—two kinds of acids, the aromatic principle, a volatile alkali, a piquant principle isolated from onions, gum, resin extract (which presumably sub-

---

17 In addition to "immediate materials," Venel also used the terms "constituting principles" (*principes constituans*, ibid. p. 319) and "immediate principles" (*principes immédiats*, ibid. p. 321).
18 Ibid. p. 320.
19 See Rouelle [n.d.] p. 279f. As Venel made clear in a footnote, and as is also revealed in part by the names of the substances, extraction by solvents was not the only way of chemically separating the proximate materials of plants. The essential salts precipitated when the expressed juices of plants were evaporated, and the distilled waters and the essential oils were procured by wet distillation. The important point in all of these cases was that the temperature used in the separation of the materials was moderate, no higher than the boiling point of water.

VENEL'S CLASS OF TWELVE IMMEDIATE
PRINCIPLES OF PLANTS (1755)

Distilled Waters
Essential Oils
Balsams
Gum-Resins
Resins
Extracts
Resino-Extractive Materials
Extracto-Resinous Materials
Oils by Decoction
Mucilages
The Essential Salts
The Coloring Parts of Plants

Figure 13.2: From Venel [1755] p. 320.

VENEL'S CLASS OF EIGHTEEN IMMEDIATE
PRINCIPLES OF PLANTS (1765)

Water (aromatic or non aromatic)
The Aromatic Principle
Spontaneous Acid (*l'acide spontané*)
Spontaneous Volatile Alkali (*l'alkali volatil
    spontané*)
The Vivacious, Piquante Priniciple
    (as contained in onions)
Essential Oil
Fatty Oils
Balsam
Resin
Gum
Gum Resin
Extract
Resin Extract
The Mucous Body
Essential Salt
Weak Acid (*acidule*)
The Green Coloring Part
Several Other Coloring Matters

Figure 13.3: From Diderot and d'Alembert [1966]
vol. XVI p. 869.

CARTHEUSER'S CLASS OF EIGHTEEN COMPONENTS
(*PRINCIPIA PRINCIPIATA*) OF PLANTS (1754)

Aromatic Spirit (*spiritus balsamicus et aromaticus*)
A Sharp, Pungent Exhaltion (*halitus acris pungens*)
A Narcotic Vapour (*vapor narcoticus*)
An Essential Fatty Oil (*oleum essentiale unguinosum*)
A Liquid Balsam (*balsamus liquidus*)
Gum (*gummi*)
Resin (*resina*)
Mucilage (*mucilago*)
Essential Acid Salt (*sal essentiale acidulum*)
Niter (*nitrum embryonatum*)
A Salt (*sal salsum, culinari analogum*)
Camphor (*camphora*, five species)
Volatile Oily Salt (*sal volatile oleosum siccum*, three species)
Wax (*cera*, six species)
Talc (*sevum*, three species)
[Natural] Soap (*sapo*, three species)
Sugar (*sacharum*, nine species)
Balsamic Acid Spirit (*spiritus balsamico acidulus*, two species)

Figure 13.4: From Cartheuser [1754] vol. 4 pp. 7–75.

VOGEL'S CLASS OF THIRTEEN
COMPONENTS OF PLANTS (1755)

A Specific Juice
Oils (essential and fatty)
Soap
Talc (*sevum*)
Mucilage
Resin
Fluid Balsam
Wax
Gum
Gum-Resins
Sugar
Manna
Camphor

Figure 13.5: From Vogel [1775]
pp. 13–25.

LEWIS'S CLASS OF "NATIVE PRINCIPLES" OF
PLANTS, BASED ON C. NEUMANN (1759)

Gums (Mucilages)
Resins (and Balsams)
Essential Oils
Camphors
Expressed Oils
Essential Salts
(A Saline Matter)
(Phlegma or Water)
(An Earthy Substance)

Figure 13.6: From Neumann [1759]
pp. 264–281.

stituted for Rouelle's materials), and the green coloring part of plants (see figures 13.1 and 13.2). Such differences in classification relied in part on new experience in the laboratory, and also in part on differences in the way varieties of plant materials were identified, named, and grouped together. For example, many chemists identified sugar as an essential salt—presumably the reason why sugar is missing from Venel's list.[20] But what all of these new classifications of proximate principles shared was, first, that both the natural raw materials (simples or juices) and the chemically extracted plant materials were unified; second, that all of these substances were separated from plants by techniques other than dry distillation; and third, that all of them could be further decomposed by dry distillation.

What conditions contributed to the change in chemists' classification and their grouping together of *simplicia* and chemical preparations, which had been clearly distinguished from each other before 1750? One reason for the transformation in the chemists' way of classifying plant materials was the accumulation of experimental experience.

Experimental experience

The pharmaceutical divide between plant materials obtained from nature and classified as *simplicia* from plant materials procured in the laboratory and classified as chemical preparations and *composita* had already become a subject of chemical reflection in the early eighteenth century. How did natural balsams, resins, gums, wax, or sugar, which separated spontaneously or almost spontaneously from barks, stems, flowers, and other plant organs, relate to the essential oils, resinous extracts, mucilages, and salts extracted from various parts of plants in the chemical laboratory? Chemists gave quite different answers to this question in the first and second halves of the eighteenth century. In the first half of the eighteenth century chemists such as Herman Boerhaave, for example, occasionally pointed to the similarities of the properties of the vegetable raw materials available as commodities to several substances extracted from plants in the laboratory. But in contrast to chemists in the second half of the eighteenth century, this comparison of similarities did not spur Boerhaave or other early eighteenth-century chemists to abolish their distinction between the vegetable raw materials and the chemically extracted plant substances.

By 1750, when the new analytical technique of solvent extraction proliferated all over Europe, experimental reports about new kinds of plant materials extracted from plants in the chemical laboratory began to accumulate. Among the designations for these new materials, terms like "resinous part" or "gummy part" had become quite common. This terminology exhibits to some extent an ongoing erosion of the traditional pharmaceutical distinction between vegetable simples, such as resins and gums purchased from merchants, and chemical preparations such as resinous and gummy parts. Nevertheless, it would be misleading to depict the factors conditioning chem-

---

20 For further differences in this kind of classification on the lowest taxonomic rank, which continued well into the nineteenth century, see also chapter 15.

ists' new mode of grouping together *simplicia* and chemically extracted substances as the results of internal developments of analytical techniques, even with respect to experimental experience. In order to understand adequately how chemists' experimental experience with substances extracted from plants by means of solvents was constituted, we must also take into account the role played by vegetable commodities in this process.

Gums, resins, balsams, wax, sugar, and so on belonged to an eighteenth-century world of objects shared by apothecaries, chemists, merchants, botanists, and various travelers to foreign countries. These materials were not, as several historians of chemistry tacitly assumed, chemical substances created in the laboratory.[21] Gums, resins, natural balsams, wax, camphor, manna, honey and sugar, and many other natural plant materials originated outside the academic chemical laboratory. Botanists studied them as the juices of plants, merchants imported them as commodities, and apothecaries sold them as *simplicia*. We argue that these vegetable commodities played a decisive role in chemists' analysis of plants by means of solvents: they were prototypes for the identification and naming of the newly extracted plant substances. The possibility of observing similarities between plant substances isolated in the laboratory and the vegetable simples, designated "gums," "wax," "resins," and so on and purchased from merchants, spurred chemists to revise their traditional mode of classification.

When chemists such as E. F. Geoffroy, his younger brother Claude-Joseph Geoffroy (1685–1752),[22] H. Boerhaave, C. Neumann, A. S. Marggraf, G. F. Rouelle, G. F. Venel, and P. J. Macquer examined vegetable substances extracted from plants and their organs by means of solvents, they observed similarities between their chemical preparations and the vegetable commodities sold in the apothecary trade as *simplicia* and described by botanists as juices of plants. Based on the comparison of the physical and chemical properties of the extracted vegetable substances with samples of well-known vegetable commodities, the former were identified as a kind of such and such material and designated as vegetable "gum" (or "gummy part"), "resin" (or "resinous part"), "balsam," or similar. That is, chemists labeled them using the same names as the familiar *simplicia* and juices of plants. A few decades later, Louis Bernard Guyton de Morveau (1737–1816) commented on the role played by knowledge derived from natural history for the identification of new chemically separated substances:

> Chemistry can absolutely not do without natural history; it is the latter which provides
> the specific names of the things we want to designate, and which determines invariably
> the ideas that must be attached to them in order to avoid confusion.[23]

Chemists also combined the familiar names of vegetable juices to designate new plant materials obtained in the laboratory, as was the case with G. F. Rouelle's resino-extractive and extracto-resinous materials. In the middle of the eighteenth century,

---

21 For this alternative historical view, see Holmes [1971], and [1989] pp. 63–83; and Löw [1977].
22 For life and work of C.-J. Geoffroy, see Gillispie [1970–1980] vol. V p. 351f.
23 Guyton de Morveau [1777–1778] p. 98.

when Rouelle performed his experiments, extracts were familiar chemical remedies obtained by the decoction of plants with water, or by the making of tinctures with spirit of wine and the subsequent evaporation of the water or spirit of wine. In his experiments Rouelle had obtained materials that displayed both the chemical properties of resins and of extracts. Both kinds of materials dissolved partly in water (like extracts) and partly in spirit of wine (like resins), but water dissolved a greater part of the extracto-resinous material, whereas spirit of wine dissolved a greater part of the resino-extractive material. This was the case, Rouelle explained, because in the extracto-resinous material the component "which participated in the nature of the extract" was predominant over the resinous part, whereas in the resino-extractive the resinous part was predominant over the extract.[24]

Vegetable juices, available as commodities, could of course play their role as prototypes for the identification of extracted vegetable substances only when the extracted laboratory materials resembled them to some extent. When from the middle of the eighteenth century onward chemists began to claim more often that they had found novel proximate principles of plants—such as the coloring principle, the bitter principle, the acrid principle, or the narcotic principle—they introduced new names to indicate the properties of or the effects caused by these newly separated components of plants. In all of these cases, the traditional mode of individuation and identification of plant substances was preserved, as is most obvious in the cases where the newly extracted substances were christened with the same names as commodities. Whereas by 1750 chemists' way of grouping together *simplicia* and juices of plants with chemically extracted plant substances diverged from apothecaries' and naturalists' classifications, their mode of individuation and identification of plant materials still largely agreed with that of artisans, merchants, and naturalists.[25]

Theoretical conditions

The gradual development and proliferation of the new analytical technique by means of solvents was by no means a sufficient condition for chemists' interest in and grouping together of proximate principles of plants. The great attention the compound proximate components of plants received by the middle of the eighteenth century was also spurred by developments in natural history and in the chemical theory of compounds and composition.

In chapters 2 and 6 we have argued that chemists around 1700 developed concepts of chemical compound, composition, and analysis that deviated from the philosophy of ultimate principles. The new concept of chemical analysis postponed the question of ultimate principles, and instead highlighted chemical substances that had proved to

---

24 Rouelle [n.d.] p. 279f. In 1789, in his *Traité élémentaire de chimie*, Lavoisier took up this terminology by Rouelle in order to establish a new chemical terminology for organic compounds, which was able to distinguish analytically among ten different kinds of organic compounds (Lavoisier [1965] p. 118).

25 The agreement is especially great with respect to the individuation of plant materials. In identification, chemists paid more attention to the chemical properties of plant substances than did merchants and ordinary craftsmen, but they agreed in this respect with apothecaries.

behave like stable building blocks in series of chemical operations. Methodologically, the new concept of chemical analysis demanded the actual experimental isolation of chemical components in a pure form and the resynthesis of the original compound from the analytical components. We have further argued that chemists developed this modern concept of chemical analysis in a specific area of chemistry that was limited to experiments with traceable or "pure" chemical substances such as acids, alkalis, and metals. It was in this specific area, mapped first by chemical affinity tables and then by the 1787 table of chemical nomenclature, that the material and practical conditions for developing the modern concept and method of chemical analysis existed.

In plant and animal chemistry, we have further argued, the material and practical conditions of chemical experimentation were quite different. Reversible chemical operations that lent themselves to interpretation as decomposition into stable building blocks and recomposition from these building blocks could not, as a rule, be performed in these two other areas of chemistry. By contrast, plant chemistry was the domain where the philosophy of ultimate principles could be evinced to some extent experimentally. It was plants in particular that yielded similar kinds of five substances when subjected to dry distillation. The analysis of plants by dry distillation, along with chemists' attempts to isolate the most simple principles of plants, had come under attack back in the seventeenth century, mainly for medical and pharmaceutical reasons: many of the substances extracted from herbal drugs by dry distillation no longer had any medical virtues. But beginning around 1700, individual chemists also questioned the traditional methods of plant analysis for quite different reasons, namely theoretical ones. In this second respect, the concepts and analytical methods that were developing in the chemistry of salts and other pure chemical substances, mapped from 1718 by chemical affinity tables, were not without consequences for other chemical domains.

The German chemist Georg Ernst Stahl, in particular, pointed out that a "pure, natural resolution" of bodies into their ultimate principles is "not easily obtainable from the Chemistry of these days, and so can hardly be come at by Art." Only "very rarely," he continued, are simple principles found "in their purity."[26] Taking into account the experimental obstacles to the actual separation of ultimate principles in a pure form, Stahl proposed a theory of a graduated order of chemical composition. Stahl's theory maintained the basic idea of the philosophy of principles—the causation of the perceptible properties of substances by irreducible qualities of a few chemical principles—and at the same time explained away the experimental difficulties of isolating the ultimate principles. According to Stahl, many natural bodies were not composed directly from the simplest elements or principles, but from more compound proximate principles, which could be further decomposed in one or more steps into the ultimate, most simple principles. He distinguished between "mixts" or "pri-

---

26 Stahl [1730] pp. 4 and 12. See also Stahl's *Zymotechnia Fundamentalis* (1697): "[es ist] bißhero unmöglich gewesen, die anfänglichen Grundwesenheiten der natürlichen Dinge, eintzeln und einfach, entweder irgendwo zu finden, oder aus ihren Mischungen zu scheiden, und dieselbe gantz bloß und rein, ausser den Stand ihrer würcklichen Verbindung mit andern Dingen auf einigerley Weise darzustellen" (Stahl [1748] p. 66).

mary mixts," which are composed immediately of ultimate principles, and "secondary mixts" or "compounds," which are composed of mixts and are resolved in chemical analysis not into the simplest, but into more compound principles.[27] "All the darkness and disputes about Principles," he stated, "arise from a neglect of that real distinction between original and secondary Mixts, or Mixts consisting of Principles and Bodies compounded of Mixts."[28] Furthermore, Stahl claimed that there was an increase in the complexity of composition from bodies belonging to the mineral kingdom up to those belonging to the kingdoms of plants and animals. Plants were more compounded bodies than minerals, and animals were even more complex compounds than plants. Hence the analysis of plants yielded not the simplest principles, as the French chemists assumed, but compound components.

Stahl's theory was made popular among European chemists from the 1740s onward, especially by the French chemists G. F. Rouelle, G. F. Venel, and P. J. Macquer. In his public chemical lectures held at the *Jardin Royal des Plantes* in Paris, G. F. Rouelle taught his students that plants and other natural mixts (*mixtions*) did not consist immediately of the most simple principles but of "principles combined together" (*les principes combinés ensemble*). It makes no sense to claim that a house is built from water and earth instead of bricks, he explained; likewise, it makes no sense to state that plants are made up from the most simple principles instead of more compound building blocks.[29] Rouelle's pupil Venel continued chemical teaching on the theory of "different orders of composition" (*les différens ordres de combinaison*);[30] and Macquer, too, argued for a distinction between "proximate principles" and "remote principles."[31]

By the middle of the eighteenth century, the theory of a graduated order of chemical composition was broadly accepted in the communities of chemists. We argue that this theory contributed considerably to the epistemic elevation of vegetable raw materials and chemically extracted plant substances that had previously been of primarily pharmaceutical interest. Seen in the light of the theory of a graduated order of chemical composition, vegetable chemical remedies such as essential oils, decoctions, and extracts, as well as *simplicia* such as resins, gum, and balsams, appeared in a new light. They were now redefined as products of plant analysis and as the more compound proximate principles of plants. The homogeneous appearance of these substances and their uniform behavior when extracted and treated with solvents convinced chemists that they were single components of plants.[32] Although it was possible to further decompose these substances by dry distillation, the theory of a graduated order of chemical composition now redefined their status.

---

27 See Stahl [1730]. In his *Zymotechnia fundamentalis* (1697), Stahl proposed an even more complex order of composition by distinguishing between *principia, mixta, composita, decomposita,* and *superdecomposita* (see Stahl [1748] p. 57f.).
28 Stahl [1730] p. 5.
29 Rouelle [n.d.] p. 110.
30 Venel [1755] p. 320; see also above.
31 See Macquer [1749] pp. 3-4, and [1766] vol. II pp. 325–331.
32 See the section on the individuation of plant materials in chapter 11.

The simultaneity of chemists' collective acceptance of the Stahlian theory of an order of chemical composition and their new interest in and classification of the compound proximate principles of plants is a coincidence that had important consequences. Stahl's theory reinforced a continuous development of chemists' analytical practice, which in the decades before had been spurred mainly by chemists' attempt to reconcile their analytical and pharmaceutical objectives. By the middle of the century, the theory of a graduated order of chemical composition lent a clear and distinctive voice to these earlier attempts. It was an important and new condition for the acceleration of an ontological shift, which may be easily overlooked when historical studies concentrate exclusively on events taking place in the laboratory.[33] But there was a further condition, external to developments in the chemical laboratory, which contributed to this accelerated ontological shift. This third condition will be examined in detail in the next section.

13.2 Plants and animals as "organized" or "organic" bodies

In the previous section we argued that the historical changes in the meaning and objectives of plant analysis in the first half of the eighteenth century neither hinged on developments internal to academic chemical experimentation nor are adequately conceptualized as a continuous "metamorphosis of a tradition" within the chemical laboratory.[34] As for plant-chemical experimentation, vegetable commodities such as gums, resins, natural balsams, sugar, wax, and camphor played an important role as prototypes for the identification of new plant materials extracted by means of solvents in the laboratory. The identification of laboratory products by comparison with well-known commodities and *simplicia* was crucial for chemists' extension of the class of proximate principles of plants. The meaning of "proximate principles" and of plant-chemical analysis further depended on the acceptance of the theory of a graduated order of chemical composition, which went back to Georg Ernst Stahl. This theory proliferated in the European chemical community around the middle of the eighteenth century, at the same time chemists began to present classes of proximate principles of plants. Although the experimental techniques for extracting compound vegetable materials developed continuously from roughly 1700 onward, as F. L. Holmes pointed out, eighteenth-century chemists' interest in and ontology of proximate principles developed discontinuously. We assert that around 1750 a new constellation of analytical practices, theoretical reasoning, and attitudes became established, which accelerated changes in chemists' ontology of plant materials. A further condition contributing to these developments, which has not yet been studied, was the discourse on "organization" in natural history and the increase in the cognitive authority of natural history.

In the early eighteenth century, most chemists were interested in plants only inasmuch they were applicable as herbal drugs and materials for the making of more

---

33 For the latter, see Holmes [1971], and [1989] pp. 63–83.
34 Holmes [1971], and [1989] p. 73.

refined chemical remedies, on one hand, and bodies that best evinced the philosophy of ultimate principles on the other. Although a few individual chemists, such as Herman Boerhaave, also linked the chemistry of plants with botanical issues and chemical reasoning concerning the generation and functions of the juices of plants, the connection between plant chemistry and natural history was not firmly established on a communal level. Most early eighteenth-century chemists' theoretical interests in plants were restricted to their function as representatives of natural bodies in general, whose analysis evinced the philosophy of ultimate principles. By contrast, in the second half of the century many chemists became interested in plants as "organized" bodies and in their differences from "unorganized" minerals. In this intellectual context, chemists also shifted their attention away from the ultimate principles, which were the same or almost the same in all bodies from the three natural kingdoms, and toward the peculiarity of the components of plants. This new interest was naturally coupled with the concept of compound proximate principles of plants. Did the proximate principles of organized bodies (that is, plants and animals) differ substantially from those of unorganized bodies? Was there a set of proximate principles that was characteristic only of plants? What were the properties of the characteristic proximate principles of plants, and how many were there? Did exactly the same kinds of proximate principles exist in all species of plants, or did the proximate principles vary in accordance with botanical variation? These were the kinds of questions which chemists tackled with increasing interest in the second half of the eighteenth century.

In France in particular, where the Parisian Royal Botanical Garden became a socially and intellectually acknowledged center for natural history by the middle of the eighteenth century,[35] chemists extended plant chemistry to botanical and physiological inquiries, or even aspired to the amalgamation of chemistry and natural history, as Jean-Baptiste Michel Bucquet (1746–1780) did in the 1770s and A. F. de Fourcroy did in the 1780s.[36] Informed by Charles Bonnet's (1720–1793), Georges-Louis Buffon's (1707–1788), and many other eighteenth-century naturalists' and philosophers' emphasis on the "organization" of plants and animals and their new classification of the former as "organized bodies," chemists, too, began to group together plants and animals as "organized" or "organic" bodies and demarcate them from mineral bodies lacking organization.[37] In the French chemical community, the concept of plants as "organized bodies" was propagated in the middle of the century by G. F. Rouelle in particular. Referring to the botanical work by M. Malpighi, N. Grew, and Stephen Hales (1677–1761), Rouelle opened his annual lectures on the chemistry of plants with an account of the anatomy of these "hydraulically organized bodies."[38] In an article entitled *"Végétale,"* published in 1765 as part of Diderot's *Encyclopédie*, his pupil G. F. Venel also defined entire plants and their organs as "organized vegeta-

---

35 On the Parisian Royal Botanical Garden, see Spary [2000].
36 See Bucquet [1773]; and Fourcroy [1782]. For life and work of Bucquet, see Gillispie [1970–1980] vol. II p. 572f.
37 On the notion of organization in eighteenth-century natural history, see Schiller [1978]; and Roger [1997]. For life and work of Bonnet and Buffon, see Gillispie [1970–1980] vol. II pp. 286f. and 576ff., respectively.
38 See Rouelle [n.d.] p. 103.

ble materials" *(les matières végétales organisées)*. By contrast, he designated the balsams, resins, gums, and other materials separated from these bodies as "unorganized" materials *(matières végétales non organisées)*.[39] A year later, in a response to Buffon's and Bonnet's criticism of the idea of different natural kingdoms, P. J. Macquer modified Venel's argument by focusing on the difference between plants and animals, on the one hand, and minerals on the other.[40] Plants grew, they were organized, and they identically reproduced themselves, Macquer emphasized, whereas stones and metals possessed neither a "germ" for their reproduction nor true organization.

But Macquer went a step further. He not only emphasized the difference between plants and minerals, but also highlighted the difference between the proximate principles *(principes prochains)* stemming from "truly living beings" *(les êtres véritablement vivans)* and substances stemming from the mineral kingdom.[41] The proximate principles of vegetables and animals, he stated, "preserve the vegetable or animal character."[42] This statement meant more than the traditional and broadly shared view of chemists, apothecaries, and physicians that the true components of a plant must share some properties with the entire plant. For according to Macquer, the origin of the compound proximate principles of vegetables and animals was expressed chemically by the fact that all of them contained an "inflammable, fatty or oily substance."[43]

Macquer and Baumé had already made this proposition nine years earlier in their plan for a public course of chemistry.[44] In 1766 Macquer further asserted that, as long as the proximate principles of plants and animals were not denatured by putrefaction or by chemical art, their oily component rendered them fermentable. Thus fermentability became another "essential character" and an empirical criterion, too, for the identification of the true proximate principles of plants and animals. At the same time, Macquer declared that the traditional criterion of classification, namely the natural origin of the materials, was no longer a certain common denominator of plant or animal substances. Bitumes, for examples, were oily materials found within the earth; therefore, they were commonly classified as minerals. Now Macquer proclaimed that they originated from vegetables because of their oily nature. Inversely, common salt and Glauber salt were extracted from several kinds of plants and animals. However since these substances "contain[ed] nothing oily" Macquer classified them as minerals, despite the fact that they were obtained from plants and animals.[45]

Macquer summarized:

---

39 See Diderot and d'Alembert [1966] vol. XVI p. 869.

40 On the tension between Buffon's questioning of the traditional tripartite distinction, along with his idea of a "chain of beings" and his demarcation of animals and plants from minerals via the notion of organization, see Roger [1997].

41 See Macquer [1766] vol. II p. 372.

42 "[…] lorsque ces principes n'ont point été dénaturés par les analyses ultérieures, & *qu'ils conservent encore par conséquent le caractere végétal ou animal* […]." (Ibid. p. 374; our emphasis.)

43 Ibid. p. 372.

44 See Macquer and Baumé [1757] pp. 60–62.

45 Macquer [1766] vol. II p. 373.

Hence we conclude, in considering chemically all natural bodies, they must be divided into two great classes. The first class is of bodies deprived of life, *unorganized*, and the principles of which have a certain degree of simplicity which is essential to them: these are the minerals. The other class contains all those bodies which not only have been *distinctly organized*, but which also contain an oily substance, which is no where found in any of the materials which have not made part of animate bodies, and which, *by combining with all of the other principles of these animate bodies, distinguishes these principles from those of minerals by a lesser degree of simplicity*. This second class contains vegetables and animals. We ought also to remark that the *oil contained in vegetable and animal substances* renders them susceptible of *fermentation*, properly so called, which cannot by any means take place in any mineral.[46]

Apart from the distinction of two classes of natural bodies rather than three—the organized plant and animals and the unorganized minerals—in his conclusion, Macquer presented a dual hypothesis about the composition of the materials extracted from plants: they were very compound materials made by combining simple principles, and they contained an oily component, which rendered them fermentable. Thus as early as 1766 Macquer developed an understanding of the chemical specificity of the proximate principle extracted from plants and animals. Although he did not use the term "organic" to designate these substances, he highlighted their difference from mineral substances.

Macquer's view on the peculiarity of the composition and properties of the proximate principles of plants and animals did not immediately find resonance in European chemical communities. As he presented his argument in a very specific discursive context, its consequence first remained restricted to that context. For Macquer's argument that plant and animal substances contained very compound proximate principles, which contained an oily component causing their fermentability, was a counterargument directed against Bonnet's, Buffon's, and other *philosophes'* claims that the distinctions between natural kingdoms were "mere ideals that imply nothing real."[47] It was not merely organization and reproduction that demarcated plants and animals from minerals, said Macquer's new argument, but their chemical components, too. But Macquer did not develop his insights further into a new, general concept of "organic substances" that would have been independent of its critical function as a counterargument against Bonnet's and Buffon's nominalism. Nor did he attempt to group together the proximate principles of plants, or even of both plants and animals, in order to represent the whole range of materials that "preserve the vegetable or animal character." In this latter respect he remained skeptical. Chemistry, he declared, was not yet advanced enough "to determine the number and the kinds (*espèces*) of the different principles (*principes principiés*) of different orders, and in particular of the more elevated orders."[48]

In the three decades that followed, most chemists studied the proximate principles of plants and animals without demarcating principles that fermented from principles

46 Ibid. p. 374; our emphasis. Here our translation largely follows the contemporary English translation; see Macquer [1771] vol. I p. 363f.
47 See Macquer [1766] vol. II p. 371f.
48 See Macquer [1766] vol. II p. 331. See also Macquer [1778] vol. II p. 301.

that did not ferment. Their interest in the proximate principles extended to all kinds of substances that could be extracted from plants, including earths, water, metals, and various kinds of mineral salts. As we will see later, in the second half of the eighteenth century many chemists engaged in a controversy about the origin of the mineral substances extracted from plants. It was only in the 1790s that French and other chemists began to sort out organic components of plants from mineral ones and to group together the former with organic substances of animal origin. Before that time the use of the term "organic" remained restricted to plants and animals.

Theories of the organic

The chemist-pharmacist Antoine Baumé, who taught courses of chemistry together with Macquer for more than fifteen years, further contributed to the new chemical study of plants and animals as "organized bodies."[49] In 1773, Baumé declared that we can "reduce all natural bodies to two big classes, namely the organized bodies and the minerals."[50] He further proposed a more comprehensive theory of composition and decomposition in nature, based on the idea that only "organized bodies are truly combustible and the nourishment of fire, since they alone contain a fatty and truly oily substance."[51] According to Baumé's theory, nature was a "vast chemical laboratory," in which all kinds of compositions and decompositions took place. Within this global natural laboratory, plants' vegetation "was the first instrument of the creator for putting nature in action."[52] Only plants were able to combine directly the four ultimate elements or principles (*principes primitives*). Plants assimilated the simple element earth and transformed it into a vitrifiable earth. Animals, which nourished themselves from plants, further transformed the vitrifiable earth into a calcareous earth. But not only the vitrifiable and calcareous earths, which entered the composition of most natural bodies as secondary principles, were products of organized beings; the "inflammable principle" (*principe inflammable*), too, could become a component of natural bodies only via its assimilation by plants. Hence all minerals and all combustible matter that existed in nature ultimately took their components from plants. Even the inflammable principle of the "*météorites ignés*" stemmed from plants. After the decay of plants, their inflammable principle was distributed in the air, transformed into vapors by the sun, and eventually taken up by meteors.[53]

A similar theory about the cycle of composition and decomposition of natural bodies spurred by the organized bodies had already been proposed in 1766 by the chemist and pharmacist Jacques François Demachy (1728–1803).[54] Rejecting "the common view of chymists that the minerals are the origin of plants and animals," Demachy declared that it was the other way around. Plants and animals, the orga-

---

49   See Baumé [1763] pp. 335–337.
50   Baumé [1773] vol. I p. 6.
51   Ibid. vol. I p. 5f.
52   Ibid. vol. I p. x.
53   Ibid. vol. I pp. x–xvii, 2–7, 125 and 130.
54   Demachy [1766] vol. II pp. 347–371. For biographical information on Demachy, see Partington [1961–1970] vol. III p. 99.

nized bodies, were "the true origin of all layers and the crust of the globe as well as of the regular minerals."[55] Demachy's plant chemistry dealt not only with the chemical analysis of plants and plant materials, but also extended to the anatomy of plants and their organized parts.[56] Hence, Demachy presented descriptions of the roots, bark, and wood of plants, as well as their envelope, parenchyma, germs, and vessels alongside his description of vegetable products like gums, mucilages, balsams, resins, gum-resins, salts, expressed and essential oils, and so on.[57]

Plant chemistry unified with natural history

The growing interest of French chemists in plants as organized botanical objects became most manifest in 1773, when Jean-Baptiste M. Bucquet published his *Introduction à l'étude des corps naturels, tirés du règne végétale*.[58] The book, which comprised two volumes dedicated exclusively to the chemistry of plants, pioneered the extended disciplinary boundaries of plant chemistry accepted in the 1790s. "I have connected (*réuni*) the preliminary notions of botany and the physics of plants," Bucquet asserted in the preface, "with knowledge stemming from plant analysis and from the different products of plants."[59] The first volume of the book started with an account of the anatomy of plants and of their roots, leaves, flowers, and other organs. An extensive second section dealt with plant analysis and the chemical components of plants. The second volume was dedicated in part to fermentation and its products, including wine, spirit of wine, ether, and tinctures—that is, chemical remedies—and to plant physiology. This last section also included theoretical considerations on the germination and the growing of plants, which became key issues in the chemistry of life established in the 1790s.

Bucquet's plant chemistry was part of a larger attempt to unify natural history and chemistry. "I think it is necessary," Bucquet had written two years before, "to link chymistry and natural history in such an intimate way that these branches of physics are only one and the same science that includes all of the most interesting [issues] possible to know about natural bodies."[60] Compared to the plant chemistry in the first half of the eighteenth century, pharmaceutical goals were still present and important in Bucquet's plant chemistry. But as chemists' agenda now also included new natural historical and physiological issues, the pharmaceutical goals were less predominant than before.

The results of Bucquet's efforts to link chemical-analytical, natural historical, and pharmaceutical points of view can be seen in the elaborate second section of his book on plant analysis.[61] In this second section, which covered more than 400 pages, Bucquet first presented twelve chapters on vegetable components separated from plants

---

55 Ibid. vol. II p. 353.
56 Ibid. vol. I pp. 155–160, 168–172, 176–178 and 186–187.
57 Ibid. vol. I pp. 168–202.
58 See Bucquet [1773].
59 Ibid. vol. I p. v.
60 Bucquet [1771] vol. I p. viiif.

by mechanical means such as expression or by means of solvents before discussing experiments on the analysis of different plants by means of fire. The introduction to this section offers a telling example of the way he superimposed botanical and chemical knowledge. Starting with the remark that the juices contained in the vessels of plants have been divided by botanists into "common juices" (*sucs common*) and "particular juices" (*sucs propres*), he further subdivided the particular juices in a chemical way, mostly based on their solubility. The result was four classes of particular juices: first, soft and sweet juices that dissolve both in water and spirit of wine; second, soft and insipid juices that do not dissolve in water; third, fat and odorless juices with varying consistency; and fourth, fat and odorous juices with varying consistency, which are more or less mixed with saline or other materials, and which dissolve more or less easily in spirit of wine.[62] Bucquet's subsequent presentation of twelve classes of particular juices (see figure 13.7), which he conceived as "compound principles" of plants, further combined botanical and chemical-analytical with pharmaceutical and commercial knowledge.[63] All of the twelve classes were subsumed under a unifying botanical category—the heading of the entire section reads "Of the humors of plants"—but materials separated from plants in the laboratory or workshop were also included in these twelve classes. Materials that flowed more or less spontaneously from plants, and hence were identified by botanists as juices of plants and by apothecaries as *simplicia*, were grouped together with chemically separated plant materials.

Bucquet's detailed experimental histories of materials regarded as compound proximate principles of plants demonstrate impressively the diverse provenance of these substances, either from chemical experiments in the chemical laboratory or from the world of trade and commerce. Many of his specific varieties of proximate principles of plants were imported commodities also described in the contemporary pharmacopoeias. For example, the class of "extractive parts of plants" (see figure 13.7) assembled various chemical extracts separated from plants in the laboratory by decoction and subsequent evaporation as well as juices "prepared on a large scale in the trade," such as the "juice of Hypocistis," "juice of Acasia," and opium.[64] Bucquet described not only the way to obtain the latter three pharmaceutical *simplicia* from

---

61 In the preface to his plant chemistry, Bucquet distinguished between analysis by solvents and analysis by fire. Like his predecessors, he asserted that the analysis by solvents enabled chemists to extract without any alteration the "materials which enter into the composition of plants," and which are "very compound principles," (Bucquet [1773] vol. I p. vi).

62 Bucquet [1773] vol. I p. 63.

63 These twelve classes are organized in twelve chapters, which are subdivided into articles presenting species of substances. Bucquet mentioned in the preface to his book that it was not yet possible to use a systematic taxonomic terminology in plant chemistry, distinguishing clearly between classes, genera, and species, as he had done before in mineral chemistry. Plants are too numerous, he explained, and the "analysis of plants is still little advanced" (Bucquet [1773] p. iiif.). Nonetheless, his organization of twelve chapters headed by the names of the "humors" of plants, and the subdivision of these chapters into articles, equally headed by substance names, actually implied a systematic classification.

64 Bucquet [1773] vol. I pp. 73–120 and xii. The latter two juices or extracts were mentioned in the part on the *Medicamenta Simplicia* of the Paris pharmacopoeia (see Boyer [1758]); the "juice of Hypocistis" was listed in the catalogue of simples contained in the Edinburgh pharmacopoeia of 1737. The fact that Bucquet knew these pharmacopoeias as well as the London dispensatory can be taken from an explicit quotation with respect to an extracto-resinous juice; see Bucquet [1773] p. 82.

plants, but also their medical virtues. With respect to the chemically separated extracts, Bucquet highlighted their pharmaceutical uses. In doing so, he presented both their traditional pharmaceutical names, such as "rob of currant," and their chemical names, such as the equivalent "mucilaginous juice." "Chemists have distinguished different kinds of extracts," he explained, "according to their nature, and according to the menstruum used to obtain them. Water, for example, extracted from certain vegetables a mucilage, which being thickened yielded a sort of mucilaginous extract" (*l'extrait mucilagineux*). Spirit of wine, he continued, extracted the "resinous extracts" (*les extracts résineux*). To these two he added the more compound "soapy extracts" (*les extraits savonneux*), among which G. F. Rouelle had distinguished the "extracto-resinous" and the "resino-extractive" matter (*l'extracto-résineux, les resino-extractifs*), based on the difference in solubility of parts of these extracts in spirit of wine and water.[65]

To give another example of how Bucquet grouped together imported pharmaceutical commodities and plant substances separated from plants in the laboratory, we examine the class of "gummy juices."[66] Bucquet divided this class into "gums flowing naturally [from plants] or by incision" and "the gummy juices extracted by water." Among the plants yielding gummy juices he mentioned, for example, the roots of marshmallow, linseed, or the seed of quinces. To the former subclass of natural gums belonged "*la gomme de pays*" (*gummi nosiras*), gum arabic, and tragacanth, which "one finds in the trade."[67] Bucquet highlighted the role of prototypes played by commercially available vegetable commodities in the identification of chemically extracted plant materials as follows: when the roots or seeds of an appropriate plant were macerated in water, the water became mucilageous, "resembling a dissolution of gum arabic;" after the slow evaporation of these mucilages, the dry residue was "a true gum."[68]

Apart from the class of "oily juices" (to which belonged the fatty oils, essential or distilled oils, wax, and butters), the other nine classes of Bucquet's "humors of plants" were either exclusively chemically separated substances—namely, the essential salts, the vegetable-animal part of cheese, the *spiritus rector* going back to Boerhaave, and the coloring parts of plants—or exclusively natural juices, including well-known imported commodities. Among the imported commodities, Bucquet highlighted, in particular, two varieties of natural balsams, benjamin and balsam of Tolu; two kinds of resins, mastic and ladanum; and two kinds of gum-resins, scammony and caoutchouc. These examples evince not only our assertion that the study of proximate principles of plants included commodities—and hence presupposed an epistemic elevation of these commodities—but also our argument that chemist's pharmaceutical interests did not decline in the late eighteenth century. The appearance of such decline is merely the superficial effect of the role played by additional

---

65   Ibid. vol. I pp. 73–75.
66   See figure 13.7 and Bucquet [1773] vol. I pp. 156–163.
67   Ibid. vol. I p. 157.
68   Ibid. vol. I p. 162.

new studies—namely of the organization of plants and plants' physiology—which did not substitute for the old pharmaceutical interests.[69]

Bucquet's attempt to unify natural history and chemistry was continued by his pupil Antoine François de Fourcroy. In the preface to his *Leçons Élémentaires d'Histoire Naturelle et de Chimie,* Fourcroy reiterated the goal set earlier by Bucquet—to "unify (*lier*) these two sciences," (that is, natural history and chemistry) and to show "how much these two sciences mutually enlighten each other."[70] Like Bucquet, Fourcroy complained about the poor state of plant chemistry. French chemists' new concern for plants and the analysis of plants was accompanied by the growing awareness that an "immense and difficult labor" was still necessary to achieve "the same progress as in the mineral kingdom."[71] Also following Bucquet, Fourcroy distinguished between the common "humors" or "sap" of plants, and particular juices.[72] He further distinguished juices separated from plants "by mechanical means" without any alteration, which he defined as their "proximate principles," and juices altered by fire and by fermentation.[73] Although he did not explain what he meant by "mechanical means," it becomes obvious from his experimental histories of the particular proximate principles of plants that the term did not refer exclusively to expression, grinding, and the like, but also to analysis by solvents and wet distillation at low temperature through the addition of water in the retort. Based on the description of their properties, Fourcroy described seventeen proximate principles of plants, which he also presented in the form of a list in the preface to his book (see figure 13.8).

Like Macquer, Baumé, Demachy, and Bucquet, Fourcroy demarcated plants and animals from minerals, as the former were "organized beings" or "organic materials."[74] Fermentation and putrefaction were highlighted as common denominators of the compound proximate principles of plants. The movement of fermentation, Fourcroy declared, "is specific to the fluid [parts of] organic bodies, and only the substances elaborated by the principle of vegetable or animal life (*le principe de la vie végétale ou animale*) are susceptible of it."[75] L. B. Guyton de Morveau was another influential French chemist who contributed to the chemistry of organized bodies in the last third of the eighteenth century. In a short "natural history" contained in his *Élémens de Chymie* of 1777–1778, Guyton de Morveau not only separated his descriptions of the minerals from those of plants and animals, which he treated in one and the same chapter, but he also asserted that the "essence" of the proximate principles of plants was conditioned by plants' organization. Furthermore, he proclaimed that chemical art was not able to imitate the "modification" linked with organization, and hence was restricted to the separation of plants' principles only.[76]

---

69 For claims to the contrary, see, in particular Simon [2005]; and Kim [2003] p. 212.
70 Fourcroy [1782] p. iv.
71 Ibid. p. xii.
72 Ibid. p. 423.
73 Ibid. p. xiv.
74 Ibid. pp. xviii and 530.
75 Ibid. p. 546.
76 Guyton de Morveau [1777–1778] vol. I p. 152.

BUCQUET'S CLASS OF TWELVE COMPONENTS ("HUMOURS") OF PLANTS (1773)	FOURCROY'S CLASS OF SEVENTEEN "PROXIMATE PRINCIPLES" OF PLANTS (1782)
Common Juices	Juices of Plants
Extractive Parts of Plants	Extracts
Essential Salts	Essential Salts
Sweet Juices	Manna
Gummy Juices	Gums
The Vegetal-Animal Part of Cheese	Mucilages
Oily Juices	Fatty Oils
Camphor	Essential Oils
Spiritus Rector	Camphor
Resinous Juices	Spiritus Rector
The Earthy Starch of Plants	Balsams
Coloring Parts of Plants	Resins
	Gum-Resins
	Elastic Gum
	Starches
	Flours
	Coloring Parts

Figure 13.7: From Bucquet [1773] vol. I pp. 63–414.

Figure 13.8: From Fourcroy [1782] p. xiv.

Studies of proximate principles of plants outside of France

Outside France, the English chemist William Lewis (1708–1781), in his translation of Caspar Neumann's chemical lectures, pointed out similar limits of chemical art.[77] Lewis' *Chemical history of the vegetable kingdom* opened with the statement that "vegetables are bodies of a regular organic structure, analogous in some degree to that of animals."[78] The aliment of plants, he continued, is changed in plants' vessels into juices peculiar to particular plants. About these peculiar juices, which he defined as gums, resins, oils, and so on, he further asserted: "No art can prepare or extract from the substances by which plants are supported, products in any respect similar to those elaborated in the bodies of vegetables themselves."[79] In Germany, as early as 1755, Rudolph A. Vogel designated plants as "natural organic bodies."[80] Referring to the work of Bonnet and Erxleben, Christian Ehrenfried Weigel (1748–1831) also divided all natural bodies into "inorganic bodies" (*unorganische Körper*), that is, minerals and "organic bodies," plants, and animals.[81] Organic bodies, Weigel stated,

---

77 For life and work of Lewis, see Gillispie [1970–1980] vol. VIII p. 297ff.
78 Neumann [1759] p. 260. We quote Lewis, rather than Neumann, since the organization of the English translation of Neumann's lectures is Lewis' (see above).
79 Ibid.
80 Vogel [1775] p. 16. We refer here to the German translation of the Latin original, published in 1755.
81 See Weigel [1777] p. 357; Bonnet [1762]; and Erxleben [1773] pp. 55–65. Erxleben designated plants and animals as "organic bodies" (ibid. p. 55). For life and work of Weigel, see Gillispie [1970–1980] vol. XIV p. 224f. For biographical information on J. Ch. P. Erxleben, see Partington [1961–1970] vol. III p. 591.

grow by internal apposition (*innere Ansetzung*)—as opposed to the external apposition (*äußere Ansetzung*) of minerals—nourished from their juices circulating in vessels.[82] Weigel further added paragraphs on the anatomy and physiology of plants as well as on the transformation of the nourishing juice (*Nahrungssaft*) into the products of plants.[83] Likewise, Johann Christian Wiegleb (1732–1800) designated plants and animals as "organized creatures," making the interesting distinction that minerals were lifeless things whereas the latter were living beings.[84] In 1789, Friedrich Albert Carl Gren (1760–1798) also presented a description of the analysis of "organic bodies."[85] Like his predecessors, Gren applied the term "organic bodies" to entire plants rather than to the substances extracted from them; the latter he designated as "proximate components."[86] Not until circa 1800 was the term "organic" also used to refer to substances extracted from plants and animals.

A controversy: Are mineral substances truly proximate principles of plants?

After 1750, the predominant objective of chemists' analysis of plants was the separation and identification of the compound components or proximate principles of plants. The ultimate simple principles did not disappear from chemists' agenda, but were no longer at the center of conceptual inquiry. Chemists' attempts to identify and order the proximate principles of plants and animals survived the upheavals of the Chemical Revolution in the 1770s and 1780s and continued well into the nineteenth century. Yet when we compare the classes of proximate principles of plants presented in the period between 1750 and 1790 and after 1790, we discover a curious difference. Between approximately 1750 and 1790 most chemists included mineral substances in their class of proximate principles of plants, whereas after 1790 they excluded mineral substances from that class. The targets of plant analysis changed during the late eighteenth century, as did chemists' conceptual networks and interests that contributed to the selection and identification of their objects of inquiry. In this section, we first discuss a controversy concerning the inclusion of mineral substances extracted from plants in the class of the proximate principles of plants, and then show how that controversy ended in the 1790s.

If we take a look at the earliest lists of proximate principles of plants published in the 1750s, we see immediately that substances were included in these lists other than those which later became identified as "organic substances" (see figures 13.1–5 earlier in this chapter). Venel entered into his list of immediate principles a material that would not still appear four decades later, namely aromatic distilled waters. The aro-

---

82 Weigel [1777] p. 357.
83 It should be noted that Weigel designated these materials not only as juices but also as "products" of plants, rather than components. Nonetheless, as chemically isolated substances were part of this class (namely, the fixed alkaline salt, spirit of wine, and essential oils), we assume that his term "product" had the same reference and meaning as "component" or "proximate principle."
84 Wiegleb [1786] vol. I p. 7f. For life and work of Wiegleb, see Gillispie [1970–1980] vol. XIV p. 332f.
85 Gren [1787–1790] p. 2. For life and work of Gren, see Gillispie [1970–1980] vol. V p. 531ff.
86 Partington's assumption that Gren in his handbook used the term "organic bodies" for organic substances is mistaken; see Partington [1961–1970] vol. IV p. 233, and Gren [1787–1790] vol. 2.

matic distilled waters of plants were well-known chemical remedies and ingredients of perfumes, procured by the distillation of plants at low temperatures. When for analytical purposes they were obtained from fresh plants without adding water to the retort, chemists assumed they were proximate principles of plants containing elemental water and essential oils or aromatic principles.[87] Cartheuser further included niter (*nitrum embryonatum*) and *sal salsum* in his class of compound principles of plants (see figure 13.4).[88]

Beginning in the 1750s, English, French, and German chemists engaged in a debate about whether the mineral salts obtained from several species of plants actually were natural compound components of plants, and hence had to be ordered into the group of native essential salts of plants. In his translation of Neumann's lectures, William Lewis argued that most mineral salts found in plants stemmed not from plants but from the earth in which plants grew. Vegetables growing in nitrous grounds were often found to contain niter, he stated, whereas vegetables growing on the coast often contained sea salt; and: "from inattention in this circumstance, some of those who called chemistry in aid for the examination of vegetables, have erroneously ascribed to all the plants of one species, saline ingredients which have been casually absorbed by an individual."[89] Macquer and Baumé shared this opinion.[90] But the question of the origin of the mineral salts separated from the juices of plants remained highly controversial.

In the 1770s and 1780s leading French and German chemists proposed that mineral salts were indeed true native components of plants, some of which presumably were generated in plants. As essential salts were obtained from the expressed juices of plants without the use of strong fire, eighteenth-century chemists agreed that they were not artifacts created by fire but natural parts of plants; this is the reason why they were designated "essential." In the pharmaceutical tradition, these salts were separated from the expressed juices of plants or from vegetable infusions (extractions of plant materials with water) by evaporating the juices and infusions and purifying the thickened juices with egg white. This procedure did not allow for easy identification of the different kinds of essential salts. But when, later in the eighteenth century, chemists such as Hilaire Martin Rouelle (1718–1779), the younger brother of Guillaume François Rouelle, Andreas S. Marggraf, and the Swedish chemist and apothecary Carl Wilhelm Scheele (1742–1786) tried to isolate and identify the different kinds of essential salts by applying different kinds of solvents and precipitating

---

87 Distilled waters were produced not only in pharmaceutical and chemical laboratories, but also on a large scale in distilleries. For non-analytic productive purposes, the distillation of these waters was always performed by adding water to the retort in order to avoid high temperatures, which would have let to the denaturation of the materials. For the exclusive purpose of analysis, chemists also carefully distilled plants without adding external water at low temperature to obtain waters regarded as components of plants. See, for example, Boerhaave [1753] vol. II p. 9.

88 Cartheuser added that several savants also considered additional mineral substances such as the fixed alkali, the volatile alkali, alum (*sal aluminosum*), *sal vitriolicum*, earth, and water to belong to the natural components of vegetables (Cartheuser [1754] p. 5). But Cartheuser did not discuss these proposals.

89 Neumann [1759] p. 263.

90 See Macquer [1766] vol. II p. 373; Baumé [1763] p. 337.

reagents, they also discovered different mineral salts in vegetable juices.[91] Based on these types of experiments, Guyton de Morveau considered selenites, niter, vitriolated tartar, marine salt, and Sylvius's salt to be mineral salts that "exist in plants." J.-B. Bucquet also reminded his readers of experiments by Hilaire M. Rouelle and Marggraf, and claimed that these experiments had shed light on the "great question" of the origin of mineral salts found in vegetables.[92] Bucquet was not only convinced that the essential salts of plants also included mineral salts—namely the deliquescent fixed alkali, the mineral fixed alkali, vitriolated tartar, Glauber's salt, Sylvius's salt and marine salt—he further proclaimed that it was "very probable that a great portion of these [mineral] salts are formed by vegetation."[93] In 1782 A. F. Fourcroy picked up on this assumption that several mineral salts must be included in the class of essential salts of plants.[94]

The debate also extended to the mineral substances found after the dry distillation or combustion of plants. But in this respect French and German chemists did not agree. Although J.-B. Bucquet was convinced that the fixed alkalis, common salt, Glauber's salt, iron, and specific earth found after the combustion of plants were natural compound components of plants, he stated that they were less compound than the principles or juices extracted mechanically and by means of solvents.[95] Accordingly, he did not group these substances together with the more compound principles or juices of plants. The order of analytical techniques—that is, the fact that the substances belonging to the class of proximate principles of plants were all extracted by means of solvents, wet distillation, and mechanical expression, and could be further decomposed by dry distillation—clearly demarcated the more compound from the more simple components. However, German chemists who engaged in the debate were less rigid in this respect. C. E. Weigel considered the fixed alkali obtained after combustion to be a product of plants, which he grouped together with the compound components extracted mechanically and by means of solvents (see figure 13.9).[96] The German chemist and apothecary Johann Friedrich A. Göttling (1753–1809) asserted that Marggraf and Wiegleb had delivered the "infallible proof" (*untrüglicher Beweis*) that the fixed alkali and the volatile alkaline salt were natural components of plants rather than products of combustion and putrefaction.[97] J. C. Wiegleb was among the most decided German proponents of the view that mineral salts were true vegetable components generated by plants. In 1775 he presented related arguments in

91 For life and work of H. M. Rouelle and Scheele, see Gillispie [1970–1980] vol. XI p. 564 and vol. XII p. 143ff., respectively.

92 See Bucquet [1773] vol. I p. 125.

93 Ibid.

94 Fourcroy [1782] p. 432f.

95 See Bucquet, [1773] vol. I pp. 443–452.

96 Weigel [1777] p. 365. Weigel spoke in a theoretically more neutral way of "products" of plants. It should be noted that in the 1770s and 1780s—that is, the time period usually designated as the Chemical Revolution—the ontological status of all of the former simple principles was reconsidered. Chemists began to tackle the question of whether alkalis and earths were indeed simple substances, even though the 1787 table of chemical nomenclature still ordered them into the class of simple substances. For the transitory ontological status of these substances, see also chapter 10.

97 Göttling [1778] pp. 51 and 98. For biographical information on Göttling, see Partington [1961–1970] vol. III p. 595; and Kopp [1966] vol. III p. 158.

WEIGEL'S CLASS OF THIRTEEN COMPONENTS ("PRODUCTS") OF PLANTS (1777)	WIEGLEB'S CLASS OF THIRTEEN "COMPONENTS" OF PLANTS (1786)
Essential Acids	Resins
Fermented Acids	Balsams
Fixed Alkali	Gums
Spirit of Wine	Dry Juices
Fermented Juices	Mucilages
Gum	Wax
Wax	Talc
Resins	Oils (fatty and essential)
Essential and Fatty Oils	Camphor
Balsams	Wine
Camphor	Spirit of Wine
Sugar	Salty Substances (Sugar, Manna,
Mucilages	Honey, Alkalis, Middle Salts …)
	Earth

Figure 13.9: From Weigel [1777] p. 365.   Figure 13.10: From Wiegleb [1786] vol. I p. 7f.

annotations to his translation of R. A. Vogel's *Institutiones*; experience demonstrated that mineral salts were "essential components of vegetables," he pointed out.[98] A decade later, in his *Handbuch der allgemeinen Chemie*, he not only gave further explanations about the "salty substances" contained in vegetables, but also included in his class of "salty substances"—apart from compounds such as tartar, vinegar, sugar, manna, and honey—volatile and fixed alkalis, saltpeter, and middle salts (see figure 13.10).[99] In so doing, Wiegleb left no doubt that this mixed class of salty substances belonged to the larger class of compound components of plants, which even contained a group of earths, namely, calcareous earth, argillaceous earth, and siliceous earth.[100]

13.3 A second ontological shift circa 1790: The coming into being of "organic substances"

It was not until the 1790s that chemists began to dismiss mineral substances from the class of proximate principles of plants. But despite their earlier debates, they did so without any explicit argumentation. Their exclusion of mineral substances from the class of proximate principles of plants is made manifest mainly through scrutiny of

---

98 See Vogel [1775] p. 26.
99 Wiegleb [1786] vol. I p. 71f. Like Weigel, Wiegleb used a theoretically neutral terminology. He designated his class of substances simply "components" of plants. Also like Weigel, Wiegleb considered fermentation a separation of the components of natural bodies rather than a transformation in which new substances were formed. Hence tartar and vinegar, which presupposed fermentation, were defined as natural components of plants.
100 Wiegleb [1786] vol. I p. 8.

the materials actually included in the class of proximate principles. It can be observed, for example, by examining the range of substances included in Fourcroy's group of "essential salts." In 1782 Fourcroy had accepted Bucquet's claim that mineral salts belonged to the group of essential salts and could be created by plants. "Naturalists have different opinions concerning the mineral salts found in plants," he stated, but he immediately added that, given the fact that different plants growing at the same place and in the same kind of earth yielded different kinds of mineral salt, it was possible that these salts were true products of vegetation.[101] In 1792 Fourcroy presented a new list of proximate principles of plants (see figure 13.11), which, like the list from 1782, still included the essential salts.[102] However, in the subsequent verbal description of the proximate principles, Fourcroy stated:

> The essential salt, *comprehending the vegetable acids*, which are made up in general from hydrogen and carbon, and which are more oxygenated than the three preceding principles [extract, mucilage and sugar]; by adding oxygen to the latter [extract, mucilage and sugar] they are transformed into acids.[103]

Without giving any explicit explanation, in 1792 Fourcroy delimited the group of essential salts to the vegetable acids, many of which had been discovered only recently by C. W. Scheele.[104] Nine years later Fourcroy published a revised list of proximate principles in which the term "essential salts" was displaced by "vegetable acid or acids" (see figure 14.3 in the next chapter).[105]

FOURCROY'S CLASS OF SIXTEEN IMMEDIATE
MATERIALS OF PLANTS (1792)

Extract
Mucilage
Sugar
Essential Salts
Fixed Oil
Volatile Oil
Aroma
Camphor
Resin
Balsam
Gum Resin
Starch
Gluten
Coloring Matter
Elastic Gum
The Ligneous Part

Figure 13.11: From Fourcroy [1792] p. 95.

101 Fourcroy [1782] p. 432f.
102 Fourcroy [1792] p. 95.
103 Ibid. p. 98; our emphasis.
104 On Scheele and his discovery of vegetable acids, see Partington [1961–1970] vol. III pp. 231–233.
105 Fourcroy [1801–1802] vol. VII p. 125.

In 1766, Macquer had already highlighted specific features of plant substances such as combustibility and fermentability. He had also proposed a theory of composition that explained the distinctive properties of typical plant substances. According to Macquer's theory, plant substances contained an oily principle. But that theory had little impact at the time, as can be seen in the controversy over the understanding of minerals found in plants. Yet in the 1790s this seemed to be no longer a question of outstanding importance. Fourcroy and other chemists then paid renewed attention to the distinctive features of plant substances highlighted by Macquer; namely, combustibility and fermentability. Why did these features acquire outstanding significance in the 1790s? What spurred Fourcroy and other chemists to sort out minerals and to use from this time onward the label "proximate principles" or "proximate materials" only for a distinctive subgroup of plant substances? In his 1792 text Fourcroy made a remark that provides a clue to these questions. Essential salts, extractive matter (*l'extractif*), mucilage, and sugar, he stated, were compounds of carbon, hydrogen, and oxygen, and in the case of the extract, also of nitrogen.[106] In his conclusion about the "immediate materials" (*matériaux immédiats*) of plants he then generalized:

> From what has been said above, one can conclude with respect to the sixteen immediate materials of plants that they can be reduced in their last analysis to three or four principles that are their primitive components, namely, hydrogen, carbon, oxygen, and sometimes nitrogen; and that these materials differ from each other only with respect to the diverse proportions of their constituting elements.[107]

This general definition of the immediate materials constituted new boundaries for the range of proper plant substances. If materials were separated experimentally from plants that did not consist of carbon, hydrogen, oxygen, and nitrogen, then they had to be excluded from the redefined class of proximate principles of plants. Fourcroy did not ignore the fact that earths and various kinds of mineral salts had been found in the analysis of juices and ashes of plants. But in light of the new Lavoisierian theory, these substances could be clearly distinguished from those plant substances that possessed outstanding properties usually not observed in minerals. By definition, the class of proximate principles of plants now no longer comprised all kinds of more compound components of plants, which had been extracted by solvents or techniques other than dry distillation, but only those compounds consisting of carbon, hydrogen, and oxygen, and sometimes nitrogen.

Lavoisier's theory and its unintended consequences

What had happened in the period between 1782 and 1792 that spurred Fourcroy to select from the former class of proximate principles of plants only those that were composed of carbon, hydrogen, and oxygen, and in some cases also of nitrogen? In 1789, A.-L. Lavoisier had proposed a new analytical definition of plant and animal substances that was to become a compelling proposition almost immediately after its

---

106 Fourcroy [1792] p. 97f.
107 Ibid. p. 102.

publication. The acids belonging to the vegetable kingdom, Lavoisier stated, "are chiefly, and almost entirely, composed of hydrogen, charcoal and oxygen."[108] Another large class of plant substances was the oxides, including sugar, mucus, gums, and starch, which contained less oxygen than the acids.[109] He concluded that "the true constituent elements of vegetables are hydrogen, oxygen, and charcoal."[110]

In the years before 1789, Lavoisier had performed analyses of four plant materials; namely, sugar, spirit of wine, olive oil, and wax. As F. L. Holmes showed, his experiments on sugar in 1779 were performed not in the context of a study of the composition of plants, but as part of his elaboration of a new theory of acidity. When Lavoisier returned to plant analysis in 1785, he added spirit of wine, olive oil, and wax to his list of plant substances. But by this time he already envisioned a general idea about the composition of plants: "he no longer doubted that plant substances were composed qualitatively of the inflammable principle [later "hydrogen"] and *matière charbonneuse* [later "carbon"], or these two together with the oxygen principle."[111] How Lavoisier became convinced that all plant substances consisted mainly of carbon, hydrogen, and oxygen is still an open question. Yet it is clear from the limited number of plant materials he used in his experiments that this conviction was not merely a summary of experiments, but rather a generalization and theory of composition. Moreover, this theory was linked with a new concept of "simple substances" (later "chemical elements") that relied on operational criteria—simple substances were those that had not yet been decomposed in chemical analysis—and largely ignored the question of ultimate elements. Carbon, hydrogen, and oxygen were defined as simple or indecomposable substances in the Lavoisierian taxonomy.

Lavoisier further elaborated his new theory of the chemical composition of plants and plant materials by adding considerations about chemical affinity. As can be seen from the eighteenth-century tables of affinity, in the early eighteenth century chemists had established regularities or "laws" of affinity between pairs of chemical substances.[112] Even when they applied the laws of chemical affinity to chemical reactions involving three or four different substances, as in the case of double salts, they always reduced the processes to chemical interactions and affinities between pairs of substances. But now, with respect to vegetable acids, Lavoisier stated, "The power of affinity *reciprocally exerted by the hydrogen, charcoal, and oxygen,* in these acids, is in a state of equilibrium only capable of existing in the ordinary temperature of the atmosphere."[113] Lavoisier hypothesized that in the case of plant substances all three of the constituting elements had a mutual relation of chemical affinity and that this relation was in a state of equilibrium only at low temperatures, such as the ordinary

108 Lavoisier [1965] p. 120.
109 Ibid. p. 116f. As to animal substances, in addition to charcoal, hydrogen, and oxygen, Lavoisier also considered nitrogen and phosphorus as components, and as an exception the "muriatic radical" as well (ibid. pp. 117, 149 and 191).
110 Ibid. p. 123.
111 Holmes [1985] p. 263.
112 See chapter 2 and Duncan [1996]; Kim [2003]; Klein [1994a], and [1995]. E. F. Geoffroy introduced the term "laws of chemical rapports" in 1718 (see Geoffroy [1718]).
113 Lavoisier [1965] p. 120f.; our emphasis.

temperature of the atmosphere, and, we may add, in the bodies of living plants. The explanatory power of this hypothesis is immediately apparent. The fact that most of the proximate principles of plants—that is, what now counted as proximate principles based on the new theory of composition—decomposed by dry distillation and combustion was now explained by their fragile equilibrium of chemical affinities. A little later Lavoisier generalized this hypothesis for all kinds of vegetable and animal matter. "The decomposition of vegetable matter, under a high temperature" and "the distillation of every species of plant substance" were portrayed as actions of affinities between charcoal, oxygen, and hydrogen as well as caloric.[114]

Fourcroy adopted Lavoisier's theory of composition and affinity wholesale. Hydrogen, carbon, and oxygen, he stated, form oils, resins, and other vegetable substances, which are in an "equilibrium of composition" as long as the temperature is not raised and contact with water and air is avoided.[115] In his *Système* he further refined the new theory. The "constituting principles" of vegetables, he wrote, are hydrogen, carbon, and oxygen, which combine into ternary compounds in which the three principles are in a state of equilibrium.[116] This "equilibrium of rest," in which all of the "attractive forces are mutually balanced," constituted "the essence or intimate nature of vegetable compounds."[117] "But the attractions that unite these three simple bodies are susceptible of frequent alterations, of diminution and augmentation, so that the equilibrium that brings them closer to each other is easily broken off." The vegetable compounds, he continued, are much more complicated than the mineral ones, allowing "a much greater variability of attractions."[118]

Of course, the extended hypothesis of a "variability of attractions" also allowed many more explanations. In principle, the many different products obtained by decomposing a proximate principle of plants through the actions of fire, water, and air, or by fermentation, all resulted from various degrees in alterations of the attraction between the three constituting principles of the material. From the moment "any agent whatsoever" was applied to a vegetable material, a great multiplicity of alterations was possible, between the two extremes of the material's equilibrium of rest, at the one end, and complete decomposition and reduction into water and carbonic acid at the other.[119] As the proportions of carbon, hydrogen, and oxygen contained in a vegetable material could also be augmented and diminished continuously and endlessly, the "nuance of composition" changed in the "passage" of a plant material from its original state to water and carbonic acid.[120]

With Lavoisier's theory of composition and affinity, the designation "proximate principles of plants" became restricted to substances that shortly afterwards became

---

114 Ibid. p. 125ff.
115 Fourcroy [1801–1802] vol. VII p. 46.
116 Ibid. p. 53.
117 Ibid. p. 54.
118 Ibid.
119 Ibid. p. 54f.
120 Ibid. p. 55. It should be noted that this phrasing highlights that Lavoisier's and Fourcroy's theory of the composition of plant and animal substances did not imply the stoichiometric law of a definite, invariant composition. We will discuss this problem in detail in chapter 14.

renamed "organic" and grouped together with organic materials stemming from the animal kingdom. We wish to propose, however, that Lavoisier was not fully aware of the consequences of his theory. It was not Lavoisier's objective to introduce a new demarcation of the proximate principles of plants.[121] Yet once this theory was elaborated, it became a most convenient tool for chemists who were focusing on life, organization, and the peculiarities of plant and animal materials to redraw the boundaries of the class of the proximate components of living beings and to exclude mineral substances. The superposition of the Lavoisierian theory of composition and affinity with the older concept of proximate principles of plants and of animals became an originally unintended precondition for a new, unambiguous definition of organic substances.

The terminological distinction between "organized" or "organic" bodies and "unorganized" bodies made from the mid-eighteenth century on by several British, German, and French chemists, such as Lewis, Vogel, Weigel, Wiegleb, Erxleben, Gren, Rouelle, Venel, Macquer, Baumé, Demachy, Guyton de Morveau, Bucquet, and Fourcroy, separated plants and animals and their organs from minerals. This distinction should not be confused with later chemists' designation of substances extracted from living beings as "organic." Although the conception of organic substances began to develop slowly from the 1750s onward, the term "organic" did not refer to substances before circa 1800. In Fourcroy's *Système*, published in 1801–1802, for example, the term "organic compound" was slippery, sometimes meaning the organized plants and animals, but sometimes also the unorganized substances extracted from plants and animals.[122] The use of the term "organic" to refer to unorganized materials rather than to living plants and animals was still very unusual around 1800.[123] Even J. J. Berzelius, who made the term "organic substances" in the modern sense more prominent, referred to substances as "organic" only from 1812 onward. In 1812, both in a paper on chemical nomenclature and in an announcement of a new volume of his chemical textbook, he unequivocally used "organic compounds" (*composita organica*) as a designation for plant and animal substances.[124]

---

121 As F. L. Holmes showed, the theory was, first of all, an essential part of the new theoretical system, whose centerpieces were the theories of combustion and acids. Furthermore, Lavoisier did not use the term "proximate" or "immediate principles," or even attempt to determine their number and to classify them. When he referred to substances such as vegetable acids, sugar, gum, and mucilages, he used the term "vegetable substances." He did not even distinguish systematically between "vegetables" and "vegetable substances" (see, for example, Lavoisier [1965] p. 123). Neither did he address the question of whether minerals found in plants belonged to the class of their proximate principles. In that latter respect he only remarked: "we can hardly entertain a doubt that the ashes, or earth which is left by vegetables in combustion, pre-existed in them before they were burnt, forming what may be called the skeleton, or osseous part of the vegetable" (ibid. p. 155). Lavoisier made similarly vague remarks about potash (ibid. p. 156).

122 In reading Fourcroy's text it is important to interpret his term "*composés organiques*" in context. For example, in the introduction to his plant chemistry the heading "*des composés organiques végétaux*" refers to entire plants (Fourcroy [1801–1802] vol. VII pp. 3–5). However in the context of his theory of composition and affinity this term refers to the proximate materials of plants (ibid. p. 54).

123 In the *Méthode de Nomenclature Chimique*, Lavoisier's *Traité*, and Fourcroy's *Philosophie Chimique* of 1792, materials were designated either "vegetable" or "animal," but not "organic." See Guyton de Morveau et al. [1787]; Lavoisier [1789]; and Fourcroy [1792].

Vitalism

Lavoisier's and Fourcroy's theories were not restricted to plant substances, but also covered the chemical composition of entire plants. In this respect it is not incidental that the two chemists also stated that "the true constituent elements of *vegetables* are hydrogen, oxygen, and charcoal."[125] Accordingly, Lavoisier and Fourcroy extended their revised theory of composition and affinity to living plants and the processes of the formation of plant substances in the organs of plants. The two chemists were convinced that the equilibrium of chemical affinities between the three or four elements making up the proximate components of plants was established and maintained only in the living bodies of plants. Hence their assertion that it was impossible for chemical art to create vegetable materials "*de toutes pièces*;" that is, from the simple elements. Chemical art, Fourcroy proclaimed, can only diminish but never augment the attraction between carbon, hydrogen, and oxygen. "Nature alone kept the power to create, reproduce, and form *de toutes pièces* from the first elements the primitive materials of vegetable compounds."[126]

This was a thoroughly materialist and chemical explanation of the synthesizing processes taking place within plants, which was by no means shared by all chemists, let alone naturalists and philosophers. Vitalistic theories of the creation of organic matter were a widespread alternative. In Germany, the chemist and physiologist Karl Friedrich Kielmeyer (1765–1844) was an influential critic of the French chemists' approach to organic matter, not least since he had adopted the Lavoisierian chemistry for inorganic substances, and, in principle, also believed in a chemical explanation of processes of life.[127] In his lectures on comparative zoology, chemistry, and natural history, held between 1790 and 1793, Kielmeyer conceded that the French chemists had made advances in the chemical analysis of plants and animals. But he added that the chemical analysis of organic materials was still in its infancy, and further that no satisfactory application of chemical methods to the general theory of organization could be expected in the near future.[128] In a famous lecture held in 1793, he then introduced the concept of vital force to explain the organic forms and the fact that at least seven million different forms of plants and animals existed on the surface of the earth.[129]

---

124 See Berzelius [1812] p. 52. In the first edition of his chemical textbook written in Swedish, Berzelius wrote in the introduction to the second part, published in 1812, that he wanted to postpone the part on "organic chemistry." As an argument for this postponement he further stated that "the laws of composition of organic compounds and the relation of their inorganic elements" had not yet been established (quoted in Hjelt [1916] p. 23). In this statement "organic compounds" referred to substances. By contrast, Partington's claim, made in Partington [1961–1970] vol. IV p. 233, that Berzelius used the term "organic" for substances (rather than plants and animals) in the 1806 Swedish edition of his *Animal Chemistry* is mistaken; see Berzelius [1813].

125 Lavoisier [1965] p. 123; our emphasis. Fourcroy stated that the "constituting principles of *vegetables*" are hydrogen, carbon, and oxygen (Fourcroy [1801–1802] vol. VII p. 53).

126 Ibid. p. 56.

127 For life and work of Kielmeyer, see Gillispie [1970–1980] vol. VII p. 366ff. The important role played by K. F. Kielmeyer in German plant chemistry has been highlighted by Löw; see Löw [1977] pp. 91–194. On Kielmeyer see also Lenoir [1989]. Lenoir has characterized Kielmeyer's approach as "vital materialism."

128 See Lenoir [1989] pp. 38–39.

Eight years later in his inaugural lecture as a professor of botany, pharmacy, and *materia medica* at the University of Tubingen, Kielmeyer extended his concept of vital force to the chemical substances constituting organized bodies. "The elements that constitute an organism," he stated, "combine into constituting parts, which have been designated by the awkward name 'proximate' (*proximae*) or 'immediate' (*immediatae*) components."[130] He then emphasized that the proximate components could only be destroyed, but not created, by chemical art, since chemical art was not in command of the "affinities" or "force" necessary for their resynthesis. From this he concluded that this vital force was "different from all kinds of affinity."[131] In 1806, he reiterated this view with a slightly different meaning:

> All mixtures in the inorganic realm are pure chemical works, capable of being explained merely by the laws of chemical affinity as products of the affinity of matter. The mixtures in the organic realm on the other hand are either contrary to the laws of affinity which are observed to hold outside of organized bodies, or at least they are not formed according to them. The only exception to this general observation occurs in cases where the material of the organic body is expelled as a dead substance as in urine and even in lifeless bones. Here in the excreted parts of the organic body the normal affinities begin to reappear.[132]

Like Lavoisier and Fourcroy, Kielmeyer explained the chemical transformations of inorganic substances using laws of chemical affinity. Yet unlike the two French chemists, who extended the laws of chemical affinities to the organic realm, he sharply demarcated the inorganic from the organic realm. According to Kielmeyer, in the organic realm the laws of affinity did not apply. At least as long as materials were part of the organized bodies, they obeyed forces that differed from chemical affinities. This may change, he added somewhat elusively, when the material is separated from the body; only when plant and animal substances were separated from the processes of life would the ordinary chemical affinities reappear.[133]

In the same year (1806), a second opponent to the French approach to organic matter uttered his views, one who would become even more influential than Kielmeyer: the Swedish chemist Jöns Jacob Berzelius. In his lectures on animal chemistry published in 1806, Berzelius proclaimed right at the beginning:

> [...] there exist processes between the unorganic constituent, or elementary particles within the animal body, which have sometimes not the least resemblance to those we see

---

129 Ibid. p. 45.
130 Quoted and translated in Löw [1977] p. 98.
131 Ibid. It should be noted that Kielmeyer distinguished in this lecture between organic and inorganic bodies. Like other chemists of the time, he applied the term "organic" to entire plants and animals rather than to substances. Löw has sometimes paraphrased the term with "organic substances," which in our view is mistaken.
132 Quoted in Lenoir [1989] p. 51.
133 Lenoir interpreted Kielmeyer's statement, quoted above, as follows: "The essential difference between organic and inorganic materials appeared to lie in a difference in the laws of chemical affinity characteristic of each" (ibid.). It should be noted that at the time bones and urine were regarded as ordinary animal substances grouped together with fatty oils, acids, and so on into the class of proximate principles of animals; and that Kielmeyer argued that the proximate principles that were extracted from the living body were no longer governed by the force of life. Hence, he distinguished between organic substances existing in the living body, on the one hand, and extracted organic *and* inorganic substances, governed by the laws of chemical affinity, on the other hand.

in unorganised matter. We may consider the whole animal Body as an instrument, which, from the nourishment it receives, collects materials for continual chemical processes, and of which the chief object is its own support. But, with all the knowledge we possess of the forms of the body, considered as an instrument, and of the mixture and mutual bearings of the rudiments to one another, yet the cause of most of the phenomena within the Animal Body lies so deeply hidden from our view, that it certainly will never be found. We call this hidden cause *vital power.*[134]

Berzelius agreed with Lavoisier and Fourcroy on the principle of a chemical approach to living beings. The living body was a chemical workshop or a large instrument that altered chemically the materials imbibed as nourishment.[135] However, when he stated that the cause of those alterations was unknown, he also attacked the Frenchmen's extension of the laws of affinity to physiological processes. Like Kielmeyer, in 1806 Berzelius contested the idea that the chemical syntheses and alterations of substances taking place in animal bodies could be explained by the ordinary laws of chemical affinities. A different, still unknown cause was at work in this case, which he designated "vital power." Tentatively, Berzelius ascribed the vital power to the nervous system. The chemical processes forming the particles of the blood, salvin, milk, and urine, he stated, act "under the influence of the nerves, which enter into these parts, and which determine as well the nature of the secreted matter as its quantity."[136]

Throughout his entire professional career, Berzelius struggled with the concepts of vital power and chemical affinity. The many different statements he made in this respect ranged from the assumption of the existence of a modified chemical affinity in living bodies to the claim that the force of life was totally different from chemical affinity. For example, in 1812 Berzelius stated that the elements contained in organic compounds obeyed "the same general law of composition as the inorganic compounds." Yet he added that organic nature allowed "innumerable variations" of that former law. And in continuation of his earlier views presented in his *Animal Chemistry,* he also asserted that the organic-chemical processes were subordinated to the nervous system, which modified the ordinary chemical affinity. Hence, he stated, "the main condition of organic creation (*organische Bildung*) seems to be an electrical-chemical modification of the elements."[137] But later he rejected this understanding of a modified chemical affinity in living beings. For example, in the third edition of his chemical textbook he proclaimed that "in living nature, the elements seem to follow *entirely different laws* as in dead nature, the products of their mutual effects are thus *totally different* from the area of inorganic chemistry."[138]

---

134 Berzelius [1813] p. 4; our emphasis. We quote from the English translation of Berzelius' *Föreläsningar i Djurkemien* of 1806.
135 For Kielmeyer's use of the workshop metaphor, see Lenoir [1989] p. 51. Berzelius also used the workshop metaphor in later publications. See Berzelius [1833–1841] vol. VI p. 3. On Berzelius' concept of vital force, see also Rocke [1992].
136 Berzelius [1833–1841] vol. VI p. 7.
137 Berzelius [1812] p. 53.
138 Berzelius [1833–1841] vol. VI p. 3; our emphasis.

The Failure of Lavoisier's Plant Chemistry

14.1 Lavoisier's analytical program for classifying plant and animal substances

In the 1780s Lavoisier and his collaborators not only articulated a new cluster of theories, but also proposed a unified analytical mode of classifying and naming substances based on knowledge of chemical composition (see part II). The *Méthode de nomenclature chimique*, the collective work by A.-L. Lavoisier, L. B. Guyton de Morveau, A. F. Fourcroy, and C. L. Berthollet published in 1787, epitomized this approach.[1] Arguing for an identification and classification of compound substances according to elemental composition, it devalued other earlier and contemporary natural historical and artisanal modes of classification. Origin from the three natural kingdoms, modes of extraction, and perceptible properties were excluded from the list of significant taxonomic criteria. What remained was chemical composition, understood in a more rigid methodological way than before and invested with new reference. Based on knowledge of elemental composition, the multifaceted objects of eighteenth-century chemistry were to be transformed into unambiguous scientific objects.

As we showed in part II, the paradigmatic core of the new analytical program was constituted by pure mineral compounds such as metal oxides, acids, and salts. Accordingly, the table of chemical nomenclature included in the *Méthode* classified mainly mineral (later called "inorganic") compounds. By contrast, plant and animal substances seemed to elude rigid analytical circumscription. The problem can be seen at a glance in the table of chemical nomenclature. Animal substances were largely omitted from the table, with the exception of acids, and most plant substances were grouped together separately in an appendix to the table. But far from a total exclusion, this ordering of plant substances in an appendix actually meant a kind of inclusion in the overall system. It meant there was a program, though a very tentative one, to create a unified system of classification and nomenclature encompassing substances from all three natural kingdoms as well as artificial chemical preparations made in the chemical laboratory.

If composition was the most significant feature of compound substances and the main criterion for classification, the older natural historical divide between vegetable, animal, and mineral bodies and substances, and the newer one between organic and inorganic bodies, had to be re-examined. It either had to be given up entirely or be redefined in a new, analytical way. Origin from plants and animals could no longer be the criterion distinguishing plant and animal substances from mineral compounds. The new analytical definition of plant and animal substances had to be independent of natural origin, not only because of the methodological rules highlighting composition, but also because of the inclusion of laboratory artifacts in the group. But what kind of substitute did the authors of the *Méthode* actually introduce for the notion of plant and animal substances? The title of the appendix—"several more compound

---

1 Guyton de Morveau et al. [1787].

substances which combine without decomposition"—was still a quite unspecific terminology. If we further study the names of the substance species included in the appendix, it is quite obvious that in most cases the old names were preserved and changed only in minor aspects. Mucus, gluten, sugar, starch, fixed oil, volatile oil, aroma, resin, extract, resinous extract, extractive resin, and fecula were quite traditional eighteenth-century names for vegetable substances.[2] They neither provided information about the composition of the substances nor designated new substances that would have been individuated and identified by examining their composition. In his *Mémoire sur le développement des principes de la nomenclature méthodique*, which comments on the table,[3] Guyton de Morveau conceded that "it is evident on the first view of our table that we have done here no more than adopting some denominations out of the number which are in use."[4] Indeed, the authors of the *Méthode* made no offer of a new nomenclature of plant and animal substances.

Two years later, in his famous *Traité*, Lavoisier still conceded that chemists were far from being able to classify plant materials in a "methodical manner," and that the chemistry of plants "still remains in some degree obscure."[5] Substances from the animal kingdom, he added, were even "less known than those from the vegetable kingdom."[6] Nevertheless, Lavoisier now presented a new analytical definition of plant substances. "The true constituent elements of vegetables," he proclaimed, "are hydrogen, oxygen, and charcoal."[7] As we showed in the previous chapter, through this redefinition of plants and plant substances elemental composition became the outstanding criterion for the identification of these substances and their demarcation from other kinds of substances also found in plants, such as mineral salts. Yet the new definition provided only a rough grid for a classification of plant substances. The problem of how to determine a genus of plant substances, or even a particular species of them, based on knowledge of elemental composition, and how to demarcate species and genera of plant substances from each other, was still unsolved.[8] Moreover, if knowledge of elemental composition was to become a working tool for the identification and demarcation of species and genera of plant substances, quantitative differentiation was necessary. But how finely could such quantitative differentiation be made? Was it possible to grasp the great variety of plant substances distinguished in

---

2   For the translation of the French terms, see the English translation of 1788 (Guyton de Morveau et al. [1788]).

3   Guyton de Morveau et al. [1787] pp. 26–74.

4   Ibid. p. 71. See also Guyton de Morveau et al. [1788] pp. 19–53.

5   Lavoisier [1965] p. 119. On the role played by plant and animal chemistry in Lavoisier's *Traité*, see also Holmes [1985] pp. 385–409. Holmes asserted that in its chapters devoted to plant and animal chemistry Lavoisier's *Traité* undergoes a "transition from polished textbook to almost a preliminary progress report" (ibid. p. 385).

6   Lavoisier [1965] p. 121.

7   Ibid. p. 123. Apart from these three elements, Lavoisier also considered *azote* and phosphorus to be elemental components of plant substances. On Lavoisier's redefinition of organic substances, see also the previous chapter. The quantitative aspects of Lavoisier's redefinition of organic substances are discussed later.

8   Therefore Lavoisier cautioned his readers not to expect too much: "As it is but of late that I have acquired any clear and distinct notions of these substances, I shall not, in this place, enlarge much upon the subject" (Lavoisier [1965] p. 117).

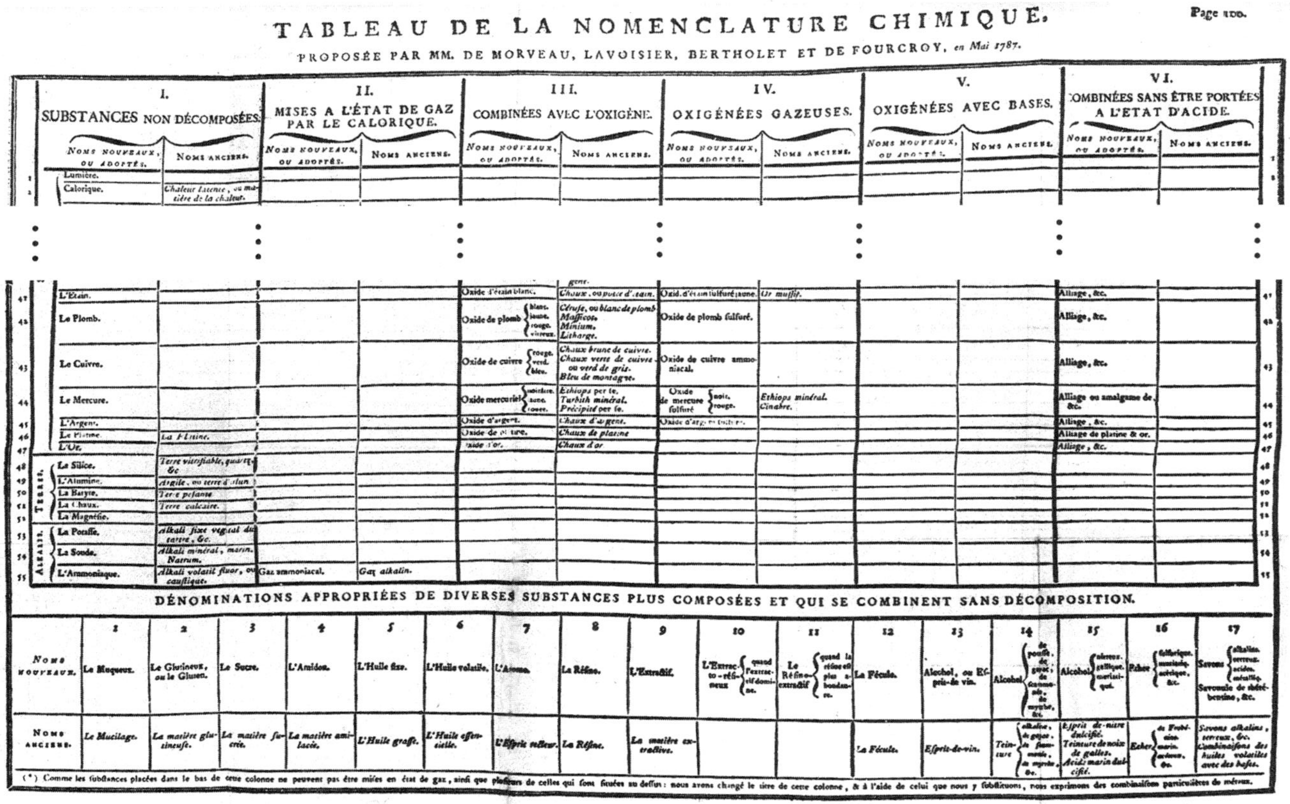

TABLEAU DE LA NOMENCLATURE CHIMIQUE,

PROPOSÉE PAR MM. DE MORVEAU, LAVOISIER, BERTHOLET ET DE FOURCROY, *en Mai 1787.*

Page 100.

	I. SUBSTANCES NON DÉCOMPOSÉES.		II. MISES A L'ÉTAT DE GAZ PAR LE CALORIQUE.		III. COMBINÉES AVEC L'OXIGÈNE.		IV. OXIGÉNÉES GAZEUSES.		V. OXIGÉNÉES AVEC BASES.		VI. COMBINÉES SANS ÊTRE PORTÉES A L'ÉTAT D'ACIDE.		
	NOMS NOUVEAUX, OU ADOPTÉS.	NOMS ANCIENS.	NOMS NOUVEAUX, OU ADOPTÉS.	NOMS ANCIENS.	NOMS NOUVEAUX, OU ADOPTÉS.	NOMS ANCIENS.	NOMS NOUVEAUX, OU ADOPTÉS.	NOMS ANCIENS.	NOMS NOUVEAUX, OU ADOPTÉS.	NOMS ANCIENS.	NOMS NOUVEAUX, OU ADOPTÉS.	NOMS ANCIENS.	
1	Lumière.												1
2	Calorique.	*Chaleur latente, ou matière de la chaleur.*											2
⋮	⋮	⋮	⋮	⋮	⋮	⋮	⋮	⋮	⋮	⋮	⋮	⋮	⋮
						gene.							
41	L'Étain.				Oxide d'étain blanc.	*Chaux, ou potée d'étain.*	Oxid. d'étain sulfuré jaune.	*Or mussif.*			Alliage, &c.		41
42	Le Plomb.				Oxide de plomb {blanc. jaune. rouge. vitreux.	*Céruse, ou blanc de plomb. Massicot. Minium. Litharge.*	Oxide de plomb sulfuré.				Alliage, &c.		42
43	Le Cuivre.				Oxide de cuivre {rouge. verd. bleu.	*Chaux brune de cuivre. Chaux verte de cuivre ou verd de gris. Bleu de montagne.*	Oxide de cuivre ammoniacal.				Alliage, &c.		43
44	Le Mercure.				Oxide mercuriel {noirâtre. jaune. rouge.	*Ethiops per se. Turbith minéral. Précipité per se.*	Oxide de mercure sulfuré {noir. rouge.	*Ethiops minéral. Cinabre.*			Alliage ou amalgame de, &c.		44
45	L'Argent.				Oxide d'argent.	*Chaux d'argent.*	Oxide d'argent sulfuré.				Alliage, &c.		45
46	Le Platine.	*La Platine.*			Oxide de platine.	*Chaux de platine*					Alliage de platine & or.		46
47	L'Or.				Oxide d'or.	*Chaux d'or*					Alliage, &c.		47
48	TERRES. La Silice.	*Terre vitrifiable, quartz, &c*											48
49	L'Alumine.	*Argile, ou terre d'alun*											49
50	La Baryte.	*Terre pesante*											50
51	La Chaux.	*Terre calcaire.*											51
52	La Magnésie.												52
53	ALKALIS. La Potasse.	*Alkali fixe vegetal du tartre, &c.*											53
54	La Soude.	*Alkali minéral, marin. Natrum.*											54
55	L'Ammoniaque.	*Alkali volatil fluor, ou caustique.*	Gaz ammoniacal.	*Gaz alkalin.*									55

DÉNOMINATIONS APPROPRIÉES DE DIVERSES SUBSTANCES PLUS COMPOSÉES ET QUI SE COMBINENT SANS DÉCOMPOSITION.

	1	2	3	4	5	6	7	8	9
NOMS NOUVEAUX.	Le Muqueux.	Le Glutineux, ou le Gluten.	Le Sucre.	L'Amidon.	L'Huile fixe.	L'Huile volatile.	L'Arome.	La Résine.	L'Extractif.
NOMS ANCIENS.	*Le Mucilage.*	*La matière glutineuse.*	*La matière sucrée.*	*La matière amilacée.*	*L'Huile grasse.*	*L'Huile essentielle.*	*L'Esprit recteur.*	*La Résine.*	*La matière extractive.*

	10	11	12	13	14	15	16	17
NOMS NOUVEAUX.	L'Extracto-résineux {quand l'extractif domine.	Le Résino-extractif {quand la résine est plus abondante.	La Fécule.	Alcohol, ou Esprit-de-vin.	Alcohol {de potasse, de gayac, de scammonée, de myrrhe, &c.	Alcohol {nitreux. gallique. muriatique.	Ether {sulfurique. muriatique. acétique, &c.	Savons {alkalins. terreux. acides. métalliq. Savonule de thérébentine, &c.
NOMS ANCIENS.			*La Fécule.*	*Esprit-de-vin.*	*Teinture* {alkaline, de gayac, de scammonée, de myrrhe, &c.	*Esprit de nitre dulcifié. Teinture de noix de galles. Acide marin dulcifié.*	*Ether* {de Frobénius. marin. acéteux, &c.	*Savons alkalins, terreux, &c. Combinaisons des huiles volatiles avec des bases.*

(*) Comme les substances placées dans le bas de cette colonne ne peuvent pas être mises en état de gaz, ainsi que plusieurs de celles qui sont situées au dessus: nous avons changé le titre de cette colonne, & à l'aide de celui que nous y substituons, nous exprimons des combinaisons particulières de métaux.

Figure 14.1: Appendix to the *Tableau* of the *Méthode*.

DÉNOMINATIONS APPROPRIÉES DE DIVERSES SUBSTANCES PLUS
COMPOSÉES ET QUI SE COMBINENT SANS DÉCOMPOSITION

	NOMS NOUVEAUX	NOMS ANCIENS
1	Le Muqueux	Le Mucilage
2	Le Glutineux, *ou* le Gluten	La matière glutineuse
3	Le Sucre	La matière sucrée
4	L'Amidon	La matière amilacée
5	L'Huile fixe	L'Huile grasse
6	L'Huile volatile	L'Huile essentielle
7	L'Arôme	L'Esprit recteur
8	La Résine	La Résine
9	L'Extractif	La matière extractive
10	L'Extracto-résineux	
11	Le Résino-extractif	
12	La Fécule	La Fécule
13	Alcohol, *ou* Esprit-de-vin	Esprit-de-vin
14	Alcohol de potasse de gayac de scammonée de myrrhe &c.	Teinture alkaline de gayac de scammonée de myrrhe &c.
15	Alcohol nitreux gallique muriatique	Esprit de nitre dulcifié Teinture de noix de galles Acide marin dulcifié
16	Ether sulfurique muriatiq. acétique &c.	Ether de Frobénius marin acéteux &c.
17	Savons alcalins terreux acides métalliq. Savonule de thérébentine, &c.	Savons alcalins, terreux, &c. Combinaisons des huiles volatiles avec des bases

Figure 14.2: Transcription of the appendix to the *Méthode's Tableau*.

pharmaceutical and commercial practices in this manner? These questions, which actually touched on the theoretical limits of the new chemistry, will be discussed later.

Apart from vexing problems that arose within the analytical program of classifying and renaming plant substances, the new approach almost inevitably stirred up conflicts with extant chemical ontologies. The first challenge of the approach consisted in the fact that it highlighted knowledge of elemental composition as the predominant, or even exclusive, criterion for the classification and nomenclature of plant materials. For centuries chemists had classified plant substances from natural historical, pharmaceutical, and artisanal points of view. As technological and commercial occupations of chemists continued to expand in the second half of the eighteenth century, the ways of handling, identifying, and classifying substances multiplied, too. In contexts of practical chemistry it made more sense to classify substances as various kinds of food or dyestuffs, for example, than as chemical entities with a specific kind of elemental composition. But even in contexts of conceptual inquiry, the taxonomic approach proclaimed in the *Méthode* was far from uncontested. Chemists' more recent concern with life directed their attention to questions other than the elemental composition of plant substances. In the 1790s, the analysis of entire plants and the separation of the more compound proximate principles of plants, the peculiarity of organic vis-à-vis inorganic materials, and the explanation of physiological functions became even more important issues than they had been in the three decades before. Once a clear demarcation of the proper plant substances had been achieved by means of the new theory of elemental composition, chemists' interest in the organization and physiology of plants shifted their attention away from the painstaking scrutiny of the elemental composition of plant substances.

Second, the Lavoisierian taxonomic program demanded the grouping together of natural plant substances with chemical preparations or artifacts created from natural plant materials in the laboratory. Accordingly, the appendix to the table of nomenclature assembled natural plant substances considered to be proximate principles of plants—mucus, gluten, sugar, starch, fixed oil, volatile oil, aroma, resin, extract, resinous extract, extractive resin, and fecula—and artificial laboratory products created from the former, such as spirit of wine and alcohol, produced by the distillation of fermented plant substances;[9] "alcohol of potash," of guaiacum, of scammony, and of myrrh;[10] "nitrous alcohol," gallic alcohol, and muriatic alcohol;[11] sulfuric ether, muriatic ether, and acetic ether;[12] and various kinds of soap made on a large scale in workshops, such as alkaline soaps, or on a small scale in chemical and pharmaceutical laboratories. The new imperative of identifying and classifying plant substances according to their elemental composition no longer privileged things given by nature. This meant a great challenge to the ongoing attempts to unify chemistry with natural history, as well as to the existing classificatory practice of grouping together only the natural proximate components of plants and animals. How European chemists met this challenge is discussed in the next section.

---

9   See Guyton de Morveau et al. [1787]. Spirit of wine and alcohol differed only in the proportion of water contained.
10   These substances were the renamed pharmaceutical "tinctures" made by digesting plants and plant materials with spirit of wine or alcohol.
11   These substances were obtained by mixing alcohol and the respective acids. In the pharmaceutical tradition, these substances were designated "dulcified nitric acid" or "dulcified muriatic acid" and tincture of gall nuts.
12   These substances were made by distilling alcohol or spirit of wine and the respective acid.

Classifications in the new chemistry of life

Whereas in the two decades after 1787 many European chemists accepted the overall definition of plant materials as compounds composed mainly of carbon, hydrogen, and oxygen, they did not set out to refine the methods of quantitative elemental analysis of plants, nor did they elaborate a new mode of identifying and classifying plant substances based on knowledge of elemental composition. In chemists' classificatory practices, the natural plant substances remained separated from their artificial counterparts. Even A. F. Fourcroy, one of the co-authors of the *Méthode*, followed traditional lines of investigation. He concurred in the definition of plant substances as compounds of carbon, hydrogen, and oxygen, which made their identification and classification potentially independent of their natural origin.[13] But in his taxonomic practice he continued to group together only the natural proximate principles or "immediate materials" of plants. The "immediate materials" of plants, he emphasized, were the combined and stable "materials that make up the tissue," and "which are contained pre-formed in the vessels and reservoirs of the plants."[14] In the context of Fourcroy's chemical-physiological occupations, the elemental-analytical definition of plant substances as compounds consisting mainly of carbon, hydrogen, and oxygen, favored by the *Méthode,* receded into the background.

In his classificatory practice Fourcroy continued the ancient tradition, examining the physical and chemical properties of the materials and their natural origin or mode of extraction. He emphasized that this mode of classification required the selection of "distinctive properties" or "characters" of plant materials.[15] Alluding to Linnaeus's taxonomic method, he asserted that it was "not impossible to treat this object in the same manner as botanists do and to select only one character or specific phrase (*phrase*) for each of these [immediate] materials."[16] In his *Système,* Fourcroy explicitly retained the naturalist division of substances according to their origin from the mineral, vegetable, and animal kingdoms.[17] In a chapter entitled "*Du dénombrement, de la classification des matériaux immédiats des végétaux*," which tackled questions concerning the classification of plant substances, he introduced elemental analysis and knowledge about composition only as an auxiliary precondition of classification.[18] Having summarized chemists' recent achievements in classifying plant substances, he drew a conclusion that distinguished four main ways of classifying plant substances.

First is classification highlighting natural origin and following the anatomical order of plants, thus dividing plant substances into materials stemming from such parts of plants as the root, bark, stem, leaves, and flowers. Second is classification highlighting the useful properties of plant materials to form classes such as vegetable nutriment, medicines, combustible materials, textiles, fermentable materials, and

---

13 Fourcroy [1792] p. 98, and [1801–1802] vol. VII p. 53f.
14 Fourcroy [1801–1802] vol. VII p. 51.
15 Fourcroy [1792] p. 96.
16 Ibid.
17 Fourcroy [1801–1802] vol. VII p. 3.
18 Ibid. pp. 120–127.

solid and permanent dyes. A third mode of classification was based on the "proper chemical characters" (*charactères vraiment chimiques*) of plant materials, distinguishing among mucousy, sugary, acid, oily, inflammable, colored, and indissoluble kinds of plant substances. Taking into account perceptible properties such as consistency, taste, and odor, Fourcroy added, the identification of plant substances by means of chemical characters could be further refined. As the most philosophical way of classifying plant materials, however, Fourcroy presented a fourth method grounded in knowledge of vegetation.

Fourcroy's chemical-physiological method grouped together only the natural immediate materials of plants, and it further ordered these materials according to their successive formation in vegetation. In a new list of twenty immediate materials of plants (see figure 14.3), he placed first the transparent, uncolored sap of young plants, followed by mucilage, which was also formed early in vegetation; then came sugar, created in the germination and maturation of fruits. Fourcroy conceded that his new way of ordering the single plant materials, which was based on the "order of successive formation," was still in its infancy. But he was quite optimistic "that some day one will find the order, succession, and period of formation of each proximate vegetable material."[19]

As all of the twenty different immediate materials of plants were products of "progressive modifications" of one initial plant material, Fourcroy further proposed that physiological inquiries might be complemented by chemical analyses. "Comparative analysis" of plant substances and "attentive examination of phenomena of vegetation" were defined as different poles of one "philosophical chain."[20] It was only in this connection with chemical physiology that chemical composition played a role for the classification of plant substances and for plant chemistry more broadly. "Since all of the most exact analyses have shown, without exception, that plant materials are made up in their first principles from carbon, hydrogen and oxygen," Fourcroy stated, "it is evident that the *problem of the natural formation of these compounds* consists in knowing from where the plants obtain these simple substances, how they appropriate them and combine them in groups of three with one another. *This problem in fact encompasses all of plant physics*."[21] In Fourcroy's chemistry of plants, knowledge about composition was subordinated to the acquisition of physiological knowledge, which was the main goal.

Fourcroy's chemical-physiological goals and interests also shaped his individuation and identification of the immediate materials of plants. For example, he defined vegetable sap (*la sève*) as an immediate material that contained all materials for plants' vegetation and was comparable with the function of animals' blood.[22] From a chemical-analytical point of view, a substance containing all materials for plants' vegetation cried out for further chemical analysis and identification of its components. Yet in the framework of Fourcroy's physiological agenda, sap was a functional

19   Ibid. p. 125.
20   Ibid.
21   Ibid. p. 57; our emphasis.
22   Ibid. p. 132.

unit of plants' economy, just like blood that acted in animal bodies as a coherent physiological entity. Seen from the perspective of a chemical physiology, vegetable sap was a coherent single substance or a chemical-physiological individual rather than a mixture of different kinds of substances. Another example is wood fibers (*le ligneux*), which Fourcroy described as final, inalterable products of vegetation contributing to the stability of plants' bodies, and which he included in his class of proximate materials of plants.[23] Throughout the eighteenth century botanists and chemists had considered wood, like cork and bast, as an organ of plants—comparable to roots, stem, bark, seeds, and so on—rather than one of their chemical components. Now, in the context of the new chemistry of life, the former demarcations between the organs and the chemical components of plants became fluid. Not only Fourcroy included wood fibers or suber (or cork) in his new list of proximate materials or plants; Thomson, Hermbstädt, Stromeyer, and Berzelius did so as well (see figure 14.3).

Fourcroy's conception of plant substances as natural products of plants' economy and his preoccupation with the classification of compound components of plants were not idiosyncratic deviations from mainstream chemistry. Rather, this way of doing plant chemistry was broadly shared in the European chemical community. Between 1790 and 1792 even Lavoisier abandoned his elemental-analytical agenda and instead turned to the large, mainly theoretical questions concerning the economy of plants and animals and the chemistry of life.[24]

Responses outside of France

In Germany, "analytical chemists" such as Sigismund Friedrich Hermbstädt (1760–1833) and Friedrich Stromeyer (1776–1835), who not only wholeheartedly agreed with the paradigmatic, inorganic core of Lavoisierian chemistry but also pursued a Lavoisierian analytical program with respect to inorganic substances, also circumscribed a plant chemistry that studied plants, processes of vegetation, and plant substances in the sense of natural proximate components of plants.[25] S. F. Hermbstädt's famous manual for the "analysis of plants" presented recipes for the separation of compound proximate components of plants (*nächste Bestandtheile*) and descriptions of these substances that retained the traditional mode of identification via physical and chemical properties.[26] The agenda of Hermbstädt's "analysis of plants" included neither a comparative elemental analysis of plant substances nor the analysis of artificial derivatives of natural plant substances. The range of his experimental objects was restricted to plants and their natural components. By contrast, Stromeyer grouped together under the label "oxygenated substances with compound bases (*Grundlagen*)" organic substances made in the laboratory—namely alcohol and four kinds

---

23 Ibid. vol. VIII p. 92.
24 See Holmes [1985] p. 411ff. Holmes designated this shift of Lavoisier's interests as a "decline in the quality" of his investigation (ibid. p. 482).
25 On the emergence of the subdiscipline of analytic chemistry, see Homburg [1999b]. For life and work of Hermbstädt, see Gillispie [1970–1980] vol. XV p. 205ff. For biographical information on Stromeyer, see Kopp [1966] vol. IV p. 124.
26 See Hermbstädt [1807].

LIST OF PROXIMATE COMPONENTS OF PLANTS

Fourcroy (1801)	Thomson (1804)	Hermbstädt (1807)[*]	Strohmeyer (1808)	Berzelius (1812)
1. Sap				
2. Mucos or Gum	Gum	Gum Mucus	Gum or mucus	Gum
3. Sugar	Sugar	Sugar	Sugar	Sugar
4. Veg. Albumen	Albumen	Albumen		
5. Veg. Acid or Acids	Acids	Acids	Acids	Acids
6. Extract	Extractive Principle	Soap	Extractive Matter	Extract
7. Tannin	Tannin	Tannin	Tannin	Tannin
8. Starch	Starch		Starch	Starch
9. Gluten	Gluten		Gluten	Gluten
10. Coloring Matter	Indigo	Coloring Principle	Coloring Matters	
11. Fixed Oil	Oils	Fat	Fatty Oil	Fatty Oil
12. Veg. Wax	Wax	Wax	Wax	
13. Veg. or Volatile Oil		Oil	Volatile Oil	Volatile Oil
14. Camphor	Camphor	Camphor	Camphor	
15. Resin	Sandracha Gum resins	Resin	Resin	Resin
16. Gum Resin	Sarcocoll Gum Resins			
17. Balsam				
18. Caoutchouc	Caoutchouc	Caoutchouc	Caoutchouc	*gutta*
19. Wood Fiber (*le ligneux*)	Wood Fiberina	Wood Fiber	Wood Fiber	Wood
20. Suber	Suber		Suber	
	Jelly			
	Bitter Principle	Bitter Principle	Bitter Principle	
	Narcotic Principle	Narcotic Principle	Narcotic Principle	
		Acrid Principle	Acrid Principle	
				cinchonum
		Picromel (Manna)		
		Chinastoff		

[*] Hermbstädt uses "Stoff" (such as Gummistoff) or "principle" (such as *principium gummosum*) throughout.

Figure 14.3: Classification of proximate principles of plants c. 1800.

of ethers—with natural proximate components of plants and three animal substances (urea, gelatin, and "*Eyweißstoff* [protein]").[27] Based on experiments performed by the French chemist Louis Jacques Thenard (1777–1857) a year before, he took into

27   See Stromeyer [1808] vol. II pp. 220–319.

account the qualitative and quantitative composition of alcohol and ethers, and he did the same with olive oil, beeswax, gum, cane sugar, and urea.[28] However, in addition to these first steps of a Lavoisierian mode of classification, Stromeyer also retained the traditional division into the two large classes of plant and animal substances. In so doing, he grouped together only the natural proximate components (*nähere Bestandteile*) of plants and of animals.[29] Moreover, in connection with this second way of classifying separately the natural plant and the animal substances, he also followed Fourcroy's program of a chemical physiology dealing with subjects such as germination and nutrition of plants.

One of the leading British chemists, Thomas Thomson (1773–1852), also described the boundaries of plant chemistry in terms of the chemical-physiological program outlined by Fourcroy.[30] Having presented a class of organic substances that grouped together only the natural compound components of plants (see figure 14.3), Thomson extended the objectives of plant chemistry to the study of vegetation: "We have now seen the different substances which are contained in plants; but we have still to examine the manner in which these substances are produced, and to endeavor to trace the different processes which constitute vegetation."[31] In the wake of Fourcroy, Thomson studied only natural substances contained in or extracted from plants, plant organs (as the natural sites of substance production and the processes of vegetation), the production of these substances in living plants, plant growth, and other types of chemically explicable transformation processes in plants. The individual chapters of the part on plant chemistry in his textbook thus include topics such as plant nutrition, the transformation of fluid nutrients in plant organs, the function of the leaves, the different plant juices, and plant decay.

The program of a plant chemistry focusing on natural plant substances, proximate principles of plants, the physiology of plants, and "life" more broadly was endorsed in particular by the Swedish chemist J. J. Berzelius, the most authoritative voice of European chemists in the first decades of the nineteenth century. "All that our research in this secretive area of chemistry can accomplish," Berzelius stated,

> is to observe chemical changes which are produced in living bodies. [...] to follow the phenomena which accompany the life process as far as possible, and then to separate the organic products from one another, study their properties, and determine their composition.[32]

Berzelius was one of the first chemists who actually began, around 1810, to fulfill the Lavoisierian dream of a quantitative analytical organic chemistry. His research contributed considerably to the development of a quantitative elemental analysis of organic compounds. Nevertheless, Berzelius was one of the most explicit proponents of an organic chemistry whose boundaries were defined by life and living beings. He considered the study of life processes so central to organic chemistry that he called it

---

28 See Thenard [1807a], [1807b], [1807c], [1807d], [1807e], [1807f], [1809a], [1809b], and [1809c].
29 Stromeyer [1808] vol. II pp. 541–574 and 575–632.
30 For life and work of Thomson, see Gillispie [1970–1980] vol. XIII p. 372ff.
31 See Thomson [1804] vol. IV p. 367.
32 Berzelius [1833–1841] vol. VI p. 27f.

the "key" to its theory.[33] Accordingly, in a proposal for a classification and nomenclature of chemical compounds, elaborated with the help of the Swedish botanist Göran Wahlenberg (1780–1851) and published in 1812, Berzelius highlighted the difference between inorganic and organic compounds including only natural extracted plant materials in the subclass "*vegetabilia*" of the class "organic compounds" (see figure 14.3).[34]

Even after 1830, when a new mode of classifying organic substances had been proposed by Jean-Baptiste Dumas and Polydore Boullay, Berzelius did not abandon the natural historical classification of plant and animal substances. As late as 1837 he wrote that a classification of plant substances on the basis of their chemical composition, analogous to the classification of inorganic substances, was "without particular advantages."[35] The revision he then made to his plant-chemical taxonomy by distinguishing between three "major classes" of plant substances—acidic, basic, and neutral—made no decisive change in his former orientation. The three "major classes" were based on shared "chemical properties," not, as in inorganic chemistry, on analytical criteria of composition. Furthermore, the subdivision of such large classes was still accomplished according to the conventional natural historical criteria. For example, the genus "gum and plant mucus" was divided according to natural origin into gum arabic; cherry tree gum; gum from roasted starch; gum from the spontaneous decomposition of starch paste; gum from the treatment of linen, wood, starch, or gum arabic with sulfuric acid; gum tragacanth; plum gum; linseed mucus; quince mucus; salep; and marigold mucus.[36] Even in the fifth edition of his textbook, Berzelius explicitly maintained his natural historical mode of classification:

> We first divide the subject according to the two kingdoms of organic nature into plant and animal substances, thus dividing organic chemistry into plant chemistry and animal chemistry. Just as the individuals in the plant and animal kingdoms can be classified according to sexual similarities in genera and species, this is also the case for the organic substances obtained from these plants or animals; in organic chemistry this situation has been used in order to group under certain divisions those bodies which are most similar to each other in terms of their properties.[37]

In accordance with this approach, Berzelius was never willing to recognize substances produced in the chemical laboratory by the chemical "metamorphosis" of extracted plant and animal substances as "organic." He never grouped these artificial organic substances together with the extracted natural ones.

---

33 Ibid. p. 3. On Berzelius' classification of organic substances, see also Klein [2003a] pp. 52–58; and Melhado [1992] p. 165.
34 See Berzelius [1812]. For life and work of Wahlenberg, see Gillispie [1970–1980] vol. XIV p. 116f.
35 Berzelius [1833–1841] vol. VI p. 107.
36 Ibid. pp. 396–409.
37 Berzelius [1843–1848] vol. IV p. 92.

14.2 Theoretical limits of Lavoisier's analytical program

Around 1800 the extraction, identification, and classification of natural proximate principles or compound components of plants were chemists' main occupations in plant chemistry. Doing plant chemistry at this time meant, first of all, performing experiments with plants in order to extract, by means of various kinds of solvents, different kinds of materials, which underwent fermentation and "denaturated" easily when heated or mixed with chemical reagents.[38] It also meant examining the properties of the extracted materials to identify them, as well as classificatory work at the writing desk. Furthermore, since the middle of the eighteenth century the scope of plant chemistry had been extended to botany and plant physiology, which were mainly subjects for teaching and theoretical work.[39] Chemists' concern with the organization of plants, plant physiology, and the extraction and identification of proximate principles of plants, along with their definition of the latter as "organic," flew in the face of the Lavoisierian analytical program for plant chemistry presented in the *Méthode* and the *Traité*. Elemental analysis and knowledge of elemental composition were not the basis for the individuation and identification of organic substances or for drawing the boundaries of classes of organic substances, but rather physical and chemical properties, along with natural origin and mode of extraction. What were constructed were not comprehensive classes of both natural and artificial organic substances, along the lines proposed in the appendix to the *Méthode*'s table, but rather classes of natural materials only. Instead of a new, systematic nomenclature for plant substances based on knowledge of elemental composition, the traditional names were largely preserved. Lavoisier's program for plant chemistry was a failure in terms of social acceptance. It also failed in another aspect: the chemical theory of composition that informed the analytical mode of identifying and classifying organic substances.

More than a decade after the propagation of the new program for plant chemistry in the *Méthode*, Fourcroy was still coping with vexing problems of classification: "One asks oneself what characters make possible to distinguish these materials [the immediate materials of plants] from each other, and what is the number of their species or rather their genera."[40] Characteristics of plant substances such as their physical and chemical properties were too slippery to allow clear demarcations between species or genera of plant substances. But why was this not all the more incentive for Fourcroy to turn to elemental composition as criterion for classification? If all plant substances consisted mainly of carbon, hydrogen, and oxygen — as had been accepted by Fourcroy and many other chemists — quantitative inquiries into the proportions of carbon, hydrogen, and oxygen contained in different kinds of plant substances should have been a promising route to settle problems of their identification. How can we explain the lack of effort by French and other chemists in the two decades after the publication of Lavoisier's *Traité* to improve the quantitative analysis of organic com-

---

38 On this experimental practice of plant chemistry around 1800, see also Tomic [2003].
39 On the theoretical outlook of physiology around 1800, see Cunningham [2002].
40 Fourcroy [1801–1802] vol. VII p. 120.

pounds? And why was the question of the number of species or genera of plant substances considered a significant taxonomic problem?

We argue that the theory of composition accepted by Lavoisier and his collaborators did not actually promise a solution to chemists' taxonomic problems in organic chemistry. The reason for this was that their theory of composition did not include a general law of definite proportions. Before the law of definite proportions was formulated by Joseph Louis Proust (1754–1826) in 1797, chemists tacitly assumed that many mineral compounds, such as oxides, acids, alkalis, salts, and sulfides, had a definite, invariable composition, based on the large differences observed in points of saturation.[41] But they also applied the concept of "chemical compound" to nonstoichiometric compounds such as alloys and solutions. This is evinced by the fact that the sixth column of the table contained in the *Méthode* includes alloys (see figures in chapter 5). Furthermore, the plant substances enlisted in the appendix to the table include tinctures, which were plant substances made by mixing herbal drugs with varying proportions of spirit of wine or alcohol. It was well known to chemists at the time that the making of tinctures, like that of alloys, did not require exact quantities of ingredients, and that the proportion of herbal extract and spirit of wine contained in one and the same kind of tincture varied.

But if such kinds of plant substances were not excluded from the range of chemical compounds—if it was allowed theoretically that the quantitative composition of carbon, hydrogen, and oxygen varied to some extent even for one and the same kind of plant substance—then the way out of the problem of identification and demarcation of organic substances by means of exact quantitative analysis was barred. For under these conditions, the fact that repeated quantitative analyses of different samples of tincture of guaiacum, for example, yielded different quantitative results would not necessarily imply that these samples represented different kinds or species of tincture of guaiacum. If definite, invariant composition was not theoretically required for one and the same kind of plant substance, it could be assumed with almost logical certainty that the comparison of quantitative composition of different kinds of plant substances would reveal overlapping composition rather than clear-cut boundaries between the different kinds. Indeed, as our following analysis demonstrates, the question of whether the constituting elements of organic compounds combined in definite proportions to constitute one specific kind of substance was as open after Proust's proclamation of the law of constant proportions as it had been before. The fact that the technique of quantitative analysis of organic compounds was still in its infancy was not the only reason for this; there were theoretical reasons as well.

It had long been assumed by chemists that the properties of a substance depended not only on the qualities of its constituting elements, but also on the proportion of

---

41 Kiyohisa Fujii has stated that the law of definite proportions was a truism of the eighteenth-century notion of affinity (see Fujii [1986]). However, this interpretation is contradicted by the fact that eighteenth-century affinity tables also included alloys and solutions. Furthermore, as Kapoor has pointed out, it was a problem for Proust to explain the general validity of his law (Kapoor [1965]). For the content of the law of definite proportions, see also Christie [1994]. On the acceptance of the law, see also Mauskopf [1976]. For life and work of Proust, see Gillispie [1970–1980] vol. XI p. 166ff.

these elements. But what did this assumption mean in the case of plant substances, which consisted mainly of carbon, hydrogen, and oxygen? What kind of variations in the proportions of these three elements did chemists assume around 1800? We assert that a careful historical examination of what the founders of the analytical program for plant and animal chemistry actually stated in this respect reveals that they assumed a continuous variation of proportions. Yet the assumption of continuous variation in the proportions of carbon, hydrogen, and oxygen contained in a great number of different kinds of plant substances excluded the definite quantitative composition of each single kind of substance. If there is a great number of different kinds of substances consisting of the same three elements, definite composition of one kind requires discontinuous variation of proportions. Fourcroy and Lavoisier took slightly different approaches to solving this vexing problem.

Different approaches by Fourcroy and Lavoisier

In his *Philosophie Chimique* Fourcroy stated that most of the sixteen immediate materials of plants, which he had clearly discerned as different kinds of materials, agreed in their elemental composition and differed only by "the diverse proportions" of these elements. If one calculated the number of different compounds resulting from the combinations of different proportions of three or four simple elements, he continued, a very great number of compounds was possible:

> But as each of these ternary or quaternary compounds of immediate materials of plants allows for a *certain latitude of proportions* (*certaine latitude de proportions*) for keeping its general nature of an extractive matter, mucilage, oil, acid, resin, etc., one sees that the *diverse proportions* of their principles that are contained in these latitudes (*renfermées dans ces lattitudes*) determine the immense, incommensurable variety of color, odor, taste, and consistency of all vegetable materials, which everybody distinguishes using several of these materials as food, garment, and construction material.[42]

In this statement, Fourcroy took into account the many distinctions made between plant materials in everyday life and in the arts and crafts. He also aimed to grasp the boundaries of his sixteen proximate components of plants, which he defined unequivocally as generic. But he did not set up exact quantitative boundaries for either of the two taxonomic ranks. The sixteen generic classes were defined by a "certain latitude" of proportions, and the great diversity of plant materials within these classes was caused by the "diverse proportions" of elements. Fourcroy did not propose a hypothesis about definite and invariant proportions. Similarly, in his *Système*, published after the pronouncement of Proust's hypothesis, he wrote that there was a "great alterability of the proportions" of carbon, hydrogen, and oxygen contained in vegetable materials, and that the "nuance of composition" changed in the "passage" of a given plant material from its original state to the latest products of decomposition, water and carbonic acid.[43] Again, these metaphors provide an image of a continuous

---

42 Fourcroy [1792] p. 104; our emphasis.
43 Fourcroy [1801–1802] vol. VII p. 55.

alteration in the proportions of elements contained in different kinds of plant materials, rather than a discontinuous one.

Lavoisier tried to grasp the boundaries of the different kinds of plant substances in a slightly different way than Fourcroy. Like Fourcroy he assumed continuous alteration in the proportions of the constituting elements of different plant substances, but unlike Fourcroy he further postulated that there was only a small number of different kinds of substances. The latter assumption allowed him to cut through the quantitative continuity by emphasizing the dominance of one element compared to the two or three other elements. If the focus was on one constituting element only, such as oxygen, and if the proportion of oxygen contained in different kinds of plant substances was compared with the entire rest of the compound, a prevalence or shortage of oxygen could be discerned. The comparison of the prevalence or shortage of oxygen contained in different plant substances enabled Lavoisier to divide these substances into different kinds of vegetable oxides and acids according to their relative content of oxygen:

> In this method we might indicate which of their *elements [of vegetable substances] existed in excess*, without circumlocution, after the manner used by Rouelle for naming vegetable extracts: he calls these extracto-resinous when the extractive matter *prevails in their composition*, and resino-extractive when they *contain a larger proportion* of resinous matter.[44]

Based on considerations like these, however, the number of chemically defined kinds of vegetable acids and oxide substances had to be very small indeed:

> Upon that plan, and by varying the terminations according to the formerly established rules of our nomenclature, we have the following denominations: Hydro-carbonous, hydro-carbonic; carbono-hydrous, and carbono-hydric oxyds. And for the acids: Hydro-carbonous, hydro-carbonic; oxygenated hydro-carbonic, carbono-hydrous, carbono-hydric, and oxygenated carbono-hydric. *It is probable that the above terms would suffice for indicating all the variations in nature.*[45]

According to Lavoisier, the entire range of vegetable acids and oxides—such as sugar, mucilage, gum and starch—comprised only ten different chemical kinds.[46] Lavoisier was quite explicit about this consequence for the number of plant substances when he stated that the ten distinctions and names "would suffice for indicating all the variations in nature."[47]

But how could this statement be reconciled with the fact that he at the same time acknowledged thirteen different vegetable acids?[48] Lavoisier did not ignore the existence of a great variety of different vegetable acids, which had been accepted by many European chemists after Scheele had published the results of his experimental inves-

---

44 Lavoisier [1965] p. 118; our emphasis.
45 Ibid.; our emphasis.
46 Ibid. p. 117.
47 Ibid. However, it should be noted that, in addition to acids and oxides, Lavoisier also mentioned the class of vegetable oils (ibid. p. 182). Hence, he did not consider the division into acids and oxides to be exhaustive.
48 Ibid. p. 120.

tigations in the 1770s and 1780s. But these acids had been identified and demarcated in a traditional chemical way by examining their physical and chemical properties as well as the properties of their salts. As names such as citric acid, malic acid, and camphoric acid show, natural origin was a further means of their identification. By contrast, from a chemical-analytical point of view focusing on elemental composition and the quantitative prevalence of one element, it was probable that the former great variety of acids and of other plant substances could be reduced to a much smaller number of chemical kinds, in a similar way as, for example, solid, fluid, and vaporous varieties of water had previously been reduced to one chemical kind or species of water. Hence Lavoisier believed that "in proportion as the vegetable acids become well understood, they will naturally arrange themselves under these [six] denominations."[49]

When Lavoisier estimated the number of different kinds of vegetable oxides and acids, he referred to his new type of *chemical* kinds of plant substances that were to be distinguished by means of comparative elemental analysis. The absence of a general law of definite proportions for plant substances, along with Lavoisier's method of estimating the quantitative dominance of one constituting element of a plant substance, resulted in a small number of analytically defined kinds of plant substances. Under the auspices of a rigid analytical program, all of the different sorts of vegetable acids and of sugar, mucilage, gum, starch, and other vegetable oxides known in the apothecary trade and the arts and crafts at the time were merely varieties of the true chemical kinds of plant substances. It did not even make sense to differentiate in these varieties between genera that could be subdivided into species and species that could only be subdivided into individual specimens, since any subdivision below the taxonomic rank of the ten analytically defined vegetable oxides and acids had to apply criteria other than composition. All of the varieties resulting from that further division were not chemical kinds in the new Lavoisierian sense.

From the point of view of Lavoisier, such varieties had to be excluded from the domain of pure scientific chemistry as non-scientific objects. Under these theoretical conditions it also becomes understandable why the names of the plant substances enlisted in the appendix of the *Méthode* were singular rather than plural (see figures 14.1 and 14.2). The main table contained in the *Méthode* used names in the plural only to designate generic classes, whereas singular names designated species of substances. Accordingly, the appendix presents only non-generic names of plant substances. "Volatile oil," for example, designates one chemical kind of plant substance rather than a generic group of different kinds of volatile oils. This was not a terminological flaw, but exactly what Lavoisier and his co-authors wanted to present. From a rigid chemical-analytical view, the varieties of mucus, gluten, volatile oil, sugar, and so on were not significant.

Most European chemists of the time did not accept Lavoisier's rigid analytical stance toward the variety and number of organic substances. When Fourcroy raised the question of the number of plant substances in his *Système*, he re-addressed the

---

49 Ibid. p. 118.

issue as a problem.[50] On the one hand, he explicitly reminded his readers of the broad variety of plant substances distinguished in pharmaceutical and artisanal practices, and described these varieties in detail in the subsequent chapters of his book.[51] But on the other hand, by 1801 he no longer stated unequivocally that his twenty immediate materials of plants were generic. Instead, he presented their names in the singular, except for the vegetable acid which he designated "*l'acide végétal ou les acides végétaux.*" Other European chemists, too, presented classes of approximately twenty proximate components of plants (see figure 14.3), but they were rarely explicit as to the question of whether or not these were generic kinds of plant substances. From the fact that in some cases they described varieties of a named plant substance—particularly in the case of vegetable acids, and sometimes for volatile oils, fatty oils, and resins as well—we may conclude that they considered these substances to be generic. But there are also examples to the contrary. Hermbstädt contended unambiguously that his class of sixteen immediate components of plants comprised non-generic kinds, with the exception of vegetable acids, and that these sixteen substances were distributed universally in the vegetable kingdom.[52] None of these chemists presented arguments for their taxonomic judgments by referring to the Lavoisierian theory of the composition of organic substances or quantitative elemental analysis. Rather, taxonomic judgments relied on the traditional identification and demarcation of plant substances according to natural origin, method of extraction, and observable properties.

---

50 See Fourcroy [1801–1802] vol. VII p. 125f.
51 Ibid. p. 120.
52 See Hermbstädt [1807].

15

Uncertainties

15.1 Ambiguities and disagreement in chemists' identification and classification

In the four decades after the publication of the *Méthode*, chemists' individuation and identification of plant substances still followed along traditional lines. European chemists continued to individuate plant substances by referring to their homogeneous appearance and their coherent behavior in extractive operations and tests of chemical properties. And they identified single substances by their natural origin, method of extraction, and a combinatorial set of properties comprising chemical properties tested experimentally, sensible properties that could be observed directly without the use of instruments (such as consistency, color, smell, and taste), and, on occasion, measured physical properties, such as specific gravity, melting point, and boiling point.

It goes without saying that the use of combinatorial sets of properties for the identification and demarcation of plant materials constituted a stark contrast to logical methods of classification like the exhaustive bifurcation used in the artificial taxonomies of eighteenth-century botany. It inevitably led to ambiguities,[1] which the British chemist Thomas Thomson commented on:

> It is necessary therefore to be well acquainted with their [the plant substances'] essential characters, that we know the marks by which they are recognized. These unfortunately are sometimes ambiguous; so that a good deal of skill and experience are necessary before we can distinguish them readily.[2]

According to Thomson, the identification of plant substances required the identification of their characteristic or "essential" properties. But which were the essential properties? Like Fourcroy, Thomson was not able to define the essential properties by any explicit rules.[3]

Not even the focus on the chemical properties of plant materials could rescue chemists from their taxonomic problems. Beginning in the mid-eighteenth century, chemists considered testing the chemical properties of plant substances the most valuable method for their identification. As they enlarged the number of solvents applied to test chemical properties and combined different kinds of solvents, new taxonomic distinctions could be made, such as the one between extracto-resinous matter and resino-extractive matter introduced by G. F. Rouelle in the 1750s. By the end of the century, chemical properties such as acidity, combustibility, and solubility in different solvents had become the most reliable characteristics for identifying and demarcating plant substances. But these elected characteristics still did not allow unambiguous classification.

---

1   For a discussion of the taxonomic ambiguities following from these intersections of properties, see also Melhado [1981] pp. 119–122.
2   Thomson [1804] vol. IV p. 358.
3   See Fourcroy [1792] p. 96 and our discussion above.

The effects produced experimentally in testing the chemical properties of plant materials with reagents and solvents allowed chemists to distinguish at best between major types of plant substances—such as the fixed oils, volatile oils, balsams, resins, and gums—but not between the otherwise observable varieties of fixed oil, for example, such as olive oil, oil of sweet almonds, and linseed oil. Chemists' experimental tools for identifying and demarcating kinds of plant materials worked to some extent on the taxonomic level of the approximately twenty kinds of proximate principles of plants, but not at all on that of their varieties. Varieties of fixed oils, volatile oils, resins, gums, balsams, essential salts, extracts, and so on were distinguished without performing experiments; that is, via directly perceptible properties such as consistency, color, smell, and taste. Chemists were used to observing all kinds of perceptible properties, and they continued that habit even at a time when they considered chemical properties the most significant ones for taxonomic purposes. Furthermore, as these varieties of the major types of plant substances were of great interest in commercial and pharmaceutical contexts, chemists were far from dismissing them. Yet their taxonomic status remained utterly unclear, as is shown by the fact that Fourcroy classified almost all of these different materials under the ambiguous heading "species or varieties."[4]

In the following we present a few examples of the ambiguities of chemists' classification of plant substances, which are characteristic of the entire period of plant chemistry. Around 1800, many European chemists agreed that volatile oils such as the distilled oils obtained from roses, aniseed, and peppermint, and fatty oils, such as olive oil, oil of almonds, and linseed oil, were different kinds of plant materials since they differed in many properties (see figure 14.3 in the previous chapter). However, volatile and fatty oils also had properties in common, even chemical properties. Like fatty oils, volatile oils became solid at low temperatures, were inflammable, dissolved sulfur and phosphorus, combined with alkalis to form soaps (though not as easily as fatty oils), absorbed oxygen from the air and thereby changed their consistency and color, were carbonized by sulfuric acid, and ignited with concentrated nitric acid. How was the significance of these similarities to be evaluated? Quite obviously, there were no reliable rules to decide the question. Fourcroy, Hermbstädt, Stromeyer, and Berzelius decided that fatty oils (or fixed oils, fat) and volatile oils were different kinds of plant substances, but Thomson grouped them together as "oils" (see figure 14.3).

Another example is the distinction between volatile oils and resins. Were the properties shared by volatile oils and resins—the fact that both were aromatic and dissolvable in spirit of wine, but not dissolvable in water—more significant than the fact that the former were liquid whereas the latter were solid? Hence, were volatile oils and resins to be grouped together, or were they different species of substances? As we can see from figure 14.3, Fourcroy, Thomson, Hermbstädt, Stromeyer, and Berzelius all agreed that volatile oils and resins were indeed different species of plant substances. In this case it seems that the difference in consistency played a crucial

---

4   See Fourcroy [1801–1802] vol. VII p. 8.

role in the chemists' classification, as the agreement in chemical properties between the volatile oils and resins could hardly be ignored. But it may also be the case that their judgment relied less on observation than on artisanal tradition, for in very similar cases chemists drew quite different conclusions. A case in point is the relationship between the aromatic natural balsams and resins. Fourcroy distinguished between natural balsams and resins based on their different consistency, but all other chemists mentioned in our table felt that consistency was not a sufficient criterion for their distinction and hence decided they were the same kind of substance (see figure 14.3).[5] The relationship between vegetable butters, such as cacao butter and coconut butter, and fatty oils is another example. Whereas in the earlier eighteenth century chemists often distinguished between butters and fatty oils, around 1800 most chemists agreed to regard butters as mere varieties of fatty oils and to group them together. Friedrich Stromeyer, for example, discussed the question and explicitly argued that they should be grouped together, since they differed only in consistency.[6] Likewise, Johann Bartholomäus Trommsdorff (1770–1837) and Berzelius grouped together fatty oils and wax despite the fact that they differed in consistency and a few other properties.[7] By contrast, Fourcroy, Thomson, and Hermbstädt did distinguish between fatty oils and wax (see figure 14.3).[8] The demarcation of volatile oils from camphor posed similar problems. As many of the chemical properties of the two materials agreed (with one exception: volatile oils absorbed oxygen from the air and thereby were transformed, whereas camphor did not alter its state when exposed to the air), Berzelius considered camphor a variety of volatile oil, though it was a solid substance.[9] Fourcroy, Thomson, and Hermbstädt, however, believed that volatile oil(s) and camphor were different kinds of materials (see figure 14.3).

Similar questions arose about the natural origin of plant materials. Was the difference between the natural origins of gums, for example, significant enough to consider gum arabic, gum Senegal, and cherry-tree gum as different species of gum, or were these gums only varieties of one and the same species of gum? Despite Fourcroy's and other chemists' efforts to select a few characteristics for the identification and classification of plant materials, early nineteenth-century chemists, like their predecessors, neither agreed on a standard set of reagents and solvents for the testing of chemical properties nor established reliable rules for selecting essential characteristics—comparable to number, figure, position, and proportion of the reproductive organs of plants in Linnaeus' taxonomic system.[10] As a consequence, eighteenth-century and early nineteenth-century chemists disagreed about the identification and demarcation of kinds of plant materials almost as often as they agreed. Not just particular questions remained unsolved, but principle taxonomic problems, too. How

5 See also Fourcroy [1801–1802] vol. VII p. 126 and vol. VIII pp. 15–26 and 43–51.
6 Stromeyer [1808] vol. II p. 267f.
7 Trommsdorff [1800–1807] vol. II p. 525; Berzelius [1812] p. 81. For life and work of Trommsdorf, see Gillispie [1970–1980] vol. XIII p. 465f.
8 For Fourcroy's arguments, see Fourcroy [1801–1802] vol. VII pp. 126, 319–338, and 339–351.
9 See Berzelius [1812] p. 87.
10 See Linnaeus [1737a], and [1737b]. See also Beretta [1993] pp. 50–61; Eriksson [1983]; and Müller-Wille [1999], and [2003].

could mere varieties of plant substances be distinguished from true species of plant substances? What kind of characteristics demarcated the border between species and genera of plant substances?

Questions like these were raised explicitly and systematically in 1812 by Berzelius, and a few years later by the French chemist Michel Eugène Chevreul (we will discuss their approaches later in the chapter). As Berzelius pointed out shortly afterward, a solution for these problems could not be expected from quantitative elemental analysis alone. In their attempts to solve taxonomic problems, both Berzelius and Chevreul pointed to the limits of quantitative elemental analysis and the theoretical underpinnings of the Lavoisierian analytical program for plant and animal chemistry—limits that some twenty years earlier had been invisible.

15.2 What were the taxonomic consequences of the Lavoisierian analytical program?

The first attempts to implement the Lavoisierian analytical program for plant and animal chemistry were made by the two French chemists Joseph Louis Gay-Lussac (1778–1850) and Louis Jacques Thenard, both of whom were members of the Society of Arcueil, as well as by the Swedish chemist Jöns Jacob Berzelius. In 1810 and 1811, Thenard and Gay-Lussac published the results of their research on the technique of quantitative elemental analysis of organic substances, along with the results of quantitative elemental analyses of fifteen different plant substances.[11] Five of these were vegetable acids; the others were sugar, gum arabic, starch, sugar of milk, oak and beech wood, resin obtained from turpentine, copal (a natural resin), wax, and olive oil. Based on their analytical results, which they expressed in percentages of carbon, hydrogen, and oxygen, the two chemists then divided vegetable substances into three large classes: acid substances, resinous or oily substances, and substances that were neither acid nor resinous.[12] They correlated this division with a comparison of the proportions of elements contained in the plant substances, using the method suggested by Lavoisier two decades before; that is, indicating "which element existed in excess" or "prevailed" in the composition.[13] In the case of vegetable acids, the proportion of oxygen was always in excess, such that the proportion of oxygen to hydrogen was greater than in water. In resinous or oily substances the proportion of oxygen to hydrogen was less than in water; and in the case of the third class, into which sugar, gum arabic, starch, sugar of milk, and wood fibers (oak and beech wood) were ordered, the proportion of oxygen to hydrogen was the same as in water. The two chemists designated these analytical regularities "laws" of composition.[14] Shortly

---

11 Gay-Lussac and Thenard [1810], and [1811]. On the development of quantitative elemental analysis see Brock [1993] p. 194 f.; Rocke [2000]; Usselman et al. [2005]; and Szabadváry [1966] pp. 287–307. For life and work of Gay-Lussac, see Gillispie [1970–1980] vol. V p. 317ff. On the institutional context of the Society of Arcueil for the development of techniques of quantitative elemental analysis, see Crosland [1967], and [2003].
12 See Gay-Lussac and Thenard [1811] vol. II p. 322.
13 See Lavoisier [1965] p. 118.
14 See Gay-Lussac and Thenard [1811] vol. II p. 321.

PREMIER TABLEAU

Contenant la proportion des principes de quinze substances végétales.

Substance analysée.	Carbone contenu dans cette substance.	Oxigène contenu dans cette substance.	Hydrog. contenu dans cette substance.	Ou en supposant que l'oxigène et l'hydrog. soient à l'état d'eau dans les substances végétales.		
				Carbone.	Eau.	Oxig. excéd.
Sucre.........	42,47	50.63	6,90	42,47	57,53	o
Gomme arabiq.	42,23	50,84	6,93	42,23	57,77	o
Amidon.......	43,55	49,68	6,77	43,55	56,45	o
Sucre de lait...	....	....	....	....	....	o
Chêne........	52,53	41,78	5,69	52,53	47.47	o
Hêtre........	51,45	42,73	5,82	51,45	48,55	o
Acid. muqueux.	33,69	62,67	3,62	33,69	30,16	36,15
Acid. oxalique..	26.57	70,69	2.74	33.57	22,87	50,56
Acid. tartareux.	24,05	69. 2	6.63	24,05	55,24	20,7
Acid. citrique..	33,81	59.86	6.33	33,81	52,75	13,44
Acid. acétique..	50,22	44,15	5,63	50,22	46,91	2,87
						Hydrog. excéd.
Résine de téréb.	75. 4	13,34	10.72	75,94	15,16	8,90
Copale........	76,81	10.6.	12.58	76,81	12,05	11,14
Cire.........	81,79	5.54	12.67	81.79	6,30	11,91
Huile d'olives..	77,21	9,43	13,36	77,21	10,71	12,08

Figure 15.1: Analytical results. From Gay-Lussac and Thenard [1811] vol. 2.

afterwards Thenard added to these three classes a class of vegetable substances containing nitrogen, a class of vegetable "coloring materials" (*matières colorantes*), and a class of substances whose existence was not yet proven.[15]

Apart from the fact that the latter two classes were not correlated with knowledge of composition, Gay-Lussac and Thenard restricted their classification to the natural plant substances. Artificial organic compounds procured from the former were not included in their classification. This was one major deviation from the taxonomic program envisioned in 1787 in the *Méthode*. With respect to taxonomic ranks below these five large classes, Gay-Lussac and Thenard also retained the traditional natural historical modes of identification, demarcation, and naming. They did not discuss the problems of demarcating plant substances, although it is obvious from a comparison of their analytical results that the quantitative composition of several different species of plant substances overlapped considerably. For example, gum (arabic) and sugar could not be clearly demarcated from one another based on experimental knowledge of their elemental composition (see figure 15.1). Moreover, in the presentation of their analytical results the two chemists ignored the well-known varieties of the different kinds of plant substances. They translated the specific names for the varieties of plant substances used in their experiments into general names—with "gum arabic," for example, becoming "gum," and "oak" and "beech wood" becoming "wood fibers"—without commenting on the shift of taxonomic rank involved in that translation. Quite obviously, Thenard and Gay-Lussac followed Lavoisier in his assumption that the number of chemical kinds of plant materials, identified by means of quantitative elemental analysis, had to be small. Nonetheless, in his *Traité,* published shortly afterward, Thenard described varieties of sugar, gum wood fibers, and so on. But he did so in the experimental-historical part of his textbook rather than in its analytical and theoretical chapters.[16] Varieties of plant materials were of interest from an artisanal point of view and from that of an experimental history, but not when the focus was on elemental composition.

Berzelius's criticism of the Lavoisierian understanding

Around the same time that Thenard and Gay-Lussac set out to endorse the Lavoisierian analytical program for plant and animal chemistry by resuming and refining the quantitative elemental analysis of organic substances, the program was attacked by another leading figure in the European chemical community, Jöns Jacob Berzelius. It was especially the Lavoisierian view that the number of plant (and animal) materials was small that provoked Berzelius's critique. In an 1812 paper on chemical nomenclature, Berzelius proclaimed that it was necessary to take into account the varieties among the plant substances that were known up to that point:

> In certain plant analyses, which I have made known, for example that of Icelandic
> mosses and other species of lichens that produce starch, as well as of cinchona bark and

---

15   Thenard [1813–1816] vol. III p. 47.
16   Ibid. pp. 58–351.

the inner bark of the pine, *I tried to prove that the plant substances which we take to be the general building blocks of the vegetables are as different from one another as the plants themselves from which they are obtained.* Thus, for example, I have shown that everything which we call tannin, although sharing certain chemical qualities with one another, nevertheless differs from one another quite dramatically according to the nature and difference of the plants by which it has been produced, and *that the tannin of the oak galls, the cachou, the uva ursi, the bark of the willow, etc., are many different species of the same substance which we call tannin. The same is true, as everyone knows, of the volatile oils, the fatty oils, the resins, and no less, as I think, of the starch, the sugar, etc.* Thus, in classifying these bodies, one must proceed in the same fashion as in the system of botany, and devise *genera* and *species*, but thereby allow each genus to retain the name which it has had up until now.[17]

According to Berzelius plant substances like tannin, volatile and fatty oils, resins, starch, sugar, and the like, were genera requiring division into different species. Berzelius did not justify this claim by referring to the quantitative composition of the substances. Rather, he proclaimed that "it would not be helpful to classify them [plant and animal substances] according to their composition, that is, according to the amount of oxygen, hydrogen, nitrogen or carbon which they contain."[18] In 1812, Berzelius still continued to argue for the varieties of plant substances in a natural historical fashion. The variations between plant substances, he stated, accorded with botanical variation. For example, the starch extracted from lichens varied according to the species of lichen, the plant substance tannin varied according to the species of tree it was extracted from, and so on. Hence, the chemical classification of plant substances followed botanical classification.

Three years later, Berzelius had already carried out numerous quantitative analyses of plant and animal substances. In addition to several vegetable acids, by 1815 he had also analyzed gum arabic, cane sugar, milk sugar, and tannin from oak galls, and in the process had obtained experimental results that he found satisfactory.[19] Based on these experiments, he once again confirmed his opinion that the known plant substances were genera still requiring division into species. "We have, then, various species of tannin, gum, sugar, &c.," he wrote. And he immediately added: "I shall not attempt at present to determine the difference of composition of the species. Such experiments appear too delicate [...]."[20] By 1815, Berzelius no longer argued that knowledge of composition was not helpful in identifying and classifying organic substances. On the contrary, by then he was fully convinced that quantitative elemental analysis was, in principle, the right way to solve the taxonomic problems in organic chemistry. Nevertheless, he argued that his quantitative analyses were too "delicate" to highlight the differences in composition of species and hence did not allow species of plant substances to be demarcated from each other. What was the problem?

After many struggles, by 1815 Berzelius no longer hesitated to extend the stoichiometric laws, as well as his new formulaic system, from inorganic to organic sub-

---

17 Berzelius [1812] 76f.; our emphasis.
18 Ibid. p. 88.
19 See Berzelius [1814], and [1815] pp. 260–275.
20 Berzelius [1815] p. 179.

stances.[21] Unlike Lavoisier and his immediate followers,[22] in 1815 Berzelius defined organic substances as compounds that have a definite, invariable composition and obey the stoichiometric laws. As E. Melhado pointed out, from that time onward Berzelius "taught the chemical profession, as no one had before, that in organic chemistry as in mineral chemistry, the distinct compound in definite proportions must be the focus of chemical practice."[23] Yet the experimental results of quantitative analysis of organic compounds often did not allow clear identification and demarcation of the different plant substances.

Like the quantitative elemental analyses of his French colleagues Gay-Lussac and Thenard, Berzelius's own quantitative analyses had shown that in some cases the quantitative composition of two different genera of plant substances agreed with one another so strongly that the difference of composition was within the margin of measurement accuracy. Berzelius described the problem by way of the following example: "We have found that sugar is made up of 6.802 pct. hydrogen, 44.115 pct. carbon, 49.083 pct. oxygen; gum of 6.792 pct. hydrogen, 41.752 pct. carbon, 51.456 pct. oxygen. It is plain to see that very good analyses of inorganic bodies often show much greater discrepancies than these two results."[24] As these numeric relations show, experiments—that is, quantitative analysis alone—did not yield the certain knowledge of composition that would allow kinds of plant substances to be identified and distinguished, such as sugar and gum, let alone varieties of sugar and gum. Based on his quantitative analyses of organic substances, Berzelius was well aware of the empirical limits of the stoichiometric laws in organic chemistry, and of the obstacles that arose when he tried to transform analytical data into small integers to build chemical formulae of organic compounds.[25]

On the other hand, by 1815 Berzelius had fully developed his theory of chemical proportions or portions, according to which chemical elements combined in definite, invariable quantitative units—designated "equivalents," "volumes," "portions," "atoms," and so on—to build inorganic compounds. In 1813 he had introduced chemical formulae in order to represent this theory and to highlight its differences from the atomic theory of John Dalton.[26] Although he was at first skeptical that his theory and his new symbolic system could be applied to organic compounds as well, in 1815 his skepticism had vanished. Whatever the reason he had for changing his mind, it was not experimental evidence alone. As Berzelius still pointed out in 1819, the very application of the stoichiometric laws in organic chemistry was questionable.

It is tempting to connect Berzelius's extension of his theory of chemical portions and chemical formulae to organic chemistry with his attempt to solve the taxonomic

21 For problems concerning the validity of stoichiometric laws in organic chemistry, see Klein [2003a] pp. 12–23.
22 See chapter 14 on the theoretical limits of Lavoisier's analytical approach to organic substances.
23 Melhado [1992] p. 135.
24 Berzelius [1973] p. 50f. We refer to the 1820 German translation of the French text by Berzelius (Berzelius [1819]).
25 For an example of such an obstacle, see Klein [2003a] p. 23.
26 For the distinction between Dalton's "physical atomism" and "chemical atomism," see Rocke [1984]. For the introduction, meanings, and uses of Berzelian formulae, see Klein [2003a].

problems of that chemical domain. Although he did not explicitly argue in this way, in his publication of 1815 he represented both issues in tandem. In 1815 Berzelius saw the principle for the solution of the taxonomic problems in organic chemistry — especially that of the identification and demarcation of the varieties of sugar, gum, resins, oils, and other proximate principles of plants and animals — in his theory of chemical portions and the application of chemical formulae. Chemical formulae proved to be excellent paper tools in organic chemistry. They cut across the continuum of proportions of elements obtained from the comparison of the analytical data for different organic substances. In so doing, chemical formulae were blueprints clearly identifying species of organic substances and demarcating them from each other. Thus, for example, the two plant genera gum and sugar were clearly distinguished by the two formulae 12 O + 13 C + 24 H for the former and 10 O + 12 C + 21 H for the latter.[27]

In 1819 Berzelius also used his theory of chemical portions for explanatory purposes. Why did the traditional identification of plant substances according to chemical and other perceptible properties work quite well in some cases whereas in other cases it did not work at all? Why was it easy to distinguish different kinds of vegetable acids, for example, but much more difficult to demarcate plant substances such as volatile oils? Berzelius asserted that, presupposing that organic compounds were made up of atoms, one could distinguish between two large classes of such compounds. The first class was of organic species like vegetable acids, whose compound atoms were made up of very few elementary atoms. In such cases, the difference in only one atom caused obvious differences on the macroscopic level, in the perceptible properties of the acids. Hence, in these cases the traditional method of identifying and demarcating a species worked quite well.[28] The second, much more extensive class of organic compounds was made up of compound atoms consisting of a far larger number of elementary atoms. In this case, Berzelius observed, "the difference in composition that arises by adding or subtracting one or more atoms of one element is utterly unimportant, and the body which is thus produced is quite similar to the original in its properties, but is distinct to the extent that one can no longer regard them as absolutely one and the same body."[29] As examples, he named the different kinds (or species) of sugar, tannin, volatile oils, and fatty oils. The different varieties or species of volatile oils, for example, deviated from each other in one or the other property, but were nonetheless grouped together under the label "volatile oils" since they agreed in their major features.

Chevreul's criticism and emphasis on stoichiometric organic compounds

Based on comparative quantitative analyses, the theory of chemical portions, and the use of Berzelian chemical formulae, attempts to create a new taxonomic system of

---

27 For these two formulae, see Berzelius [1815] pp. 271 and 265, respectively.
28 Berzelius [1973] p. 46.
29 Ibid. p. 47.

organic substances from the bottom up to the highest taxonomic rank were no longer out of the question. But Berzelius never undertook more comprehensive taxonomic efforts of this kind. He never turned his theoretical insights into a working resource for a new taxonomic system in organic chemistry. In France, another outstanding chemist tackled questions similar to Berzelius's about the species of organic substances. In 1823, Michel Eugène Chevreul published a treatise on fats that also discussed taxonomic problems concerning the rank of organic species.[30] Like Berzelius, Chevreul complained first, that most of the plant and animal substances chemists considered to be proximate principles of all plants and animals actually were generic groups requiring division into different species. Second, even as late as 1823—thirty-six years after the publication of the *Méthode* and eight years after Berzelius's extension of the stoichiometric laws and chemical formulae to organic compounds—Chevreul still found it necessary to demand that an organic species must contain its elements in definite proportions. If the latter was not the case, he argued, a substance was not a single chemical species but rather a mixture of different species. Fatty oils extracted from animals, for example, were not single pure compounds, as common belief had it, but mixtures.

In numerous experiments Chevreul had analyzed fatty oils of different natural origins. He had isolated from these apparently homogeneous proximate components of animals a whole range of formerly unknown substances, which actually contained their elements in definite proportions. Yet the original natural oils, which had long been identified as chemical species and proximate principles of animals, did not contain their elements in definite proportions. Hence, Chevreul concluded, they were not chemical species but rather mixtures of different substance species.

"Following my outlook on things," Chevreul summed up, it is necessary to consider "sugars, [and] gums as genera formed of several species; and the oils, resins and so on as compounds consisting of several immediate principles, whose union [into oils, resins] *is not subject to definite proportions*, although the latter is the case with respect to the elemental composition of each of them."[31] In this condensed statement Chevreul referred to two taxonomic problems at once: first, the necessity of considering the existing proximate principles of plants as genera; and second, the necessity of taking into account that many of the traditional genera actually comprised substances that were not stoichiometric compounds, or chemical species, but rather mixtures. A year later, he formulated a "first principle" for the classification and study of organic compounds: "Only definite compounds are to be admitted as organic species."[32]

Like Berzelius's theory of chemical portions and chemical formulae, Chevreul's principle was significant not only for the identification of organic substances, but also for the mode of their individuation. As in inorganic chemistry, the boundaries of a single organic substance were no longer to be drawn only by reference to its homogeneous appearance and its coherent behavior when tested with chemical reagents.[33]

---

30 See Chevreul [1823].
31 Ibid. p. 2, our emphasis.
32 Chevreul [1824] p. 157.
33 See chapter 11.2.

Rather, Chevreul highlighted invariant definite composition, qualitative and quantitative—that is, chemical purity in the modern sense—as an unambiguous mark of chemical individuals in organic chemistry as well. This meant that substances such as fatty oils and resins had to disappear from the list of organic species.[34] As Chevreul was eager to emphasize—presumably with an eye on Fourcroy and his followers—chemical individuals and species differed from those of physiology.[35] However, despite his criticism of the extant mode of individuation, identification, and classification of organic substances, Chevreul, like Berzelius, did not set out to elaborate a new classificatory order. His outline for a new taxonomy remained elusive: it had to be based on the natural origin and the chemical properties of the organic substances—that is, on traditional criteria—as well as on knowledge of quantitative elemental composition.[36] As Chevreul stated repeatedly, the time was not yet ripe for a "natural classification" above the rank of organic species.[37]

34 Chevreul [1824] p. 158.
35 Ibid. pp. 27 and 29f.
36 Ibid. p. 184.
37 Ibid. pp. VI, X, and 166.

A Novel Mode of Classifying Organic Substances and an Ontological Shift around 1830

In 1828, only four years after Chevreul formulated his taxonomic principle, the two French chemists Jean-Baptiste Dumas and Polydore Boullay proposed a novel mode of identifying and classifying organic species, which was based on knowledge of the qualitative and quantitative elemental composition of organic compounds as well as on their chemical constitution.[1] Based on the work on the binary constitution of alcohol and ordinary ether by Nicolas Theodore de Saussure (1767–1845) and Gay-Lussac,[2] and Chevreul's work on the binary constitution of fats, as well as on their own experiments with ethers and other alcohol derivatives, Dumas and Boullay made the daring suggestion that all of the organic substances they had grouped together — two kinds of sugar, ethal (isolated from fat), alcohol, and ethers — were not directly composed of the elements carbon, hydrogen, and oxygen, but had a binary constitution, and were made up from two more proximate components.[3] The common denominator of the new class of binary compounds, which was given the name "compounds of bicarbonated hydrogen," was the component "bicarbonated hydrogen." The differences between the species were explained by the differences in the second component. This novel mode of classification assimilated the classification of organic substances to the classification of inorganic ones.[4] Dumas and Boullay increased the plausibility of the implied analogy between the constitution of organic compounds and inorganic ones by constructing formula models displaying the specific analogy between their new class of "compounds of bicarbonated hydrogen" and a special class of inorganic compounds, the salts of ammonia.

Dumas and Boullay presented their new class of organic substances in the form of a table at the end of an article, in which they primarily reported the results of new experiments exploring the formation reactions of ethers (see figure 16.1). Their novel mode of classification presupposed an extended series of experiments coupling quantitative elemental analysis with the exploration of the chemical reactions of organic substances. As was common at the time, from their experimental studies of chemical reactions chemists drew conclusions concerning the more proximate components of substances and their constitution. But the new classificatory system did not just rely on experiments. It also required additional work on paper with Berzelian chemical formulae. Using chemical formulae as paper tools, Dumas and Boullay were able to model chemical reactions along with the binary constitution of the organic substances

---

1   Dumas and Boullay [1828]. For a more detailed analysis of the work of Dumas and Boullay and its historical context, see Klein [2003a]. On support for Dumas among leading figures of the Arcueil circle, which established links between his approach and Lavoisier's program for organic chemistry, see Crosland [1967] and [2003].
2   For life and work of de Saussure, see Gillispie [1970–1980] vol. XII p. 123f.
3   See de Saussure [1807], and [1814]; Gay-Lussac [1815]; Chevreul [1823]; and Dumas and Boullay [1827]. On the conception of binary constitution of organic compounds, which was broadly shared by chemists at the time, see Klein [2003a] and the primary and secondary literature quoted there.
4   See also Kapoor [1969a], who pointed out the special relevance of analogy in the chemical classifications of organic substances. The general significance of analogies with inorganic compounds was also emphasized by Brooke; see Brooke [1973], and [1987].

NOM DU COMPOSÉ.	BASE.	ACIDE.	EAU.
Hydro-chlorate d'ammoniaque......	$Az\,H^3$	$2\,H\,Ch$	
Hydro-chlorate d'hydrogène bi-carboné (éther hydro-chlorique)......	$2\,H^2\,C^2$	$2\,H\,Ch$	
Hydriodate d'ammoniaque.....	$Az\,H^3$	$2\,H\,I$	
Hydriodate d'hydr. bi-carb. *(éther hydriodique)*......	$2\,H^2\,C^2$	$2\,H\,I$	$\dot{H}H$
Hypo-nitrite d'ammoniaque hydraté......	$2\,Az\,H^3$	$\ddot{A}\,Az$	$\dot{H}H$
Hypo-nitrite d'hydr. bi-carb. hydraté *(éther nitrique)*......	$4\,H^2\,C^2$	$\ddot{A}\,Az$	$\dot{H}H$
Acétate d'ammoniaque hydraté......	$2\,Az\,H^3$	$H^6\,C^4\,O^3$	$\dot{H}H$
Acétate d'hydr. bi-carb. hydraté *(éther acétique)*.....	$4\,H^2\,C^2$	$H^6\,C^4\,O^3$	$\dot{H}H$
Benzoate d'ammoniaque hydraté......	$2\,Az\,H^3$	$H^{12}\,C^{30}\,O^3$	$\dot{H}H$
Benzoate d'hydr. bi-carb. hydraté *(éther benzoïq.)*.....	$4\,H^2\,C^2$	$H^{12}\,C^{30}\,O^3$	$\dot{H}H$
Oxalate d'ammoniaque cristallisé et desséché.........	$2\,Az\,H^3$	$C^4\,O^3$	$\dot{H}H$
Oxalate d'hydr. bi-carb. hydraté *(éther oxalique)*.....	$4\,H^2\,C^2$	$C^4\,O^3$	$\dot{H}H$
Bi-sulfate d'ammoniaque........	$2\,Az\,H^3$	$2\,\dot{S}$	
Bi-sulfate d'hydr. bi-carb. *(Acide sulfo-vinique)*.........	$4\,H^2\,C^2$	$2\,\dot{S}$	
Binoxalate d'ammoniaque......	$2\,Az\,H^3$	$2\,C^4\,O^3$	
Binoxalate d'hydr. bi-carb. *(acide oxalo-vinique)*.........	$4\,H^2\,C^2$	$2\,C^4\,O^3$	
Bi-carbonate d'ammoniaque hydraté......	$2\,Az\,H^3$	$4\,\dot{C}$	$\dot{H}H$
Bi-carbonate d'hydr. bi-carb. hydraté *(sucre de cannes)*..	$4\,H^2\,C^2$	$4\,\dot{C}$	$\dot{H}H$
Bi-carbonate d'hydr. bi-carb. bi-hydraté *(sucre de raisins)*......	$4\,H^2\,C^2$	$4\,\dot{C}$	$2\,\dot{H}H$
Hydrate d'hydr. bi-carb. octo-basique *(éthal)*......	$16\,H^2\,C^2$	....	$\dot{H}H$
Hydrate d'hydr. bi-carb. bi-basique *(éther sulfurique)*......	$4\,H^2\,C^2$	....	$\dot{H}H$
Hydrate d'hydr. bi-carb. *(alcool)*......	$4\,H^2\,C^2$	....	$2\,\dot{H}H$
Ammoniaque liquide......	$Az\,H^3$	....	$2\,\dot{H}H$

Figure 16.1: Classification of organic substances by Dumas and Boullay [1828].

undergoing the reactions. Their novel mode of classification was embedded in an experimental culture coupling experimentation with the modeling of invisible experimental processes on paper and with chemical theory.[5] Here chemical theory was built into the paper tools—that is, the chemical formulae applied for modeling. Chemical classification thus became at once experimentalized and theorized.[6]

The perceptible properties of organic substances, their natural origin, and the mode of their extraction were the dominant criteria for classifying plant substances until the 1820s. By contrast, in Dumas and Boullay's new classificatory system the composition and constitution of organic compounds, investigated by way of series of experiments and construction of formula models on paper, became the most important taxonomic resources. Moreover, the new class comprised both natural plant substances and their derivatives, which were experimentally produced artificial organic compounds. Only a few years later the German chemist Justus Liebig followed their example, building new classes of organic substances based on experiments and the construction of formula models of their binary constitution.[7] The French chemists Auguste Laurent and Charles Frédéric Gerhardt (1816–1856) continued these early attempts at a new mode of classification by developing the first more comprehensive classification systems in the 1840s.[8]

Later in the century, Dumas and Boullay's new mode of classifying organic substances was celebrated in the European community of chemists as a breakthrough. Justus Liebig, who was otherwise often critical of Dumas's work, recognized its importance: "This work was the first in the field of the new organic chemistry, where it was attempted to link with a shared tie one series of bodies with another; in this light, it marked the beginning of the current era, and is important for the history of chemistry."[9] Friedrich August Kekulé (1829–1896), one of the most prominent proponents of synthetic carbon chemistry in the 1860s,[10] stated that Dumas and Boullay's classification was "the first grouping of a large number of derivatives" in organic chemistry.[11] Looking back on the history of classifying organic substances, Aleksandr Butlerov (1828–1886) pointed out:

> As long as our familiarity with organic bodies was merely superficial and *limited itself primarily to external properties*, while transformations and mutual relations remained almost entirely unknown, the classifications of organic compounds, considered from the standpoint of chemistry, *could not be natural*. At that time, bodies were grouped according to their origin, color, consistency, etc., divided into volatile oils, resins, dyestuffs, etc. A closer, but not yet complete familiarity with the chemical properties of a part of the organic compounds led to the distinction between acids, alkalis, and indifferent bodies,

5   On the notion of "experimental culture" used here, see Klein [2003a].
6   For a more detailed analysis of the table and of the function of Berzelian formulae as paper tools for modeling and classification, see Klein [1999], and [2003] pp. 130–148.
7   See Liebig [1832–1834], and [1834].
8   On these further attempts after 1828 to classify organic compounds based on composition and constitution (or structure), see Fisher [1973a], [1973b], and [1974]; Kapoor [1969b]; and Mauskopf [1976]. For life and work of Gerhardt, see Gillispie [1970–1980] vol. V p. 369ff.
9   Liebig [1837] p. 26.
10 For life and work of Kekulé, see Gillispie [1970–1980] vol. VII p. 279ff.
11 Kekulé [1861–1862] vol. I p. 63.

and allowed the first to be divided into volatile and non-volatile, the second into bodies without oxygen and containing oxygen, the third into bodies containing nitrogen and free of nitrogen, etc. *The emergence of theoretical views made it possible*—at least in terms of the substances to which these viewpoints were applied—*to move towards more scientific systems*. Each theory was reflected in classification [...].[12]

Carbon chemists of the 1860s defined their way of classifying organic compounds, based on knowledge of composition, constitution, and chemical structure (from the 1860s onward) and acquired by experimentation and work on paper, as both natural and scientific. They clearly demarcated this experimental and theoretical mode of classification from previous ones, which relied in particular on perceptible properties. Whereas earlier chemists' modes of classifying organic substances shared basic attitudes with naturalists and craftsmen, carbon chemists' analytical approach was emblematic of a culture of experts and the difference of that culture from the mundane world.

16.1 The ontological shift in the 1830s: stoichiometric substances

Not only the mode of classifying organic substances was transformed in the period after the late 1820s; the mode of their individuation and identification changed as well.[13] In 1824, Chevreul had claimed that the boundaries of chemical individuals and species of plant substances must differ from the boundaries of substances defined from a physiological point of view (see the previous chapter). In 1835, Dumas continued this line of argument when he wrote in the fifth volume of his *Traité*, concerned with organic chemistry, "The main goal of this book is a familiarization with the organic substances from a purely chemical standpoint."[14] "From a purely chemical standpoint" meant first that the study of physiological processes and of "organized substances" or "complex products" had to be excluded.[15] "The history of those substances," Dumas stated, "which are only organs or part of organs, like wood fibers (*le ligneux*), fibers (*la fibrine*), corn starch (*l'amidon*) and all the other complex products, which interest the chemist only as a starting point for his operations, must be left to physiology."[16] A "purely chemical standpoint" toward organic substances meant, second, that they had to be individuated and identified in an experimental, analytical, and theoretical way. "The most general characteristic regarding organic materials," Dumas wrote, "is everything that concerns their analysis, the determination of their atomic weights and the search for the rational formulae which represent their nature and their properties."[17] This statement left no doubt that the chemical individuals of carbon chemistry were exclusively stoichiometric carbon compounds, represented by chemical formulae and molecular weights. Moreover, a pure single

---

12 Butlerov [1868] p. 82f.; our emphasis.
13 The following account of the new mode of individuation and identification of organic compounds in carbon chemistry provides only a rough summary of the detailed historical studies in Klein [2003a].
14 Dumas [1828–1846] vol. V p. 1.
15 Ibid. p. 2.
16 Ibid. p. 77f.
17 Ibid. p. 3.

carbon compound was further identified by its "rational" formula, which represented the binary constitution of the substance.

The experimental and theoretical individuation and identification of organic compounds also extended to organic chemical artifacts, such as ethers and other derivatives of natural carbon compounds. Whereas around 1800 chemists had grouped together only the natural compound components of plants, Dumas stated: "The boundary drawn in this book lies in the claim that as long as an organic material is not transformed into carbon, carbon oxide, carbonic acid, carbonated hydrogen, nitrogen, ammonia, nitrogen oxide, or water, it must remain in the array of organic matter."[18] With this statement, Dumas emphasized that "organic" meant only a distinctive kind of composition—which led to the distinctive products of analysis mentioned by him—but not origin from organized bodies. It was no longer synonymous with "organized," but instead with carbon compounds—compounds consisting, as a rule, of carbon, hydrogen, oxygen, and nitrogen—independent of their provenance, be it natural or artificial. From Dumas's perspective, which highlighted elemental composition and binary constitution, the provenance of organic compounds became entirely insignificant. Dumas even went a step further, questioning the demarcation between organic and inorganic chemistry. "In my opinion," he proclaimed, "there are no organic materials."[19] Similar statements were made several years later by his French colleague Charles F. Gerhardt: "Since all organic materials, without exception, contain carbon, one can say that this is the chemistry of carbon. It only includes organic materials in their purely chemical relations, without accounting for the role that they play in living organization."[20]

Chemists' new ontology of stoichiometric carbon compounds was embedded in a deep structural transformation of organic chemistry taking place between the late 1820s and the early 1840s.[21] As a result of these transformations, not only the meaning of "organic substance" changed, but also the material world of referents. First, chemists' improved extraction techniques enabled them to increase the number of pure plant and animal substances extracted from organic tissues and juices. Better techniques of quantitative analysis allowed them to control stoichiometric purity and to dismiss all substances lacking constant quantitative composition. Studies of chemical reactions, and the modeling of these reactions on paper by means of chemical formulae, further contributed to the control of stoichiometric purity as well as to the identification of the pure compounds. In this way, organic substances became wholly experimentally defined objects, from the bottom (that is, their individuation and identification), up to their classification. As a consequence of the new mode of individuation and identification, many of the plant materials that were emblematic for eighteenth-century plant chemistry no longer belonged to the scientific objects of the new carbon chemistry. Not only wood fibers (*le ligneux*), fibers (*la fibrine*), and corn starch (*l'amidon*), explicitly mentioned by J.-B. Dumas, left the stage of carbon

---

18 Ibid. p. 2.
19 Ibid. p. 78.
20 Gerhardt [1844–1845] vol. I p. 1.
21 On this transformation see Klein [2003a].

chemistry, but also resins, balsams, gums, and extracts, which now were excluded as non-stoichiometric mixtures. Likewise, all chemical-pharmaceutical preparations lacking invariable definite quantitative composition, such as decoctions and tinctures, which had still been included in Lavoisier's system and the *Méthode*, now were dismissed.

At the same time, the world of organic substances was also considerably enlarged by the introduction of two new types of organic compounds. Chemists now had access to coal tar, a waste product of the gas industry that proved to be an unexpectedly rich reservoir of carbon compounds. An even richer source of new carbon compounds was experimental studies of chemical reactions of organic compounds in the chemical laboratory. Studies of chemical reactions yielded new reaction products, often even entire series of new reaction products, which chemists examined not merely as material experimental traces of imperceptible reactions, but also as perceptible substances.[22] Organic substances, as a rule, undergo a large number of reactions—"metamorphoses," as the historical actors called them—through oxidations, substitution reactions, and many other types of reactions, which yield ever-extending chains and trees of derivatives. Once chemists were able to trace these metamorphoses in the 1830s—by means of quantitative analysis of reaction products, subsequent comparison of their composition with that of the original substances, and the modeling of the reaction on paper by means of chemical formulae[23]—the number of new organic substances grew exponentially. Whereas by 1800 the number of stoichiometric organic compounds was at best 100, in 1872 Pierre E. Marcellin Berthelot (1827–1907) already reported the impressive number of more than 10,000 pure organic compounds.[24] Some thirty years later, the number had increased to approximately 90,000.[25] The bulk of these new substances were artificial carbon compounds, products of the laboratory procured during the study of chemical reactions as reaction products or synthesized for their own sake.

The carbon compounds individuated and identified in carbon chemistry—the remaining stoichiometric plant and animal substances and the pure carbon compounds isolated from coal tar, as well as the artificial carbon compounds created in the laboratory—were nested in extended networks of experiments and work on paper. In the late 1840s, when the culture of carbon chemistry was firmly established, the individuation and identification of carbon compounds required quantitative elemental analysis, control of stoichiometric purity by studies of the chemical properties and reactions of a substance, experimental examination of their proximate components or "constitution" (later "structure"), and work on paper with chemical formulae to demarcate the substances and to model their constitution and chemical reactions. Analysis of composition (qualitative and quantitative), control of purity, studies of reactions, and modeling on paper allowed chemists to draw ever more sophisticated

22 On the sequence of experiments procured by studies of reaction products, and the creation of new substances in this way, see Klein [2005b].
23 See Klein [2003a].
24 See Berthelot [1872] p. v.
25 See Schummer [1997].

boundaries of single substances. Sometimes these substances even agreed in elemental composition and merely differed in constitution, as was the case with isomers.

By 1840 the substitution products—differing in elemental composition by just one portion or atom of an element—and isomers—agreeing in elemental composition and differing only with respect to their proximate components—had become emblems of the new ontology of stoichiometric substances in the culture of carbon chemistry. Embedded in a sophisticated network of experiments, theories, and model building, they were the first emblematic pure laboratory substances of chemistry that were individuated and identified in ways utterly different from the existing chemical arts and crafts. It was only with the emergence of the new synthetic-dye industry in the late 1850s that this gap was bridged again.

16.2 The trajectory of ontological shifts in plant chemistry

Eighteenth-century plant chemistry has been depicted as a continuously evolving laboratory tradition culminating in the experimental successes of plant chemistry in the early nineteenth century, when pure plant substances like morphine and strychnine were isolated.[26] According to this historical interpretation, chemists concerned with plant chemistry in the eighteenth century and with organic chemistry in the nineteenth century always studied the same kinds of objects given by nature. What changed, according to this view, were the techniques of access to pure organic substances and the number of pure organic substances. "Chemical analysis" and "purity," it further appears from this perspective, always had the same meaning and the same experimental objective.

Contrary to this widespread opinion, we argue that chemists' ontology of plant materials shifted in the eighteenth and early nineteenth centuries in quite discernible ways, as did their modes of classifying plant materials. Our comparison of eighteenth-century chemists' ways of classifying plant substances in contexts of conceptual inquiry brings to the foreground a trajectory of ontological shifts and changes in classification punctuated in the 1750s, 1790s, and 1830s. Beginning in the 1750s, along with naturalists' and philosophers' increased concern for organization and phenomena of life, chemists became interested in plants as living organized beings and plant materials as products of life. This move was less visible in the laboratory than at the writing desk and in the lecture hall. It spurred chemists' sorting out and grouping together of compound components or proximate principles of plants, as well as their separation of plants and animals as organized bodies from unorganized minerals, without altering the traditional artisanal modes of individuating and identifying plant materials. The new classification of the compound proximate principles was further conditioned by the accumulation of experimental experience, and the proliferation and collective acceptance of the Stahlian theory of a graduated order of chemical composition. The latter theory focused chemists' attention more closely on the com-

---

26   See Holmes [1971], and [1989]; Löw [1977]; and Tomic [2003].

pound proximate components of plants. Further, it underpinned ongoing developments in the laboratory, which culminated in the distinction between the analysis of plants—which separated the more compound proximate components of plants—and the further analysis of the proximate principles of plants into simpler elements or principles.

In the 1790s, when chemists clearly defined "organic substances" and dismissed mineral substances from the class of proximate principles of plants and animals, a second ontological shift became visible. This second ontological shift was reinforced by Lavoisier's new theory of the elemental composition of organic substances from carbon, hydrogen, oxygen, and nitrogen. Yet, far from promoting a new classificatory practice and ontology of both organic and inorganic substances based on knowledge of chemical composition—as envisioned in the *Méthode de nomenclature chimique* published in 1787—Lavoisier's theory contributed to a theoretical underpinning of the ongoing separation of organic from inorganic substances. In the two decades after 1787, chemists largely ignored the quantitative elemental analysis of organic compounds, and instead continued their studies of the natural proximate principles of plants and animals, processes of life, and the specificities of organic matter.

What remained untouched by these two ontological shifts was the overall naturalistic approach to classification and the artisanal mode of individuation of plant materials. Well into the nineteenth century, chemists ordered materials into three large classes according to their natural origins from the mineral, vegetable, and animal kingdoms. On the lowest taxonomic rank of generic plant materials—such as resins, gums, essential oils, and essential salts—eighteenth- and early nineteenth-century chemists grouped together varieties of plant materials according their natural origin from plant species, their mode of extraction, their perceptible properties, and their practical, in particular pharmaceutical, uses. Furthermore they individuated plant substances according to homogeneous appearance and coherence in chemical extraction and tests of chemical properties.

It was not until the late 1820s that chemists began to individuate, identify, and classify organic substances in an experimental and theoretical way based on knowledge of their invisible elemental composition and constitution (later "structure"). The ontology of carbon compounds, constituted in the new expert culture of carbon chemistry between the late 1820s and the early 1840s, allowed only substances that obeyed the stoichiometric law of definite proportions and could be represented by chemical formulae. As a consequence, many substances individuated and identified in eighteenth-century plant and animal chemistry, such as balsams, resins, gums, fats, extracts, wood, milk, blood, bile, and bones, were excluded from the list of scientific objects. In addition to the remaining stoichiometric organic substances stemming from plants and animals, in the 1830s substances extracted from coal tar and a plethora of novel artificial organic compounds, such as chlorinated hydrocarbons, entered the chemical laboratory. Based on knowledge of composition and constitution, the new classes of carbon compounds grouped together pure plant and animal substances, pure substances extracted from coal tar, and the new kind of artificial carbon compounds procured in the laboratory.

If we compare the materials handled and classified in the plant (and animal) chemistry of the eighteenth century with the substances studied in the type of organic chemistry that began to thrive in the 1830s, the ontological difference is striking. The plant substances studied by early eighteenth-century chemists were natural materials applied in the apothecary trade and other arts and crafts. Balsams, resins, gums, fats, extracts, vinegar, spirit of wine, distilled waters, tinctures, composite elixirs, and the like were commodities imported by merchants or procured by apothecaries, distillers, and chemists. Chemists' individuation of plant materials, which was based on their homogeneous appearance and their coherent behavior in extraction and tests of chemical properties, as well as their mode of identification and classification of plant materials using natural origin, method of extraction, perceptible properties, and practical uses, coincided with the points of view of artisans and craftsmen. When chemical properties became more important for the identification of plant substances in the second half of the eighteenth century, individuation, identification, and classification were linked more closely to experimentation and the use of extended and refined sets of chemical reagents. But this link did not transform the artisanal and natural historical individuation and identification of plant materials. The testing of chemical properties was an observation of effects procured by applying chemical reagents to the substance to be tested. It did not include the separation of components for analytical purposes or for the exploration of the invisible regrouping among components taking place in chemical reactions. As the term "chemical properties" almost reveals, the testing of chemical properties was an extension and refinement of artisanal methods of identifying objects by observing their properties.

In the new experimental culture of carbon chemistry that emerged after the late 1820s, the material world of organic substances changed profoundly. Many of the substances chemists grouped together under the label "carbon compounds" in the full-fledged culture of carbon chemistry of the late 1840s were organic chemical preparations, which existed neither in nature nor in the eighteenth- and early nineteenth-century arts and crafts. At the same time, many substances individuated, identified, and classified as plant or animal substances in the eighteenth century were no longer accepted as chemical kinds in carbon chemistry. Carbon compounds were objects constituted, materially and epistemically, in a scientific expert culture. They were nested in extended networks of experiments and work on paper. The individuation and identification of these stoichiometric compounds required quantitative elemental analysis, control of purity by studies of the chemical properties and reactions of a substance, experimental examination of their proximate components or constitution (later called "structure"), as well as work on paper with chemical formulae to demarcate substances and to model their constitution and chemical reactions.

Conclusion: Multidimensional Objects and Materiality

Materials have long been marginalized by historians of science. Whereas most of the recent historical studies of the material culture of the sciences have highlighted instruments, bodily techniques, and sites, very few studies have been devoted to stuff, materials, and the materiality of the objects of scientific inquiry. The quiet inventions involving materials seem to be unworthy of historical scrutiny. This view stands in stark contrast to our familiar periodization of civilization into the Stone Age, the Bronze Age, and so on. It is also at odds with the fact that in the history of science, technology, and art prior to the social differentiations of the nineteenth century, learned experimenters, engineers, artisans, and artists invested time and energy to the study and improvement of materials. The high social esteem of these endeavors is manifest not least in the fact that among the new experimental sciences of the early modern period the science of materials—chemistry—was the first to be acknowledged and established in academies and other scientific institutions. Materials are indeed so central to technology and society that they were objects of learned inquiry centuries before the establishment of materials science in the twentieth century. In the eighteenth century they were studied both by practitioners and learned men, especially chemists. The chemistry of this time, we have argued, was the science of laboratory substances and materials applied in industry and everyday life. Comparable to the integrated worlds of "high" and "low" mathematics and mechanics in the early modern period, early modern academic chemistry was embedded in artisans' and merchants' world of materials.[1]

We have further argued that materials have a history. By contrast, most historians have assumed that early modern chemists were always concerned with the same type of material substances, given by nature and discovered in the academic laboratory. What changed in history, according to this common view, were the techniques of access to natural substances, the number of known natural substances, and their theoretical understanding. Chemical analysis, it further appears from this perspective, always had the same meaning and the same practical, experimental goal, namely, to isolate and discover pure single substances whose boundaries were determined by nature. Our study of the chemical practices of producing, individuating, identifying, and classifying substances from the late seventeenth century to the early nineteenth century challenges this familiar view. The upshot of it is that chemists' substances were not universal immutable objects given by nature, but objects shaped in human practice and having the same historicity and contingency as human practice.

Our concept of a historical ontology of materials tries to capture the historicity and contingency of material substances. A historical ontology of materials is concerned with the questions of what type of materials were produced, manipulated, and studied by the human actors of the past, how these types of materials changed in history, and what kind of collective resources conditioned their material boundaries and meaning. In part III we have examined in fine detail the dynamics of change of plant

---

1   See Bennett [1986]; and Lefèvre [2000].

materials along with the ontological shifts that reconfigured the place of those materials in nature and society. In the early eighteenth century chemists studied plant materials as material goods, especially chemical remedies, and some of them also as the simplest principles of natural bodies; in the middle of the eighteenth century they strengthened the connection of their studies of plant materials with natural history by means of new methods of plant analysis, which were embedded in revisions of the theory of chemical composition and the concept of components of plants; by the end of the century this latter approach led to a selection of objects of inquiry that chemists conceived as the true products of organic life or "organic substances;" in the 1830s chemists redefined organic substances as stoichiometric carbon compounds in the context of a comprehensive change of the entire culture of organic chemistry. Especially the latter shift of chemists' ontology of organic materials highlights an important issue. Our historical ontology is not merely concerned with changes in the meaning of objects. Rather material production, ways of individuation, and the related inclusions in or exclusions of material objects from the sphere of inquiry are at stake too. We have shown that the experimental production of substances by chemists and the way they determined pure single substances changed considerably in the 1830s, and we consider production and individuation of material objects, and the instruments, skills, and knowledge involved in these activities, to be no less a part of the constitution of an object of inquiry than theories, beliefs, social interests and power.[2]

The list of material substances studied by eighteenth-century chemists excluded the philosopher's stone, philosophical mercury, the elixir of life, or other secret substances of medieval and early modern transmutational alchemy as well as the pure stoichiometric compounds of nineteenth-century chemistry, which were beyond the horizon of their practice. In the decades around 1700 the chemists' concern with the transmutation of metals and the secret substances triggering transmutation began to decline. Instead, chemists paid increased attention to chemical reactions that can be reversed and to those substances that underwent reversible decompositions and recompositions. E. F. Geoffroy's table of chemical affinities, published in 1718, is a mark of this ontological shift. When we consider the substances of interest to eighteenth-century chemists from a broad comparative perspective, one feature stands out clearly. Far from belonging to the ancient world of transmutational alchemy, and far also from belonging exclusively to the new domain of academic chemistry, most of these substances were also applied in the arts and crafts and wider society. The vast majority of the eighteenth-century chemical substances were materials as familiar in apothecary's shops, foundries, assaying laboratories, arsenals, potteries, dye-works, distilleries, coffee houses, and market-places as they were in academic laboratories, studies, and lecture halls. This is particularly true for the plant and animal substances, almost all of which were applied as remedies, food, or dyestuffs, but also for many of the eighteenth-century mineral substances. Even those substances that were true

---

2   Our concept of historical ontology differs in this respect from Hacking's concept of historical ontology as well as from Daston's approach; see Hacking [2002]; Daston [2000], and [2004a].

inventions or discoveries of the academic laboratory, such as the different kinds of gases, became quickly transferred to the mundane world to be applied as remedies, or even used in such brand-new enterprises as ballooning. Circulating between different social sites, the eighteenth-century chemical substances were boundary objects that linked together academic chemists, apothecaries, distillers, miners, metallurgists, assayers, dyers, and potters as well as state commissioners, merchants, and polite connoisseurs of the expanding luxury market.[3]

But these substances were more than boundary objects. We have argued that they were also multidimensional objects that were both commodities and objects of learned inquiry, amalgamating perceptible and imperceptible, useful and philosophical, and technological and scientific features. Chemists frequently shifted their inquiries from the perceptible to the imperceptible dimension of these things. They observed their color, smell, taste, consistency, density, boiling point, combustibility, solubility, acidity and other perceptible properties with the same intensive attention as they analyzed their invisible composition and the regroupings of their components in chemical reactions. Experimental history, technological inquiry, and analytical experimental philosophy were the three main types of investigative practices in eighteenth-century chemistry that evolved around the many faces of chemical substances.

Multdimensional objects of inquiry

Our argument that eighteenth-century chemical substances were multidimensional objects of inquiry stands not only in stark contrast with the common divide of the history of chemistry into a history of chemical technology and a history of chemical science, but also with some larger historical pictures of eighteenth-century science. In *Les mots et les choses* Michel Foucault characterized the eighteenth century as an age which privileged the visible surface of things.[4] What counted most in observation were visible mechanical qualities — "lines, surfaces, forms, reliefs" — that entailed the possibility of linking perceptible objects to mathesis.[5] Botany and botanical taxonomy were important occupations in the eighteenth century "not because there was a great interest in botany," but rather "because it was possible to know and to say only within a taxonomic area of visibility."[6] In other words, botany was a prominent discipline in the eighteenth century because it provided a space for the realization of the episteme of that period.[7] Linnaean botany in particular served as a model, as it reduced the taxonomic characters to "four variables only: the form of the elements, the quantity of those elements, the manner in which they are distributed in space in relation to each other, and the relative magnitude of each element."[8] According to

---

3   For the notion of boundary object see Star and Griesemer [1989].
4   See Foucault [1966]. In the following we quote from the English translation (Foucault [1970]).
5   See Foucault [1970] pp. 133 and 136.
6   Ibid. p. 137.
7   For the notion of episteme in this context, see ibid. p. 128. The Hegelian character of Foucault's concept of episteme becomes particularly manifest in statements like those quoted above concerning the realization of the episteme of the classical age.
8   Ibid. p. 134.

Foucault, this kind of natural history came to an end by the early nineteenth century, when learned men such as Cuvier "substituted" "organism for structure, [and] internal subordination for visible character."[9] Whereas the naturalists of the classical age had "opposed historical knowledge of the visible to philosophical knowledge of the invisible" the epistemic regime of the new age privileged studies of hidden structure.[10]

How far are these views corroborated by our studies of eighteenth- and early nineteenth-century chemistry? As chemical substances resemble in many ways the objects of natural history—both with respect to their perceptible dimension and their broad variety—we should expect that the episteme of Foucault's classical age manifested itself not only in eighteenth-century botany, but in chemistry, too. John Pickstone recently spelled out these consequences when he highlighted the Lavoisierian chemistry of the late eighteenth century as a new model of the analytical way of knowing for both chemistry and other sciences. Adopting Foucault's theory of stages of different epistemes, Pickstone characterized the earlier eighteenth century as a period in the development of science, technology, and medicine in which the descriptive natural historical mode of knowing predominated over analysis.[11]

Our studies have shown that chemical analysis in its modern meaning and methodology already existed at the beginning of the eighteenth century, though only in a comparatively small area, largely mapped by E. F. Geoffroy's table of affinities of 1718. This modern practice and concept of analysis pursued the separation by chemical means of pre-existing components of compounds, with the additional methodological requirement that the original compound could be recomposed from its substance components. The fact that this modern practice, concept, and method of chemical analysis could not be applied seamlessly in the other chemical domains, especially not in plant and animal chemistry, does not imply that analysis played no role in these other fields. On the contrary, analysis played an important role in plant and animal chemistry as well as in the chemistry of raw minerals. What divided the different chemical domains in the eighteenth century was not the absence or presence of analysis, but the objectives, meaning, methodology, and techniques of analysis.

Beginning shortly after the establishment of the Paris Academy of Sciences in 1666, chemists were already attempting to analyze plants and to separate their most simple principles. By the middle of the eighteenth century, chemists agreed on the objective of separating from plants and animals the more compound proximate principles. Additionally, they continued their attempts to further analyze these proximate components and to identify their simple elements or principles as well as the proportions of them. But whereas they regarded their analysis of the proximate principles as a success, they took a skeptical stance toward the further elemental analyses of the proximate principles. This began to change when Lavoisier introduced the analytical technique of complete combustion. In other words, Lavoisier did not introduce

9   Ibid. p. 138.
10   Ibid.
11   See Pickstone [2000].

organic elemental analysis from scratch, but rather proposed a new technique along with new conceptual resources built into that technique. The Lavoisierian analytical approach gained momentum on the communal level of chemistry only around 1810, when chemists such as Jöns Jacob Berzelius, Louis Jacques Thenard, and Joseph Louis Gay-Lussac further developed the technique of combustion analysis. In the three decades prior to 1810, chemists' collective agenda was tuned to different issues, such as chemical physiology, the characteristics of organic versus inorganic matter, and the separation, identification, and classification of the proximate principles of plants and animals. These issues required analysis and theoretical considerations no less than the analysis of the elemental composition of substances.[12] Again, it was the meaning, objectives, and methods of analysis which differed from Lavoisier's elemental analysis, rather than the analytical approach per se.

Throughout the entire eighteenth century, analyses of the imperceptible dimension of natural objects and substances were persistent issues in chemistry. The analysis of the composition of plants, animals, raw minerals, and chemical substances and the objective of gaining philosophical knowledge were not opposed to, but rather paired with the observation of their perceptible dimension and inquiries into their usefulness. Instead of concentrating exclusively on the visible surface of things, chemists approached substances as multidimensional objects: as useful materials, as perceptible objects of nature, and as things consisting of imperceptible parts (substance components) and imperceptible affinities. Even in areas of artisanal practical chemistry, chemical analysis played a role throughout the eighteenth century, especially in mineralogy, assaying, pharmacy, the making of porcelain and ceramics, the manufacture of gunpowder, and agriculture. Foucault's concept of an overarching episteme in the eighteenth century, which focused the historical actors' attention to the visible surface of things and opposed historical knowledge of the visible to philosophical knowledge of the invisible, is untenable with respect to the science of materials of the period, chemistry. Instead we argue, eighteenth-century chemists as well as nineteenth-century chemists approached chemical substances—the most promising candidates for Foucault's theory—as multidimensional objects and switched between their perceptible and imperceptible dimensions. Analysis, or the search for simple elements of objects, existed in all domains of eighteenth-century chemistry, albeit with differences of meaning, objectives, methods and techniques.

Materiality

Another central argument of our book is that historical analyses of practices of identifying and classifying objects are a most promising method to study ontological structures and ontological shifts of the past. When we describe the ontology of materials assumed by eighteenth-century chemists from a wider perspective, a distinctive pattern becomes visible: there was a comparatively small group of substances that chem-

---

12   As we showed in part III, the new concept of "organic substances" was even unintentionally reinforced by Lavoisier's theory of the elemental composition of plant and animal substances.

ists identified and classified according to their chemical composition, and a much larger group they identified and classified according to their natural origin, mode of extraction, perceptible properties, and practical uses.

At the beginning of the eighteenth century there were no more than four groups of substances classified according to chemical composition—salts, metal alloys, compounds of sulfur, and watery solutions—to which in the course of the century a few groups were added such as the gases (as compounds of calorique) and new compounds of acidifiable bases (such as carbon and phosphorus). By contrast, there was a plethora of plant and animal materials—ranging from simple extracted substances to fermented and chemically altered materials and all the way to composite materials applied as chemical remedies. These were identified and classified according to natural origin, mode of extraction or preparation, perceptible properties, and practical uses, even in contexts of conceptual inquiry. Resins, gums, balsams, waxes, fatty oils, camphors, sugar, and honey imported from overseas; distilled waters, volatile oils, mucilages, extracts, and essential salts extracted from plants in the chemical laboratory; not to mention blood, milk, urine, bile, and fat obtained from animals, did not fit the chemical order established in the domain of traceable pure chemical substances and mapped by chemical affinity tables and the table of chemical nomenclature of 1787. Laws of affinity, theories of chemical composition, and causal philosophical explanation of chemical composition and affinity, Newtonian or otherwise, were introduced and firmly established only in the domain ordered by chemical tables, even though in the course of the eighteenth-century chemists aimed at more unified chemical practices and theories.

Why was this so? Why were the eighteenth-century substances split into two groups, one of which contained substances identified and classified according to chemical composition, while the other contained substances identified and classified according to provenance and perceptible properties? The apparently natural answer to this question, spurred by an old cliché of pre-Lavoisierian chemistry, would be that before the Chemical Revolution, plant chemistry was not yet a science but merely an art and craft. Yet our analyses demonstrate that this image is mistaken. In addition to their persistent naturalistic and artisanal identification and classification of plant substances on both the lowest and highest taxonomic ranks, chemists also classified plant substances from philosophical and theoretical perspectives. In so doing, they grouped together the simple principles of plants in the early eighteenth century, the proximate principles of plants from the middle of the eighteenth century onward, and organic substances by 1790. These three ways of classification relied on theories of the composition of plants as well as their chemical analysis, although the analysis of plants did not satisfy chemists' established methodology of analysis that demanded the resynthesis of the analyzed compound. Far from intellectually blind cookery and individual trial and error, eighteenth-century plant chemistry was a collective enterprise that implemented chemical analysis, pursued collective objectives oriented toward broadly shared theories, and established collective ways of classification linked with analysis and theory. This still leaves us with the question of how to account for the different pattern of chemical order established around the group of substances mapped by the eighteenth-century chemical tables, on the one hand, and the plant and animal substances, on the other. Specificities of the intellectual agenda and social forms of interactions in plant chemistry cannot convincingly explain this divide. We argue that its explanation must take into account the role played by the materiality of objects.

By "materiality" we mean not merely the sensory stimuli provided by substances, but, in particular, their potential for transformation in series of chemical experiments. Metals, alkalis, earths, acids, salts, alloys, sulfur and compounds of sulfur, and watery solutions—the substances included in E. F. Geoffroy's table of affinities of 1718 as well as in the 1787 table of chemical nomenclature—could be subjected to cycles of decomposition and recomposition. In replacement reactions they displayed recurrent patterns of electivity between reaction partners. Moreover, these substances combined one by one, and became separated from compounds in such a way that only two products of analysis occurred. That is, with respect to this group of substances, traceability and simplicity existed on the material level itself. Our notion of "purity" here refers not to chemists' concept of purity, which changed considerably from the seventeenth to the early nineteenth century, but to traceability and the absence of "background noise" in chemical experiments. We designate as "pure" those distinctive kinds of substances that could be traced in chemical transformations according to a building block model, supplemented by the concept of affinity. Metals, alkalis, earths, and acids disappeared when chemists transformed them into alloys or salts; but they reappeared in subsequent experiments interpreted as the analysis of salts or alloys.

By contrast, plant and animal materials did not display these features. Reversible decompositions and recompositions and replacement reactions were not found in chemical operations performed with chemically extracted plant and animal substances. For the entire eighteenth century the resynthesis of plant and animal substances from their analytical products seemed to be out of the question. Similarly impenetrable to chemists was the question of whether and how chemical affinities governed the chemical changes of plant and animal substances. With the exception of vegetable and animal acids and alkalis, patterns of elective reactivity were hard to observe in chemical transformations of plant and animal substances. Instead of two products, plant and animal substances often yielded cascades of reaction products that could not be systematized along lines developed in the area of substances which we have designated "chemistry of pure substances."

Moreover, the eighteenth century chemists' attempts to further analyze plant substances encountered technical obstacles from the start. These substances yielded not only considerably more than two analytical products. Most of these products were also liquid or volatile, and hence technically difficult to separate from each other—far more difficult than the spontaneous separations by precipitation and crystallization that were ubiquitous in the chemistry of pure substances. As dry distillation allowed the analytical products to be collected through a slow increase in temperature and repeated change of the distillate receiver, it seemed to be the most appropriate method for performing analyses of plant substances. Yet the identification and clear demarcation of the various distillates posed many problems, since these materials had very similar properties. In hindsight all of these obstacles can be easily explained. Natural organic compounds consist of the same kinds of chemical elements—mostly carbon, hydrogen, oxygen, and nitrogen; thus they yielded similar products of decomposition. When Lavoisier introduced the new analytical technique by combustion, a reliable way to elucidate the composition of organic compounds seemed to be found. However, the Lavoisierian approach entailed the new problem of how to determine the quantitative differences in the proportions of the elements. Again, in hindsight we can explain easily why these obstacles did not arise in the chemistry of pure substances. The vast majority of substances belonging to that latter domain were inorganic compounds, which display strong differences in their elemental composition as well as their saturation points. With respect to the latter quantitative aspect, it was not coincidental that stoichiometry developed in the chemistry of inorganic substances. The agency of substances—their actions and reactions—could be triggered by chemists' instruments, manipulations, and objectives, but not conditioned or directed at will. Successful chemical transformations in some cases and resistance in others, observation of electivity and recurrent patterns of transformation or chaos, the simplicity of combining and separating only two substances or the confusing complexity of handling multiple analytical products—all of this contributed to the divide in eighteenth-century chemists' modes of classification and ontology.

Considering the enormous impact the group of pure substances had on the emergence of a systematized conceptual chemical investigation in the early eighteenth century, the following aspect must be highlighted. The bulk of the pure substances that lent themselves to the systematization of early modern chemistry and the conceptual network of compound, composition and affinity were not natural raw materials, but materials processed in the arts and crafts and areas of alchemy tuned to the arts and crafts. These substances were in the epistemic horizon of and accessible to chemists, since they existed in and were available from artisanal practices. The developments in the arts and crafts, especially pharmacy and metallurgy, prior to roughly 1700 played an important role for the range of materials studied by chemists, and hence the range and kind of experiments they were able to make.

The persistence of multidimensional objects

Not all materials lent themselves equally to systematic experimentation and conceptual inquiry, as did the substances ordered in eighteenth-century chemical tables. It was this highly selective group of pure or traceable substances, mostly inorganic, that stood at the center of the Chemical Revolution in the last third of the eighteenth century. The changes in chemistry spurred by Lavoisier and his collaborators have long been a favorite occupation of historians' of science, as they have been regarded as a revolutionary overthrow of pre-Lavoisierian chemistry. Taking Thomas Kuhn's proposition seriously that scientific revolutions go hand in hand with deep revisions of the taxonomic structure of a given scientific domain, we have scrutinized the classificatory order in a key document of the Chemical Revolution, the taxonomic table contained in the *Méthode* of 1787, and compared it with chemists' earlier modes of classification. The result of this scrutiny was unambiguous: the Chemical Revolution did not engender an ontological rupture, and chemists did not live in a different world of objects of inquiry before and after the Chemical Revolution.

Lavoisier and his collaborators created a comprehensive and systematic taxonomy of pure chemical substances, based on chemical analysis and knowledge of elemental composition. In so doing, they clearly distinguished between chemical compounds and simple indecomposable substances, and they further extended the web of pure substances to new groups. Yet they neither established classification according to chemical composition and the grouping together of pure chemical compounds—in 1787 that practice had existed for almost a century—nor altered the main cluster of classes that existed in pre-Lavoisierian chemistry. The "new chemists" built on existing classificatory distinctions and their predecessors' ways of ordering. Several of their classes that seem to be novelties at first glance are in fact mere anti-phlogistic mirror images of classes already constructed in the phlogistic system. Moreover, the Chemical Revolution did not change the existing mode of individuation and identification of substances. Non-stoichiometric compounds, such as alloys and tinctures, still belonged to its range of scientific objects, and the bulk of plant and animal substances was excluded from their reform of taxonomy and nomenclature. It was only some decades after the Chemical Revolution, and not in inorganic but in organic chemistry, that chemists rigorously applied quantitative analysis and stoichiometric laws to all chemical substances, demarcated pure stoichiometric compounds from non-stoichiometric compounds and mixtures, and dismissed classification according to natural origin. This deep ontological shift, which accelerated in the 1830s, consisted not merely in a new meaning of organic substances, and of chemical compounds more broadly, but also entailed changes in the experimental production and individuation of substances along with the creation of novel types of carbon compounds and a reconfiguration of the range of included and excluded objects of inquiry.

How shall we conceptualize the transitions in chemistry around 1830? In our brief discussion of Foucault's assertion that studies of the visible surface of things were replaced by studies of their internal structure in the decades around 1800, we have argued that the eighteenth-century chemists studied both the perceptible dimension of substances, including their usefulness, and their imperceptible dimension. Did this change when nineteenth-century chemists redefined their area of scientific objects to comprise only stoichiometric compounds, embedded in systems of experiments, precision measurement, and work on paper with chemical formulae? Were the stoichiometric compounds unambiguous "scientific" objects, whereas the multidimensional eighteenth-century materials were not yet proper scientific objects? The French philosopher Gaston Bachelard argued that the objects constituted by a "scientific mind" are separated from ordinary objects through an "epistemic rupture," and that this separation is engendered by accuracy, precision instruments, mathematical representation, and permanent rigorous criticism that replaces ordinary "primary experience."[13] Drawing upon Bachelard, Lorraine Daston recently demarcated "quotidian objects"—"the solid, obvious, sharply outlined, in-the-way things of quotidian experience"—from "scientific objects." "In contrast to quotidian objects," she observed, "scientific objects are elusive and hard-won." Whereas scientific objects are subtle, evanescent, and hidden entities that require ingenuity and instruments to detect them, quotidian objects "possess the self-evidence of a slap in the face."[14]

Does this distinction between scientific and quotidian objects help to clarify what was going on in the ontological transition in chemistry around 1830? Stoichiometric compounds were indeed hard-won objects that required experimental intervention, precision measurement, and chemical formulae, which embodied a quantitative chemical theory that overlapped with chemical atomism. Their modes of individuation, identification and classification, based on systems of experiments and knowledge about their invisible elemental composition and constitution, differed profoundly from chemists' earlier practices, especially with respect to eighteenth-century plant and animal materials. Yet the new stoichiometric carbon compounds were also manifest in-the-way things that slapped the experimenter in the face when overheated, spoiled his laboratory with stinking poisonous vapors, and damaged his retorts in explosions.

Attempts to demarcate the nineteenth-century stoichiometric carbon compounds from the eighteenth-century chemical substances along the axes of the visible (or manifest) and invisible (hidden) do not work. In classical chemical analysis the most elusive—molecular chemical structure—was entwined with the most quotidian—dirt and smell. As in eighteenth-century chemistry, nineteenth-century chemical investigation of the invisible composition and structure of stoichiometric compounds required the identification and study of their isolated perceptible substance components. Their study of the perceptible surface of substances was not external to chemical analysis; it was not an activity that chemists did in addition to chemical analysis for pursuing goals and interests different from chemical analysis. Rather it was an intrinsic part of the method of classical chemical analysis to let the substance to be analyzed interact with other substances and to examine the material products of that reaction. Well into the twentieth century, when new types of physical apparatus such as IR and NMR spectrometers were introduced for chemical analysis, chemists made their hands dirty when they analyzed substances, and in preparative chemistry this is still the case today. Moreover, chemists never totally abandoned the ancient method of identifying substances through smell, taste, color, shape of crystals and similar quotidian properties. This was true even with the advent of the new techniques of quantitative analysis, the precision measurement of physical properties, and the experimental study of structure. The stabilization of quantitative analysis and the use

13 Bachelard [1996].
14 Daston [2000] p. 2.

of chemical formulae by nineteenth-century chemists privileged knowledge of composition and structure over sensible properties, but also extended the chemists' tools box for identifying and classifying chemical substances. Even in today's chemistry, the characteristic smell of a chemical substance is often a significant step in the process of its identification.

The chemists' focus on pure stoichiometric compounds that arose in the nineteenth century was a deep ontological shift that reconfigured the range of substances accepted as objects of inquiry in organic and inorganic chemistry. Yet chemists' dismissal of ancient materials that did not fit into their new ontology of stoichiometric compounds, and their introduction of novel types of carbon compounds unknown in artisanal traditions and the everyday world, did not entail the abolition of inquiries into the multiple dimensions of chemical substances. It neither drove a wedge between the perceptible and imperceptible dimension of substances nor between their scientific and technological side. What distinguished the stoichiometric carbon compounds from their counterparts in the eighteenth-century, the plant and animal materials, was their embeddedness in a new extensive network of experiments, precision measurement, and work on paper with chemical formulae. These types of network neither existed in natural historical cultures nor in the traditional arts and crafts that implemented chemical knowledge and techniques. The enormous material productivity of the new culture of carbon chemistry led to an exponential growth of pure stoichiometric compounds in the nineteenth century, and many of the new carbon compounds never left the academic laboratory. Others did leave it to be applied in industry and society. Chloral, chloroform, acetyl chloride, benzene, aniline, and toluidine are only a few examples of the newly synthesized carbon compounds that were utterly unfamiliar in the arts and crafts and everyday life of the 1830s but became applied a few decades later in pharmacy and the emerging synthetic dye industry. Like their predecessors in the eighteenth century and the early modern period, the stoichiometric chemical substances of the nineteenth century were applicable or potentially applicable materials in industry and society. In this respect they were multidimensional objects too.

References

Abbri, Ferdinando and Marco Beretta. 1995. "Bibliography of the 'Méthode de nomenclature chimique' and of the 'Traité élémentaire de chimie' and their European Translations (1787–1800)." In *Lavoisier in European Context*, edited by Bernadette Bensaude-Vincent and Ferdinando Abbri, 279–291. Canton, MA: Watson Publishing International.

Agricola, Georg. 1556. *De re metallica libri XII*. Basel: Froben & Bischoff.

———. 1557. *Vom Bergkwerck 12 Bücher*. Basel: Froben & Bischoff.

———. 1950. *Georgius Agricola: De re metallica*. Translated by H.C. and L.H. Hoover. New York: Dover.

Albury, W.R. 1972. *The Logic of Condillac and the Structure of French Chemical and Biological Theory, 1780–1801*. Ph.D. thesis: Johns Hopkins University.

Anderson, Wilda C. 1984. *Between the Library and the Laboratory: The Language of Chemistry in Eighteenth-Century France*. Baltimore: Johns Hopkins University Press.

Applebaum, Wilbur, ed. 2000. *Encyclopedia of the Scientific Revolution: From Copernicus to Newton*. New York and London: Garland Publishing.

Bachelard, Gaston. 1996. *La formation de l' esprit scientifique: contribution à une psychanalyse de la connaissance* (1938). Paris: Vrin.

Bacon, Francis. 1986–1994. *The Works of Francis Bacon*. 14 vols. Stuttgart-Bad Canstatt: Frommann-Holzboog. Reprint of the edition by Spedding, Ellis, and Heath (1857–1874).

Baird, Davis. 1993. "Analytical Chemistry and the 'Big' Scientific Instrumentation Revolution." *Annals of Science* 50: 267–290.

———. 2004. *Thing Knowledge: A Philosophy of Scientific Instruments*. Berkeley: University of California Press.

Barlet, Annibal. 1653. *Le vray et methodique cours de la physique resolutive, vulgairement dite chymie*. Paris: N. Charles.

Barnes, Barry. 1983. "Social Life as Bootstrapped Induction." *Sociology* 17(4): 524–545.

———. 2005. "Elusive Memories of Technoscience." *Perspectives on Science* 13(2): 142–165.

Barnes, Barry, David Bloor, and John Henryl. 1996. *Scientific Knowledge: A Sociological Analysis*. Chicago: University of Chicago Press.

Baumé, Antoine. 1762. *Élémens de pharmacie theorique et pratique*. Paris: Damonneville, Didot, De Hansy.

———. 1763. *Manuel de chymie, ou exposé des opérations et des produits d'un cours de chymie*. Paris: Didot, Musier, De Hansy, Panckoucke.

———. 1773. *Chymie expérimentale et raisonnée*. 3 vols. Paris: Didot.

Beguin, Jean. 1624. *Les elemens de chymie*. Geneva: Jean Celerier.

Bennett, Jim A. 1986. "The Mechanics' Philosophy." *History of Science* 4/1: 1–28.

Bensaude-Vincent, Bernadette. 1983. *A propos de 'Méthode de nomenclature chimique.' Esquisse historique suivie du texte de 1787*. Edited by Jacques Roger, Cahiers d'Histoire et de Philosophie. Paris: Centre National de la Recherche Scientifique.

———. 1994. "Une charte fondatrice, Introduction." In *Guyton de Morveau, Lavoisier, Berthollet, Fourcroy: Méthode de nomenclature chimique*, 9–60. Paris: Èditions du Seuil.

———. 1995. "Introductory Essay: A Geographical History of Eighteenth-Century Chemistry." In *Lavoisier in European Context: Negotiating a New Language for Chemistry*, edited by Bernadette Bensaude-Vincent and F. Abbri, 1–17. Canton, MA: Watson Publishing International.

————. 1998. *Éloge du mixte: matériaux nouveaux et philosophie ancienne.* Paris: Hachette Littératures.

————. 2001a. "Graphic Representations of the Periodic Systems of Chemical Elements." In *Tools and Modes of Representation in the Laboratory Sciences*, edited by Ursula Klein, 133–162. Dordrecht: Kluwer.

————. 2001b. "The Construction of a Discipline: Materials Science in the United States." *Historical Studies in the Physical and Biological Sciences* 31(2): 223–248.

Bensaude-Vincent, Bernadette and Ferdinando Abbri. 1995. *Lavoisier in European Context: Negotiating a New Language for Chemistry.* Canton, MA: Watson Publishing International.

Bensaude-Vincent, Bernadette and Agustí Nieto-Galan. 1999. "Theories of Dyeing: A View on a Longstanding Controversy through the Works of Jean François Persoz." In *Natural Dyestuffs and Industrial Culture in Europe, 1750–1880*, edited by Robert Fox and Augstí Nieto-Galan, 3–24. Canton, MA: Science History Publications.

Bensaude-Vincent, Bernadette, Antonio García-Belmar, and José Ramón Bertomeu-Sánchez, eds. 2003. *L'émergence d'une science des manuels: Les livres de chimie en France (1789–1852).* Paris: Editions des archives contemporaines.

Beretta, Marco. 1993. *The Enlightenment of Matter: The Definition of Chemistry from Agricola to Lavoisier.* Canton, MA: Watson Publishing International.

————. 1997. "Humanism and Chemistry: The Spread of Georgius Agricola's Metallurgical Writings." *Nuncius* 12(1): 17–48.

————. 2000. "The Grammar of Matter. Chemical Nomenclature during the 18th Century." In *Sciences et langues en Europe*, edited by Roger Chartier and Pietro Corsi, 109–125. Luxembourg: Office for Official Publ. of the European Communities.

Bergman, Torbern. 1782. *Sciagraphia regni mineralis.* Leipzig and Dessau: Bibliopolio Eruditorum.

————. 1783. *An Essay on the Usefulness of Chemistry, and its Application to the Various Occasions of Life.* London: Murray.

————. 1784. "On the Investigation of Truth [= Introitus de indagando vero 1779]." In *Torbern Bergman: Physical and Chemical Essays*, edited by Edmund Cullen, xxi–xliv. London: Murray.

————. 1784–1791. *Physical and Chemical Essays.* Translated by Edmund Cullen. 3 vols. London: Murray.

————. 1785. *A Dissertation on Elective Attractions [= De attractionibus electivis disquisitio 1775].* Translated by T. Beddoes, London: J. Murray.

————. 1791. "Thoughts on a Natural System of Fossils [= Meditationes de systemate fossilium naturali 1784]." In *Torbern Bergman: Physical and Chemical Essays*, xxi–xliv. Edinburgh: Mudie.

Berthelot, Pierre Eugène Marcellin. 1872. *Traité élémentaire de chimie organique.* Paris: Dunod.

Bertomeu-Sánchez, José Ramón, Antonio García-Belmar, and Bernadette Bensaude-Vincent. 2002. "Looking for an Order of Things: Textbooks and Chemical Classifications in Nineteenth-Century France." *Ambix* 49(3): 227–250.

Berzelius, Jöns Jacob. 1812. "Versuch einer lateinischen Nomenclatur für die Chemie, nach electrisch-chemischen Ansichten." *Annalen der Physik* 42: 37–89.

————. 1813. *A View of the Progress and Present State of Animal Chemistry.* London: Hatchard, Johnson, and Boosey.

————. 1814. "Experiments to Determine the Definite Proportions in Which the Elements of Organic Nature are Combined." *Annals of Philosophy* 4: 323–331, 401–409.

————. 1815. "Experiments to Determine the Definite Proportions in Which the Elements of Organic Nature Are Combined." *Annals of Philosophy* 5: 93–101, 174–184, 260–275.

————. 1819. *Essai sur la théorie des proportions chimiques et sur l'influence chimique de l'électricité*. Paris: Méquignon-Marvis.

————. 1833–1841. *Lehrbuch der Chemie*. Translated by F. Wöhler. 3 ed. 10 vols. Dresden: Arnold.

————. 1843–1848. *Lehrbuch der Chemie*. 5ʹed., 5 vols. Dresden: Arnold.

————. 1973. *Versuch über die Theorie der chemischen Proportionen und über die chemischen Wirkungen der Electricität; nebst Tabellen über die Atomgewichte der meisten unorganischen Stoffe und deren Zusammensetzungen*. Hildesheim: Gerstenberg. Reprint of the edition Dresden: Arnold, 1820.

Boerhaave, Herman. 1732. *Elementa Chemiae, quae anniversario labore docuit, in publicis, privatisque, scholis*. 2 vols. Leyden: Severinus.

————. 1753. *A New Method of Chemistry; Including the History, Theory, and Practice of the Art: Translated from the Original Latin of Dr. Boerhaave's Elementa Chemiae, as Published by Himself. To which are Added Notes; and an Appendix, Shewing the Necessity and Utility of Enlarging the Bounds of Chemistry*. Translated by Peter Shaw. 3rd. ed. London: T. Longman.

————. 1782. *Anfangsgründe der Chymie praktischer Theil, aus dem Lateinischen übersetzt von Johann Christian Wiegleb*. Berlin and Stettin: Friedrich Nicolai.

Bonnet, Charles. 1762. *Considérations sur les corps organisés: où l'on traite de leur origine, de leur développement, de leur réproduction, etc. et où l'on a rassemblé et abgrégé tout ce que l'histoire naturelle offre de plus certain et de plus intéressant sur ce sujet*. 2 vols. Amsterdam: Marc-Michel Rey.

Boullay, Pierre François Guillaume. 1807. "Observations sur l'éther sulfurique et sa préparation." *Annales de Chimie* 62: 242–247.

Bowker, Geoffrey C. and Susan Leigh Star. 1999. *Sorting Things Out: Classification and its Consequences*. Cambridge, MA: The MIT Press.

Boyer, Joanne-Baptista. 1758. *Codex medicamentarius, seu: Pharmacopoea Parisiensis*. 5 ed. Paris: Petrum-Guillelmum Cavelier.

Boyle, Robert. 1999. *The Works of Robert Boyle*. Edited by Michael Hunter and Edward B. Davis. 14 vols. London: Pickering & Chatto. Reprint of the London edition 1772.

Brakel, Jaap van. 2000. *Philosophy of Chemistry. Between the Manifest and the Scientific Image*, *Louvain Philosophical Studies* 15. Leuven: Leuven University Press.

————. 2005. "On the inventors of XYZ." *Foundations of Chemistry* 7: 57–84.

Bret, Patrice. 1994. "La vie des sciences. Lavoisier à la Régie des Poudres: le savant, le financier, l'administrateur et le pédagogue." *Comptes rendus de l'Académie des Sciences. Série générale* 11: 297–317.

Brock, William H. 1993. *The Norton History of Chemistry*. New York: W. W. Norton & Company.

————. 1997. *Justus von Liebig: The Chemical Gatekeeper*. Cambridge: Cambridge University Press.

Brooke, John Hedley. 1973. "Chlorine Substitution and the Future of Organic Chemistry: Methodological Issues in the Laurent-Berzelius Correspondence (1843–1844)." *Studies in History and Philosophy of Science* 4(1): 47–94.

———. 1987. "Methods and Methodology in the Development of Organic Chemistry." *Ambix* 34(3). 147–155.

Buchwald, Jed Z. 1992. "Kinds and the Wave Theory of Light." *Studies in History and Philosophy of Science* 23(1): 39–74.

Buchwald, Jed Z. and George E. Smith. 1997. "Thomas S. Kuhn, 1922–1996." *Philosophy of Science* 64: 361–376.

Bucquet, Jean Baptiste Michel. 1771. *Introduction à l'étude des corps naturels, tirés du règne minéral*. Paris: Jean-Thomas Herissant.

———. 1773. *Introduction à l'étude des corps naturels, tirés du règne végétal*. 2 vols. Paris: Jean-Thomas Herissant.

Butlerov, Aleksandr M. 1868. *Lehrbuch der Organischen Chemie; zur Einführung in das specielle Studium derselben*. German ed., rev. and enl., transl. from the Russian. Leipzig: Quandt & Händel.

Cartheuser, Friedrich. 1753. *Elementa Chymiae dogmatico-experimentalis*. Frankfurt ad Viadrum: Ioann. Christian Kleyb.

———. 1754. *Dissertatio Chymico-Physica de genericis quibusdam plantarum principiis hactenus plerumque neglectis*. Frankfurt ad Viadrum: Ioann. Christian Kleyb.

Cassebaum, Heinz and George B. Kauffman. 1976. "The Analytical Concept of a Chemical Element in the Work of Bergman and Scheele." *Annals of Science* 33: 447–456.

Chalmers, Alan. 1993. "The Lack of Excellency of Boyle's Mechanical Philosophy." *Studies in History and Philosophy of Science* 24: 541–64.

Chevreul, Michel Eugène. 1823. *Recherches chimiques sur les corps gras d'origine animale*. Paris: Levrault.

———. 1824. *Considérations générales sur l'analyse organique et sur ses applications*. Paris: Levrault.

Christie, Maureen. 1994. "Philosophers versus Chemists Concerning 'Laws of Nature.'" *Studies in History and Philosophy of Science* 25(4): 613–629.

Clave, Etienne de. 1641. *Nouvelle lumière philosophique des vrais principes et élemens de la nature, et qualité d'iceus, contre l'opinion commune*. Paris: Olivier de Varennes.

———. 1646. *Le cours de chimie*. Paris: Olivier de Varennes.

Clericuzio, Antonio. 2000. *Elements, Principles and Corpuscles. A Study of Atomism and Chemistry in the Seventeenth Century*. Dordrecht: Kluwer.

Clow, Archibald and Nan Clow. 1952. *The Chemical Revolution: A Contribution to Social Technology*. London: Batchworth Press.

Cohen, Benjamin R. 2004. "The Element of the Table: Visual Discourse and the Preperiodic Representation of Chemical Classification." *Configurations* 12(1): 41–75.

Cole, Arthur H. and George B. Watts. 1952. *The Handicrafts of France as Recorded in the Descriptions des Arts et Métiers 1761–1788*. Boston: Kress Library of Economics.

Condillac, Étienne Bonnot. 1948. *Oeuvres philosophiques II*. Paris: Presses Universitaires de France.

Cowen, David L. 2001. *Pharmacopoeias and Related Literature in Britain and America, 1618–1847*. Aldershot: Ashgate.

Crosland, Maurice P. 1962. *Historical Studies in the Language of Chemistry*. London: Heinemann.

———. 1967. *The Society of Arcueil: A View of French Science at the Time of Napoleon I*. Cambridge, MA: Harvard University Press.

————. 2003. "Research Schools of Chemistry from Lavoisier to Wurtz." *British Journal for the History of Science* 36(3): 333–361.

————. 2005. "Early Laboratories c. 1600–c. 1800 and the Location of Experimental Science." *Annals of Science* 62(2): 233–253.

Cunningham, Andrew. 2002. "The Pen and the Sword: Recovering the Disciplinary Identity of Physiology and Anatomy before 1800. I: Old Physiology—the Pen." *Studies in History and Philosophy of Biological and Biomedical Sciences* 33: 631–665.

Dagognet, François. 1969. *Tableaux et langages de la chimie*. Paris: Èditions du Seuil.

Darmstaedter, Ernst. 1926. *Berg-, Probir- und Kunstbüchlein*. Munich: Verlag der Münchner Drucke.

————. 1998. *Georg Agricola 1494–1555: Leben und Werk*. Reprint of the original edition Munich 1926. Munich: Verlag der Münchner Drucke.

Daston, Lorraine. 1991. "The Ideal and Reality of the Republic of Letters in the Enlightenment." *Science in Context* 4(2): 367–386.

————. 2000. "Introduction: The Coming into Being of Scientific Objects." In *Biographies of Scientific Objects*, edited by Lorraine Daston, 1–14. Chicago: University of Chicago Press.

————. 2004a. "Introduction: Speechless." In *Things that Talk: Object Lessons from Art and Science*, edited by Lorraine Daston, 9–24. New York: Zone Books.

————. 2004b. "Type Specimens and Scientific Memory." *Critical Inquiry* 31: 153–181.

Debus, Allen G. 1966. *The English Paracelsians*. New York: Franklin Watts.

————. 1967. "Fire Analysis and the Elements in the Sixteenth and Seventeenth Centuries." *Annals of Science* 23: 127–147.

————. 1977. *The Chemical Philosophy: Paracelsian Science and Medicine in the Sixteenth and Seventeenth Centuries*. 2 vols. New York: Science History Publications.

————. 1991. *The French Paracelsians: The Chemical Challenge to Medical and Scientific Tradition in Early Modern France*. Cambridge: Cambridge University Press.

Demachy, Jacques François. 1766. *Instituts de chymie, ou principes élémentaires de cette science, présentés sous un nouveau jour*. Paris: Lottin Le Jeune.

————. 1774. *Recueil de dissertations physico-chymiques*. Paris: Monory.

Diderot, Denis J. and Jean LeRond d'Alembert. 1966. *Encyclopédie ou dictionnaire raisonné des sciences, des arts, et des métiers*. 35 vols. Stuttgart: Frommann. Reprint of the first edition Paris 1751–1780.

Di Meo, Antonio. 1984. "Théories et classifications chimiques au XVIII. siècle." *History and Philosophy of the Life Sciences* 5(1): 159–185.

Dodart, Denis. 1731. "Memoires pour servir à l'histoire des plantes." *Mémoires de l'Académie Royale des Sciences, 1666–1699* 4: 121–242.

Donovan, Arthur. 1996. *Antoine Lavoisier—Science, Administration, and Revolution*. Cambridge: Cambridge University Press.

Douglas, Mary and David Hull, eds. 1992. *How Classification Works: Nelson Goodman among the Social Sciences*. Edinburgh: Edinburgh University Press.

Duhamel du Monceau, Henri-Louis. 1736. "Sur la base du sel marin." *Histoire de l' Académie Royale des Sciences: avec les mémoires de mathématique et de physique pour la même année*: 215–32.

Dumas, Jean-Baptiste André. 1828–1846. *Traité de chimie appliqué aux arts*. 8 vols. Paris: Béchet jeune.

Dumas, Jean-Baptiste André and Polydore Boullay. 1827. "Mémoire sur la formation de l'ether sulfurique." *Annales de Chimie et de Physique* 36: 294–310.

————. 1828. "Mémoire sur les ethers composés." *Annales de Chimie et de Physique* 37: 15–53.

Duncan, Alistair M. 1970. "The Function of Affinity Tables and Lavoisier's List of Elements." *Ambix* 17(1): 28–42.

————. 1996. *Laws and Order in Eighteenth-Century Chemistry*. Oxford: Clarendon Press.

Dupré, John. 1993. *The Disorder of Things: Metaphysical Foundations of the Disunity of Science*. Cambridge, MA: Harvard University Press.

————. 2001. "In Defence of Classification." *Studies in History and Philosophy of Biological and Biomedical Sciences* 32(2), Special issue: History and Philosophy of Taxonomy, edited by Nicolas Jardine: 203–213.

Eamon, William. 1994. *Science and the Secrets of Nature: Books of Secrets in Medieval and Early Modern Culture*. Princeton: Princeton University Press.

Earles, M. P. 1985. *The London Pharmacopoeia Perfected, The Durham Thomas Harriot Seminar—Occasional Papers No. 3*. London: The Chameleon Press.

Eklund, John. 1975. *The Incompleat Chymist: Being an Essay on the Eighteenth-Century Chemist in his Laboratory, with a Dictionary of Obsolet Chemical Terms of the Period*. Washington: Smithonian Institution Press.

Ercker, Lazarus. 1960. *Beschreibung der allervornehmsten mineralischen Erze und Bergwerksarten (Großes Probierbuch 1574). Hg. u. in heutiges Deutsch übertr. v. Paul Reinhard Beierlein*. Berlin: Akademie-Verlag.

————. 1951. *Lazarus Ercker's Treatise on Ores and Assaying*. Translated from the German Edition of 1580 by Annelise Grünhaldt Sisco and Cyril Stanley Smith. Chicago: The University of Chicago Press.

Eriksson, Gunnar. 1983. "Linnaeus the Botanist." In *Linnaeus: The Man and his Work*, edited by Tore Frängsmyr, 63–109. Berkeley: University of California Press.

Erxleben, Johann Christian Polykarp. 1773. *Anfangsgründe der Naturgeschichte*. Göttingen: Dietrich.

————. 1784. *Anfangsgründe der Chemie*. Göttingen: Johann Christian Dietrich.

Exner, Alfred. 1938. *Der Hofapotheker Caspar Neumann (1683–1737). Ein Beitrag zur Geschichte des ersten pharmazeutischen Lehrers am Collegium-medico-chirurgicum in Berlin*. Berlin: Triltsch & Huther.

Fachs, Modestin. 1678. *Probier Büchlein (1595)*. Leipzig: Henning Grossen.

Fester, Gustav. 1923. *Die Entwicklung der chemischen Technik bis zu den Anfängen der Grossindustrie: ein technologisch-historischer Versuch*. Berlin: Springer.

Fisher, N. W. 1973a. "Organic Classification before Kekulé." *Ambix* 20(1): 106–131.

————. 1973b. "Organic Classification before Kekulé." *Ambix* 20(3): 209–233.

————. 1974. "Kekulé and Organic Classification." *Ambix* 21(1): 29–52.

Fors, Hjalmar. 2003. *Mutual Favours: The Social and Scientific Practice of Eighteenth-Century Swedish Chemistry*. Uppsala: Universitetstryckeriet.

Foucault, Michel. 1966. *Les mots et les choses*. Paris: Gallimard.

————. 1970. *The Order of Things: An Archaeology of the Human Sciences*. London: Routledge.

Fourcroy, Antoine-François de. 1782. *Leçons élémentaires d'histoire naturelle et de chimie; dans lesquelles on s'est proposés, 1e. de donner un ensemble méthodique des connaissances chimiques acquises jusqu'à ce jour; 2e. d'offrir un tableau comparé de la doctrine de Stahl et de celle de quelques modernes: pour servir de résumé à un cours complet sur ces deux sciences*. Paris: Société Royale de Medicine.

————. 1792. *Philosophie chimique, ou vérités fondamentales de la chimie moderne, disposées dans un nouvel ordre.* Paris: Simon.

————. 1801–1802. *Systême des connaissances chimiques, et de leurs applications aux phénomènes de la nature et de l'art.* 11 vols. Paris: Baudouin.

Fourcroy, Antoine-François de and Nicolas Louis Vauquelin. 1797. "De l'action de l'acide sulfurique sur l'alcool, et de la formation de l'ether." *Annales de Chimie* 23: 203–215.

Frängsmyr, Tore. 1974. "Swedish Science in the Eighteenth Century." *History of Science* 12: 29–42.

Frängsmyr, Tore, John Lewis Heilbron, et al., eds. 1990. *The Quantifying Spirit in the Eighteenth Century.* Berkeley: University of California Press.

Fujii, Kiyohisa. 1986. "The Berthollet-Proust Controversy and Dalton's Chemical Atomic Theory." *British Journal for the History of Science* 19: 177–200.

Gay-Lussac, Joseph Louis. 1815. "Lettre de M. Gay-Lussac à M. Clément, sur l'analyse de l'alcool et de l'éther sulfurique, et sur les produits de la fermentation." *Annales de Chimie* 95: 311–318.

Gay-Lussac, Joseph Louis and Louis Jacques Thenard. 1810. "Extrait d'un mémoire sur l'analyse végétale et animal." *Annales de Chimie* 74: 47–64.

————. 1811. *Recherches physico-chemiques.* 2 vols. Paris: Deterville.

Gee, Brian. 1989. "Amusement Chests and Portable Laboratories: Practical Alternatives to the Regular Laboratory." In *The Development of the Laboratory: Essays on the Place of Experiment in Industrial Civilization*, edited by Frank A. J. L. James, 37–59. New York: American Institute of Physics.

Gehler, Johann Samuel Traugott. 1787–1796. *Physikalisches Wörterbuch oder Versuch einer Erklärung der vornehmsten Begriffe und Kunstwörter der Naturlehre mit kurzen Nachrichten von der Geschichte der Erfindungen und Beschreibungen der Werkzeuge begleitet in alphabetischer Ordnung.* 4 vols. Leipzig: Schwickert.

Gellert, Christlieb Ehregott. 1776. *Anfangsgründe zur metallurgischen Chymie: in einem theoretischen und practischen Theile, nach einer in der Natur gegründeten Ordnung.* 2. ed. Leipzig: Caspar Fritsch.

Geoffroy, Etienne-François. 1718. "Table des differents rapports observés en Chimie entre differentes substances." *Histoire de l'Académie Royale des Sciences: avec les mémoires de mathématique et de physique pour la même année*: 202–212.

————. 1736. *A Treatise of the Fossil, Vegetable, and Animal Substances that are Made Use of in Physick. Containing the History and Description of them; with an Account of their Several Virtues and Preparations. To which is Refixed an Enquiry into the Constituent Principles of Mixed Bodies, and the Proper Methods of Discovering the Nature of Medicines.* London: Douglas.

————. 1996. "Table of the Different Relations Observed in Chemistry between Different Substances—27 August 1718." *Science in Context* 9(3), Special Issue: Fundamental Concepts of Early Modern Chemistry, edited by Wolfgang Lefèvre: 313–319.

Gerhardt, Charles Fréderic. 1844–1845. *Précis de chimie organique.* 2 vols. Paris: Fortin, Masson & Cie.

Gillispie, Charles Coulston. 1957. "The Discovery of the Leblanc Process." *Isis* 48: 152–170.

————, ed. 1970–1980. *Dictionnary of Scientific Biography.* New York: Scribner.

————. 1980. *Science and Polity in France at the End of the Old Regime.* Princeton: Princeton University Press.

———. 2004. *Science and Polity in France—The Revolutionary and Napoleonic Years.* Princeton: Princeton University Press.

Glaser, Christopher. 1676. *Traité de la chymie, enseignant par une briéve & facile methode toutes ses plus necessaires preparations.* Lyon: Pierre and Binoist Bailly.

———. 1677. *The Compleat Chymist, or, A New Treatise of Chymistry [1663].* London: John Starkey.

Glauber, Johann Rudolph. 1646–1649. *Furni novi philosophici oder Beschreibung einer Newerfundener Distillir-Kunst.* 5 parts. Amsterdam: Johann Fabel.

———. 1651. *A Description of New Philosophical Furnaces, or, a New Art of Distilling, Divided into Five Parts. Whereunto is Added a Description of the Tincture of Gold or the True Aurum Potabile; also the First Part of the Mineral Work, Set Forth and Published for the Sakes of Them that are Studious of the Truth.* London: T. Williams.

Golinski, Jan. 1988. "Utility and Audience in Eighteenth-Century Chemistry: Case Studies of William Cullen and Joseph Priestley." *British Journal for the History of Science* 21: 1–31.

———. 1992. "The Chemical Revolution and the Politics of Language." *Eighteenth Century: Theory and Interpretation* 33: 238–251.

———. 2000. "'Fit instruments': Thermometers in Eighteenth-Century Chemistry." In *Instruments and Experimentation in the History of Chemistry*, edited by Frederic L. Holmes and Trevor H. Levere, 185–210. Cambridge, MA: The MIT Press.

Goodman, Nelson. 1978. *Ways of Worldmaking.* Indianapolis: Hackett Publishing Company.

Gordin, Michael D. 2004. *A Well-Ordered Thing: Dimitrii Mendeleev and the Shadow of the Periodic Table.* New York: Basic Books.

Göttling, Johann Friedrich August. 1778. *Einleitung in die pharmaceutische Chymie für Lernende.* Altenburg: Richter.

Goupil, Michelle. 1992. "Claude-Louis Berthollet, collaborateur et continuateur (?) de Lavoisier." In *Lavoisier et la révolution chimique*, edited by Michelle Goupil, 35–53. Paris: SABIX.

Gren, Friedrich Albrecht Carl. 1787–1790. *Systematisches Handbuch der gesammten Chemie. Zum Gebrauch seiner Vorlesungen entworfen*, vol. 3. Halle: Waisenhaus Buchandlung.

Grew, Nehemiah. 1965. *The Anatomy of Plants. With an Idea of a Philosophical History of Plants and Several Other Lectures Read before the Royal Society.* New York and London: Johnson Reprint Corporation. Reprint of the edition London: Rawlins, 1682.

Guerlac, Henry. 1961. *Lavoisier—The Crucial Year.* Ithaca: Cornell University Press.

———. 1977. "Some French Antecedents of the Chemical Revolution." In *Essays and Papers in the History of Modern Science*, edited by Henry Guerlac, 340–374. Baltimore and London: The Johns Hopkins University Press.

———. 1981. "Antoine-Laurent Lavoisier—Chemist and Revolutionary." In *Dictionary of Scientific Biography*, edited by Charles Coulston Gillispie, 8: 66–91. New York: Scribner.

Gustin, Bernhard Henry. 1975. *The Emergence of the Chemical Profession 1790–1867.* Ph.D. thesis: The University of Chicago.

Guyton de Morveau, Louis-Bernard. 1777–1778. *Élémens de chymie, théorique et pratique, rédigés dans un nouvel ordre, d'après les découvertes modernes, pour servir aux cours publics de l'Académie de Dijon.* 3 vols. Dijon: Frantin.

———. 1782. "Mémoire sur les dénominations chimiques, la necessité d'en perfectionner le système et les règles pour y parvenir." *Observations sur la Physique* 19: 370–382.

Guyton de Morveau, Louis-Bernard, Antoine-Laurent Lavoisier, et al. 1787. *Méthode de nomenclature chimique. On y a joint un nouveau systême de caractères chimiques, adaptés à cette nomenclature, par MM. Hassenfratz et Adet*. Paris: Cuchet.

——. 1788. *Method of Chymical Nomenclature. To which is Added, a New System of Chymical Characters, Adapted to the Nomenclature, by Mess. Hassenfratz and Adet. Translated from the French, and the New Chymical Names adapted to the Genius of the English Language, by James St. John. M.D*. London: G. Kearsley.

——. 1793. *Methode der chemischen Nomenklatur für das antiphlogistische System*. Translated by Karl von Meidinger. Wien: Wappler.

——. 1994. *Méthode de nomenclature chimique*. Paris: Èditions du Seuil.

Hacking, Ian. 1983. *Representing and Intervening: Introductory Topics in the Philosophy of Natural Science*. Cambridge: Cambridge University Press.

——. 1991. "A Tradition of Natural Kinds." *Philosophical Studies* 61: 109–126.

——. 1992. "'Style' for Historians and Philosophers." *Studies in History and Philosophy of Science* 23(1): 1–20.

——. 1993. "Working in a New World: The Taxonomic Solution." In *World Changes: Thomas Kuhn and the Nature of Science*, edited by Paul Horwich, 275–310. Cambridge, MA: The MIT Press.

——. 2002. *Historical Ontology*. Cambridge, MA: Harvard University Press.

Hannaway, Owen. 1975. *The Chemists and the Word: The Didactic Origins of Chemistry*. Baltimore: Johns Hopkins University Press.

Harré, Rom, ed. 2005a. *Philosophical Problems of Chemical Kinds*, Foundations of Chemistry 7, Special issue.

——. 2005b. "Chemical Kinds and Essences Revisited." *Foundations of Chemistry* 7: 7–30.

Heilbron, John Lewis. 1990. "Introductory Essay." In *The Quantifying Spirit in the 18th Century*, edited by Tore Frängsmyr, John Lewis Heilbron, et al., 1–23. Berkeley: University of California Press.

——, ed. 2003. *The Oxford Companion to the History of Modern Science*. Oxford: Oxford University Press.

Hermbstädt, Sigismund Friedrich. 1807. *Anleitung zur Zergliederung der Vegetabilien*. Berlin: Ferdinand Oehmigke.

Hickel, Erika. 1978. "Der Apothekerberuf als Keimzelle naturwissenschaftlicher Berufe in Deutschland." *Medizinhistorisches Journal* 13: 259–276.

Hjelt, Edvard. 1916. *Geschichte der Organischen Chemie von ältester Zeit bis zur Gegenwart*. Braunschweig: Vieweg & Sohn.

Holmes, Frederic L. 1971. "Analysis by Fire and Solvent Extractions: The Metamorphosis of a Tradition." *Isis* 62(2): 129–148.

——. 1985. *Lavoisier and the Chemistry of Life: An Exploration of Scientific Creativity*. Madison: University of Wisconsin Press.

——. 1989. *Eighteenth-Century Chemistry as an Investigative Enterprise*. Berkeley: University of California Press.

——. 1995. "Beyond the Boundaries: Concluding Remarks on the Workshop." In *Lavoisier in European Context: Negotiating a New Language for Chemistry*, edited by Bernadette Bensaude-Vincent and F. Abbri, 267–278. Canton, MA: Watson Publishing International.

——. 1996. The Communal Context of Etienne-François Geoffroy's 'Table des rapports.' *Science in Context* 9(3): 289–311.

―――. 2004. *Investigative Pathways: Patterns and Stages in the Careers of Experimental Scientists*. New Haven: Yale University Press.

Holmes, Frederic L. and Trevor H. Levere. 2000. *Instruments and Experimentation in the History of Chemistry*. Cambridge, MA: The MIT Press.

Homberg, Wilhelm. 1702. "Essays de chimie." *Histoire de l' Académie Royale des Sciences: avec les mémoires de mathématique et de physique pour la même année*: 33–52.

Homburg, Ernst. 1993. *Van beroep 'Chemiker': De opkomst van de industriële chemicus en het polytechnische onderwijs in Duitsland (1790–1850)*. Delft: Delftse Universitaire Pers.

―――. 1999a. "From Colour Maker to Chemist: Episodes from the Rise of the Colourist, 1670–1800." In *Natural Dyestuffs and Industrial Culture in Europe, 1750–1880*, edited by Robert Fox and Augustí Nieto-Galan, 219–257. Canton, MA: Science History Publications.

―――. 1999b. "The Rise of Analytical Chemistry and its Consequences for the Development of the German Chemical Profession (1780–1860)." *Ambix* 46(1): 1–31.

Houghton, Walter E., Jr. 1941. "The History of Trades: Its Relation to Seventeenth-Century Thoughts: As Seen in Bacon, Petty, Evelyn, and Boyle." *Journal of the History of Ideas* 2(1): 33–60.

Hufbauer, Karl. 1982. *The Formation of the German Chemical Community (1720–1795)*. Berkeley: University of California Press.

Hunter, Michael. 1992. *Science and Society in Restoration England*. Aldershot: Gregg.

―――. 1995. *Science and the Shape of Orthodoxy: Intellectual Change in Late Seventeenth-Century Britain*. Woodbridge: Boydell Press.

Jacques, Jean. 1985. "Le Cours de chimie de G.-F. Rouelle recueilli par Diderot." *Revue d'histoire des sciences* 38: 43–53.

Jardine, Nicholas and Emma C. Spary. 1996. "The Natures of Cultural History." In *Cultures of Natural History*, edited by Nicholas Jardine, J. Anne Secord, and Emma C. Spary, 3–13. Cambridge: Cambridge University Press.

Kapoor, Satish C. 1965. "Berthollet, Proust, and Proportions." *Chymia* 10: 53–110.

―――. 1969a. "Dumas and Organic Classification." *Ambix* 16(1, 2): 1–65.

―――. 1969b. "The Origins of Laurent's Organic Classification." *Isis* 60(4): 477–527.

―――. 1981. "Article: Berthollet." In *Dictionnary of Scientific Biography*, edited by Charles Coulston Gillispie, 2: 73–82. New York: Scribner.

Kekulé, Friedrich August. 1861–1882. *Lehrbuch der organischen Chemie, oder der Chemie der Kohlenstoffverbindungen*, vol. 1 and 2. Erlangen: Ferdinand Enke, 1861 and 1866, vol. 3. Stuttgart: Ferdinand Enke, 1882.

Kersaint, Georges. 1966. *Antoine Francois de Fourcroy, sa vie et son oeuvre*. Paris: Museum National d'Histoire Naturelle.

Kim, Mi Gyung. 2001. "The Analytic Ideal of Chemical Elements: Robert Boyle and the French Didactic Tradition of Chemistry." *Science in Context* 14(3): 361–395.

―――. 2003. *Affinity, That Elusive Dream: A Genealogy of the Chemical Revolution*. Cambridge, MA: The MIT Press.

Klein, Ursula. 1994a. *Verbindung und Affinität. Die Grundlegung der neuzeitlichen Chemie an der Wende vom 17. zum 18. Jahrhundert*. Basel: Birkhäuser.

―――. 1994b. "Origin of the Concept of Chemical Compound." *Science in Context* 7(2): 163–204.

―――. 1994c. "Robert Boyle: der Begründer der neuzeitlichen Chemie?" *Philosophia Naturalis* 31(1): 63–106.

————. 1995. "E. F. Geoffroy's Table of Different 'Rapports' Observed between Different Chemical Substances—A Reinterpretation." *Ambix* 42(2): 79–100.

————. 1996a. "The Chemical Workshop Tradition and the Experimental Practice—Discontinuities within Continuities." *Science in Context* 9(3): 251–287.

————. 1996b. "Experiment, Spiritus und okkulte Qualitäten in der Philosophie Francis Bacons." *Philosophia Naturalis* 33(2): 289–315.

————. 1998a. "Paving a Way through the Jungle of Organic Chemistry—Experimenting within Changing Systems of Order." In *Experimental Essays—Versuche zum Experiment*, edited by Michael Heidelberger and Friedrich Steinle, 251–271. Baden-Baden: Nomos Verlagsgesellschaft.

————. 1998b. "Nature and Art in Seventeenth-Century French Chemical Textbooks." In *Reading the Book of Nature: The Other Side of the Scientific Revolution*, edited by Allen G. Debus and Michael T. Walton. 239–250. Kirksville: Thomas Jefferson University Press.

————. 1999. "Techniques of Modelling and Paper Tools in Classical Chemistry." In *Models as Mediators: Perspectives on Natural and Social Sciences*, edited by Mary Morgan and Margaret Morrison, 146–167. Cambridge: Cambridge University Press.

————. 2003a. *Experiments, Models, Paper Tools: Cultures of Organic Chemistry in the Nineteenth Century.* Stanford: Stanford University Press.

————. 2003b. "Experimental History and Herman Boerhaave's Chemistry of Plants." *Studies in History and Philosophy of Biological and Biomedical Sciences* 34: 533–567.

————. 2005a. "Experiments at the Intersection of Experimental History, Technological Inquiry, and Conceptual Analysis: A Case Study from Early Nineteenth-Century France." *Perspectives on Science* 13(1): 1–48.

————. 2005b. "Technoscience avant la lettre." *Perspectives on Science* 13(2): 227–266.

Kopp, Hermann Franz Moritz. 1966. *Geschichte der Chemie*. 4 vols. Hildesheim: Olms.

Kragh, Helge. 1998. "Afterword: The European Commonwealth of Chemistry." In *The Making of the Chemist: The Social History of Chemistry in Europe, 1789–1914*, edited by David M. Knight and Helge Kragh, 329–341. Cambridge: Cambridge University Press.

Kremers, Edward and George Urdang. 1951. *History of Pharmacy: A Guide and Survey*. Philadelphia, London, Montreal: J. B. Lippincott Company.

Kripke, Saul A. 1971. "Identity and Necessity." In *Identity and Individuation*, edited by Milton K. Munitz, 135–164. New York: New York University Press.

————. 1980. *Naming and Necessity*. Oxford: Basil Blackwell.

Kuhn, Thomas S. 1989. "Possible Worlds in History of Science." In *Possible Worlds in Humanities, Arts, and Sciences: Proceedings of Nobel Symposium 65*, edited by Sture Allén, 9–32, 49–51. Berlin: de Gruyter.

————. 1993. "Afterwords." In *World Changes: Thomas Kuhn and the Nature of Science*, edited by Paul Horwich, 311–341. Cambridge, MA: The MIT Press.

————. 1996. *The Structure of Scientific Revolutions*. 3rd ed. Chicago: University of Chicago Press.

Lasswitz, Kurd. 1984. *Geschichte der Atomistik vom Mittelalter bis Newton*. 2 vols. Hildesheim: Olms. Reprint of the edition Hamburg and Leipzig: L. Voss, 1890.

Lavoisier, Antoine-Laurent. 1789. *Traité élémentaire de chimie: Présenté dans un ordre nouveau et d'après les découvertes modernes*. 2 vols. Paris: Cuchet.

————. 1965. *Elements of Chemistry, in a New Systematic Order, Containing all the Modern Discoveries*. Translated by Robert Kerr. New York: Dover. Reprint of the editon Edinburgh 1790.

Le Febvre, Nicaise. 1664. *A Compendious Body of Chymistry: Teaching the Whole Practice thereof by the most Exact Preparation of Animals, Vegetables and Minerals, Preserving their Essential Virtues*. 2 vols. London: Ratcliffe.

Lefèvre, Wolfgang. 2000. "Galileo Engineer: Art and Modern Science." *Science in Context* 13/3–4: 281–297.

———. 2001. "Natural or Artificial Systems?—The Eighteenth-Century Controversy on Classification of Animals and Plants and its Philosophical Contexts." In *Between Leibniz, Newton, and Kant*, edited by Wolfgang Lefèvre, 191–209. Dordrecht: Kluwer.

———. 2004. *Picturing Machines 1400–1700*. Cambridge, MA: The MIT Press.

Le Grand, H.E. 1975. "The 'Conversion' of C.-L. Berthollet to Lavoisier's Chemistry." *Ambix* 22(1): 58–70.

Lemery, Nicolas. 1675. *Cours de chymie*. Paris: L'Auteur.

———. 1677. *A Course of Chemistry: Containing the Easiest Manner of Performing those Operations that are in Use in Physick, Illustrated with Many Curious Remarks and Useful Discourses upon each Operation*. Translated by Walter Harris. London: W. Kettilby.

———. 1713. *Cours de chymie*. 10. ed. Paris: Delespine.

Lenoir, Timothy. 1989. *The Strategy of Life: Teleology and Mechanics in Nineteenth-Century German Biology*. Chicago: University of Chicago Press.

Lesch, John E. 1990. "Systematics and the Geometrical Spirit." In *The Quantifying Spirit in the 18th Century*, edited by Tore Frängsmyr, John L. Heilbron, and Robin E. Rider, 73–111. Berkeley: University of California Press.

Levere, Trevor H. 1994. *Chemists and Chemistry in Nature and Society: 1770–1878*. Aldershot: Variorum.

Libavius, Andreas. 1597. *Alchemia*. Frankfurt am Main: Kopff.

———. 1964. *Die Alchemie des Andreas Libavius*. Weinheim: Verlag Chemie.

Liebig, Justus. 1832–1834. *Laborbücher, Liebigiana I, C. 2*. Munich: Bayerische Staatsbibliothek.

———. 1834. "Ueber die Constitution des Aethers und seiner Verbindungen." *Annalen der Pharmacie* 9: 1–39.

———. 1837. "Ueber die Aethertheorie, in besonderer Rücksicht auf die vorhergehende Abhandlung Zeise's." *Annalen der Chemie und Pharmacie* 23: 12–42.

Lindquist, Svante. 1984. *Technology on Trial: The Introduction of Steam Power Technology into Sweden, 1715–1736*. Uppsala: Almqvist and Wiksell International.

Linnaeus, Carl. 1737a. *Genera plantarum*. Lugduni Batavorum: Wishoff.

———. 1737b. *Critica botanica*. Lugduni Batavorum: Wishoff.

Löw, Reinhard. 1977. *Pflanzenchemie zwischen Lavoisier und Liebig*, vol. 1, *Münchner Hochschulschriften: Reihe Naturwissenschaften*. Straubing: Donau-Verlag.

Long, Pamela O. 2001. *Openness, Secrecy, Authorship: Technical Arts and the Culture of Knowledge from Antiquity to the Renaissance*. Baltimore: Johns Hopkins University Press.

Lüthy, Christoph, John E. Murdoch, and William R. Newman, eds. 2001. *Late Medieval and Early Modern Corpuscular Matter Theories*. Leiden, Boston, Köln: Brill.

Macquer, Pierre Joseph. 1749. *Elemens de chymie theorique*. Paris: Jean-Thomas Herissant.

———. 1751. *Elemens de chymie-pratique, contenant la descriptions des opérations fondamentales de la chymie, avec des explications et des remarques sur chaque opération*. 2 vols. Paris: Jean-Thomas Herissant.

————. 1766. *Dictionnaire de chymie, contenant la théorie et la pratique de cette science, son application à la physique, à l'histoire naturelle, à la médicine et à l'economie animale.* 2 vols. Paris: Lacombe.

————. 1771. *A Dictionary of Chemistry. Containing the Theory and Practice of that Science; its Application to Natural Philosophy, Natural History, Medicine, and Animal Economy. With full Explanations of the Qualities and Modes of Acting of Chemical Remedies, and the Fundamental Principles, of the Arts, Trades, and Manufactures Dependent on Chemistry.* Translated by James Keir. 2 vols. London: T. Cadell and P. Emsly, J. Robson.

————. 1778. *Dictionnaire de chymie, contenant la théorie et la pratique de cette science, son application à la physique, à l'histoire naturelle, à la médicine et aux arts dépendans de la chymie.* 2nd. ed. 4 vols. Paris: L'Imprimerie de Monsieur.

Macquer, Pierre Joseph and Antoine Baumé. 1757. *Plan d'un cours de chymie expérimentale et raisonnée avec un discours historique sur la chymie.* Paris: Jean-Thomas Herissant.

Malpighi, Marcello. 1901. *Die Anatomie der Pflanzen. I. und II. Theil. London 1675 und 1679. Ostwalds Klassiker der exakten Wissenschaften Nr. 120.* Leipzig: Wilhelm Engelmann.

Mauskopf, Seymour H. 1976. "Crystals and Compounds: Molecular Structure and Composition in Nineteenth-Century French Science." *Transactions of the American Philosophical Society* 66(3): 1–82.

————. 1995. "Lavoisier and the Improvement of Gunpowder Production." *Revue d'histoire des sciences* 48: 95–121.

Mayr, Ernst. 1982. *The Growth of Biological Thought.* Cambridge, MA: Belknap.

————. 1992. "A Local Flora and the Biological Species Concept." *American Journal of Botany* 79: 222–238.

McOuat, Gordon R. 1996. "Species, Rules and Meaning: The Politics of Language and the Ends of Definitions in Nineteenth-Century Natural History." *Studies in History and Philosophy of Science* 27(4): 473–519.

Meinel, Christoph. 1983. "Theory or Practice? The Eighteenth-Century Debate on the Scientific Status of Chemistry." *Ambix* 30(3): 121–132.

————. 1985. "Reine und angewandte Chemie. Die Entstehung einer neuen Wissenschaftskonzeption in der Chemie der Aufklärung." *Berichte zur Wissenschaftsgeschichte* 8: 25–45.

————. 1988. "Early Seventeenth-Century Atomism. Theory, Epistemology, and the Insufficiency of Experiments." *Isis* 79: 68–103.

Melhado, Evan M. 1981. *Jacob Berzelius: The Emergence of his Chemical System.* Madison: University of Wisconsin Press.

————. 1992. "Novelty and Tradition in the Chemistry of Berzelius (1803–1819)." In *Enlightenment Science in the Romantic Era: The Chemistry of Berzelius and Its Cultural Setting*, edited by Evan M. Melhado and Tore Frängsmyr, 132–170. Cambridge: Cambridge University Press.

Merton, Robert K. 1970. *Science, Technology and Society in Seventeenth-Century England.* New York: Harper & Row.

Metzger, Hélène. 1923. *Les doctrines chimiques en France du début du XVIIe à la fin du XVIIIe siècle.* Paris: Presses Univ. de France.

Mittelstraß, Jürgen, ed. 1980–1996. *Enzyklopädie Philosophie und Wissenschaftstheorie.* 4 vols. Stuttgart: Metzler.

Moran, Bruce T. 2005. *Distilling Knowledge. Alchemy, Chemistry, and the Scientific Revolution.* Cambridge, MA: Harvard University Press.

Morris, Peter J. T., ed. 2002. *From Classical to Modern Chemistry: The Instrumental Revolution*. Cambridge: Royal Society of Chemistry [et al.].

Müller-Wille, Staffan. 1999. *Botanik und weltweiter Handel: Zur Begründung eines natürlichen Systems der Pflanzen durch Carl von Linné (1707–78)*. Berlin: Verlag für Wissenschaft und Bildung.

———. 2003. "Joining Lapland and the Topinambes in Flourishing Holland: Center and Periphery in Linnean Botany." *Science in Context* 16(4): 461–488.

Multhauf, Robert P. 1965. "Sal Ammoniac: A Case History in Industrialization." *Technology and Culture* 4: 569–586.

———. 1966. *The Origins of Chemistry*. London: Oldbourne.

———. 1972. "A Premature Science Advisor: Jacob A. Weber (1732–1792)." *Isis* 63(3): 356–369.

———. 1984. *The History of Chemical Technology. An Annotated Bibliography*. New York: Garland.

———. 1996. "Operational Practice and the Emergence of Modern Chemical Concepts." *Science in Context* 9(3): 241–249.

Munitz, Milton K., ed. 1971. *Identity and Individuation*. New York: New York University Press.

Neumann, Caspar. 1740. *Herrn D. Caspar Neumanns, Praelectiones chemicae seu chemia medico-pharmaceutica experimentalis & rationalis, oder gründlicher Unterricht der Chemie. Von Johann Christian Zimmermann*. Berlin: Johann Andreas Rüdiger.

———. 1756. *D. Caspar Neumanns Chymia Medica dogmatico-experimentalis. Tomus primus. Das ist, gründliche und mit Experimenten erwiesene Medicinische Chymie, Erster Band, welcher die Lehre von nassen und trocknen Artzeneyen und die chymische Untersuchung des Pflantzenreichs in sich fasset*. Züllichau: Johann Jacob Dendeler.

———. 1759. *The Chemical Works of Caspar Neumann, Professor of Chemistry at Berlin, F. R. S. etc. Abridged and Methodized. With Large Additions, Containing the Later Discoveries and Improvements Made in Chemistry and the Arts Depending thereon. By William Lewis*. London: W. Johnston, G. Keith, A. Linde, P. Davey, B. Law, T. Field, T. Caslon, and E. Dilly.

Newman, William R. 1994. *Gehennical Fire: The Lives of George Starkey, an American Alchemist in the Scientific Revolution*. Cambridge, MA: Harvard University Press.

———. 2000. "Alchemy, Assaying, and Experiment." In *Instruments and Experimentation in the History of Chemistry*, edited by Frederic L. Holmes and Trevor H. Levere, 35–54. Cambridge, MA: The MIT Press.

———. 2004. *Promethean Ambitions: Alchemy and the Quest to Perfect Nature*. Chicago: University of Chicago Press.

Newman, William R. and Lawrence M. Principe. 1998. "Alchemy vs. Chemistry: The Ethymological Origins of a Historiographic Mistake." *Early Science and Medicine* 3: 32–65.

———. 2002. *Alchemy Tried in the Fire: Starkey, Boyle, and the Fate of Helmontian Chymistry*. Chicago: University of Chicago Press.

Nieto-Galan, Agustí. 2001. *Colouring Textiles: A History of Natural Dyestuffs in Industrial Europe*, Boston Studies in the Philosophy of Science, vol. 217. Dordrecht: Kluwer.

Nummedal, Tara E. 2002. "Practical Alchemy and Commercial Exchange in the Holy Roman Empire." In *Merchants and Marvels: Commerce, Science and Art in Early Modern Europe*, edited by Pamela H. Smith and Paula Findlen, 201–222. New York and London: Routledge.

Ochs, Kathleen H. 1985. "The Royal Society of London's History of Trade Programme: An Early Episode in Applied Science." *Notes and Records of the Royal Society of London* 39(2): 129–158.

Olby, R.C., G. N. Cantor, J. R. R. Christie, and M. J. S. Hodge. eds. 1990. *Companion to the History of Modern Science*. London and New York: Routledge.

Oldroyd, D.R. 1974a. "Some Phlogistic Mineralogical Schemes, Illustrative of the Evolution of the Concept of 'Earth' in the 17th and 18th Centuries." *Annals of Science* 31(4): 269–305.

———. 1974b. "A Note on the Status of A. F. Cronstedt's Simple Earths and his Analytical Methods." *Isis* 65: 506–512.

———. 1975. "Mineralogy and the 'Chemical Revolution.'" *Centaurus* 19: 54–71.

Olschki, Leonardo. 1965. *Geschichte der neusprachlichen wissenschaftlichen Literatur*. 3 vols. 1919–1927. Vaduz: Kraus Reprint.

Pagel, Walter. 1982. *Paracelsus: An Introduction to Philosophical Medicine in the Era of the Renaissance*. Basel: Karger.

Paracelsus. 1996. *Sämtliche Werke*. Hildesheim: Olms. Reprint of the edition Munich: Barth, 1922–1931.

Partington, James R. 1961–1970. *A History of Chemistry*. 4 vols. London: Macmillan.

Peirce, Charles Sanders. 1931–1958. *Collected Papers of Charles Sanders Peirce*. Edited by Charles Hartshorne and Paul Weiss. 8 vols. Cambridge, MA: Harvard University Press.

Perrin, Carleton E. 1973. "Lavoisier's Table of the Elements: A Reappraisal." *Ambix* 20(1): 95–105.

———. 1981. "The Triumph of the Antiphlogistians." In *The Analytic Spirit*, edited by Harry Woolf, 40–63. Ithaca: Cornell University Press.

Pharmacopoea. 1732. *Pharmacopoea Leidensis*. Lugduni Batavorum: Samuelem Luchtmans.

———. 1737. *Pharmacopoeia Edinburgensis: or, the Dispensatory of the Royal College of Physicians in Edinburgh. Translated and Improved from the Third Edition of the Latin, and Illustrated with Notes, by Peter Shaw*. 3 ed. London: W. Innys and R. Manby.

———. 1741. *Pharmacopoea Wirtenbergica*. Stuttgardiae: Christophori Erhardi Bibliopolae.

———. 1748. *The Dispensatory of the Royal College of Physicians, London, Translated into English with Remarks, etc. by H. Pemperton*. London: T. Longman and J. Nourse.

———. 1781. *Dispensatorium Regium et Electorale Borusso-Brandenburgicum*. Berolini: Christiani Sigismundi Spener.

Pickering, Andrew. 2005. "Decentring Sociology: Synthetic Dyes and Social Theory." *Perspectives on Science* 13(3): 352–405.

Pickstone, John V. 2000. *Ways of Knowing: A New History of Science, Technology, and Medicine*. Manchester: Manchester University Press.

Poirier, Jean-Pierre. 1993. *Lavoisier—Chemist, Biologist, Economist*. Philadelphia: University of Pennsylvania Press.

Pomata, Gianna and Nancy G. Siraisi, eds. 2005. *Historia: Empiricism and Erudition in Early Modern Europe*. Cambridge, MA: The MIT Press.

Porter, Roy, ed. 2003. *The Cambridge History of Science*, vol. 4. *Eighteenth-Century Science*. Cambridge: Cambridge University Press.

Porter, Theodore M. 1981. "The Promotion of Mining and the Advancement of Science: the Chemical Revolution of Mineralogy." *Annals of Science* 38: 543–570.

Priesner, Claus. 1986. "Spiritus Aethereus—Formation of Ether and Theories on Etherification from Valerius Cordus to Alexander Williamson." *Ambix* 33(2/3): 129–152.

Principe, Lawrence M. 1998. *The Aspiring Adept. Robert Boyle and his Alchemical Quest. Including Boyle's "Lost" Dialogue on the Transmutation of Metals*. Princeton: Princeton University Press.

———. 2001. "Wilhelm Homberg: Chymical Corpuscularism and Chrysopoeia in the Early Eighteenth Century." In *Late Medieval and Early Modern Corpuscular Matter Theories*, edited by Christoph Lüthy, John E. Murdoch, and William R. Newman, 535–556. Leiden, Boston, Köln: Brill.

Putnam, Hilary. 1975a. *Mind, Language, and Reality*. Cambridge: Cambridge University Press.

———. 1975b. "The Meaning of 'Meaning.'" In *Language, Mind, and Knowledge*, edited by Keith Gunderson, 131–193. Minneapolis: University of Minnesota Press.

———. 1990. "Is Water Necessarily H_2O?" In *Realism with a Human Face*, edited by James Conant, 54–79. Cambridge, MA: Harvard University Press.

Quine, Willard V. 1969. "Natural Kinds." In *Ontological Relativity and Other Essays*, edited by Willard V. Quine, 114–138. New York and London: Columbia University Press.

Rappaport, Rhoda. 1960. "G. F. Rouelle: An Eighteenth-Century Chemist and Teacher." *Chymia* 6: 68–101.

Reinhardt, Carsten. 2006. *Shifting and Rearranging. Physical Methods and the Transformation of Modern Chemistry*. Sagamore Beach: Science History Publications.

Rheinberger, Hans-Jörg. 1997. *Toward a History of Epistemic Things: Synthesizing Proteins in the Test Tube*. Stanford: Stanford University Press.

Riskin, Jessica. 1998. "Rival Idioms for a Revolutionized Science and a Republican Citizenry." *Isis* 89: 203–232.

Roberts, Lissa. 1991. "Setting the Table: The Disciplinary Development of Eighteenth-Century Chemistry as Read Through the Changing Structures of its Tables." In *The Literary Structure of Scientific Argument: Historical Studies*, edited by Peter Dear, 99–132. Philadelphia: University of Pennsylvania Press.

———. 1992. "Condillac, Lavoisier, and the Instrumentalization of Science." *Eighteenth Century: Theory and Interpretation* 33: 252–271.

Rocke, Alan J. 1984. *Chemical Atomism in the Nineteenth Century: From Dalton to Cannizzaro*. Columbus: Ohio State University Press.

———. 1992. "Berzelius's Animal Chemistry: From Physiology to Organic Chemistry (1805–1814)." In Enlightenment Science in the Romantic Era: The Chemistry of Berzelius & its Cultural Setting, edited by Evan M. Melhado & Tore Frängsmyr, 107–131. Cambridge: Cambridge University Press.

———. 2000. "Organic Analysis in Comparative Perspective: Liebig, Dumas, and Berzelius, 1811–1837." In *Instruments and Experimentation in the History of Chemistry*, edited by Frederic L. Holmes and Trevor H. Levere, 273–310. Cambridge, MA: The MIT Press.

Roger, Jacques. 1997. *Buffon: A Life in Natural History*. Ithaca and London: Cornell University Press.

Rossi, Paolo. 1968. *Francis Bacon: From Magic to Science*. London: Routledge & Kegan Paul.

———. 1970. *Philosophy, Technology, and the Arts in the Early Modern Era*. New York: Harper & Row.

———. 2001. *The Birth of Modern Science*. Oxford and Malden: Blackwell.

Roth, Wolff-Michael. 2005. "Making Classifications (at) Work." *Social Studies of Science* 35(4): 581–621.

Rouelle, Guillaume François. 1744. "Memoire sur les sels neutres." *Histoire de l' Académie Royale des Sciences: avec les mémoires de mathématique et de physique pour la même année*: 353–364.

———. 1754. "Mémoire sur les sels neutres." *Histoire de l' Académie Royale des Sciences: avec les mémoires de mathématique et de physique pour la même année*: 572–588.

———. 1759. *Cours d'experience chymique*. Paris: de Grangé.

———. n.d. *Le cours de chimie de Rouelle*. Paris: Bibliothèque Interuniversitaire de Médicine, MS 5021.

Sadoun-Goupil, Michelle. 1977. *Le chimiste Claude-Louis Berthollet: Sa vie, son oeuvre*. Paris: Vrin.

Sage, Balthazar Georges. 1773. *Mémoires de chimie*. Paris: De l'imprimerie Royale.

Saussure, Nicolas Theodore de. 1807. "Mémoire: Sur la composition de l'alcohol et de l'éther sulfurique." *Journal de Physique, de Chimie, d'Histoire Naturelle et des Arts* 64: 316–354.

———. 1814. "Nouvelles observations sur la composition de l'alcool et de l'éther sulfurique." *Annales de Chimie* 89: 273–305.

Schiller, Joseph. 1978. *La notion d'organisation dans l'histoire de la biologie*. Paris: Maloine.

Schmauderer, Eberhardt. 1969. *Entwicklungsformen der Pharmakopöen*. Düsseldorf: VDI-Verlag.

Schneider, Wolfgang. 1968–1975. *Lexikon zur Arzneimittelgeschichte*. 7 vols. Frankfurt am Main: Govi-Verlag.

———. 1972. *Geschichte der pharmazeutischen Chemie*. Weinheim: Verlag Chemie.

Schummer, Joachim. 1997. "Scientometric Studies on Chemistry I: The Exponential Growth of Chemical Substances, 1800–1995." *Scientometrics* 39(1): 107–123.

Sellars, Wilfried. 1963. *Science, Perception and Reality*. London: Routledge & Kegan Paul.

Shaw, Peter. 1734. *Chemical Lectures Publickly Read at London, in the Years 1731, and 1732, and since at Scarborough, in 1733: for the Improvement of Arts, Trades, and Natural Philosophy*. London: J. Shuckburgh and T. Osborne.

Siegfried, Robert. 1982. "Lavoisier's Table of Simple Substances." *Ambix* 29(1): 29–48.

———. 2002. *From Elements to Atoms. A History of Chemical Composition*. Philadelphia: American Philosophical Society.

Siegfried, Robert and Betty Jo Dobbs. 1968. "Composition, a Neglected Aspect of the Chemical Revolution." *Annals of Science* 24(4): 275–293.

Simon, Jonathan. 2005. *Chemistry, Pharmacy and Revolution in France, 1777–1809*. Aldershot: Ashgate.

———. 1998. "The Chemical Revolution and Pharmacy: A Disciplinary Perspective." *Ambix* 45(1): 1–13.

———. 2002. "Authority and Authorship in the Method of Chemical Nomenclature." *Ambix* 49(3): 206–26.

Smeaton, W.A. 1954. "The Contributions of P.-J. Macquer, T.O. Bergman, and L.B. Guyton de Morveau to the Reform of Chemical Nomenclature." *Annals of Science* 10(2): 87–106.

———. 1957. "L. B. Guyton de Morveau (1737–1816). A Bibliographical Study." *Ambix* 6(1): 18–34.

———. 1962. *Fourcroy: Chemist and Revolutionary, 1755–1809*. Cambridge: Heffer.

Smith, John Graham. 1979. *The Origins and Early Development of the Heavy Chemical Industry in France*. Oxford: Clarendon Press.

Smith, Pamela H. 1994. *The Business of Alchemy: Science and Culture in the Holy Roman Empire*. Princeton: Princeton University Press.

————. 2004. *The Body of the Artisan: Art and Experience in the Scientific Revolution.* Chicago: The University of Chicago Press.

Smith, Pamela H. and Paula Findlen, eds. 2002. *Merchants and Marvels: Commerce, Science and Art in Early Modern Europe.* New York and London: Routledge.

Spary, Emma C. 2000. *Utopia's Garden. French Natural History from Old Regime to Revolution.* Chicago: University of Chicago Press.

Stahl, Georg Ernst. 1728. *Fundamenta pharmaciae chymicae manu methodoque Stahliana posita.* Budingae: J. F. Regelein.

————. 1730. *Philosophical Principles of Universal Chemistry: Or, the Foundation of a Scientifical Manner of Inquiry into Preparing the Natural and Artificial Bodies for the Uses of Life.* Translated by Peter Shaw. London: John Osborne and Thomas Longman.

————. 1748. *Zymotechnia fundamentalis: oder Allgemeine Grund-Erkenntniß der Gaehrungs-Kunst. Aus dem Lateinischen ins Teutsche übersetzt.* 2. ed. Stettin and Leipzig: Kunckelsche Handlung.

————. 1765. *Ausführliche Betrachtung und zulänglicher Beweis von den Saltzen, daß dieselbe aus einer zarten Erde, mit Wasser innig verbunden, bestehen [1. Aufl. 1723].* 2. ed. Halle: Verl. des Waisenhauses.

Star, Susan Leigh and James R. Griesemer. 1989. "Institutional Ecology, 'Translations' and Boundary Objects: Amateurs and Professionals in Berkeley's Museum of Vertebrate Zoology, 1907–39." *Social Studies of Science* 19(3): 387–420.

Stewart, Larry. 1992. *The Rise of Public Science: Rhetoric, Technology and Natural Philosophy in Newtonian Britain, 1660–1750.* Cambridge: Cambridge University Press.

Strawson, P. F. 1979. *Individuals: An Essay in Descriptive Metaphysics.* London: Methuen.

Stromeyer, Friedrich. 1808. *Grundriss der theoretischen Chemie.* 2 vols. Göttingen: Johann Friedrich Röwer.

Stroup, Alice. 1990. *A Company of Scientists: Botany, Patronage, and Community at the Seventeenth-Century Parisian Royal Academy of Science.* Berkeley: University of California Press.

Szabadváry, Ferenc. 1966. *Geschichte der analytischen Chemie.* Braunschweig: Vieweg.

Taylor, John K. 1986. "The Impact of Instrumentation on Analytical Chemistry." In *The History and Preservation of Chemical Instrumentation*, edited by John T. Stock and Mary V. Orna, 1–10. Dordrecht: Reidel.

Teich, Mikuláš. 1975. "Born's Amalgamation Process in the International Metallurgic Gathering at Skleno in 1786." *Annals of Science* 32: 305–340.

Thenard, Louis Jacques. 1807a. "Mémoire sur les éthers." *Mémoires de Physique et de Chimie de la Société d'Arcueil* 1: 73–114.

————. 1807b. "Deuxième mémoire sur les éthers: De l'éther muriatique." *Mémoires de Physique et de Chimie de la Société d'Arcueil* 1: 115–135.

————. 1807c. "Sur la découverte de l'éther muriatique." *Mémoires de Physique et de Chimie de la Société d'Arcueil* 1: 135–139.

————. 1807d. "Troisième mémoire sur les éthers: Des produits qu'on obtient en traitant l'alcool par les muriates métalliques, l'acide muriatique oxigéné et l'acide acétique." *Mémoires de Physique et de Chimie de la Société d'Arcueil* 1: 140–160.

————. 1807e. "Deuxième mémoire sur l'éther muriatique." *Mémoires de Physique et de Chimie de la Société d'Arcueil* 1: 337–358.

————. 1807f. "Nouvelles observations sur l'éther nitrique." *Mémoires de Physique et de Chimie de la Société d'Arcueil* 1: 359–369.

————. 1809a. "De l'action des acides végétaux sur l'alcool, sans l'intermède et avec l'inter-mède des acides minéraux." *Mémoires de Physique et de Chimie de la Société d'Arcueil* 2: 5–22.

————. 1809b. "Sur la combinaison des acides avec les substances végétales et animales." *Mémoires de Physique et de Chimie de la Société d'Arcueil* 2: 23–41.

————. 1809c. "Note sur la combinaison des matières végétales et animales avec les acides." *Mémoires de Physique et de Chimie de la Société d'Arcueil* 2: 492–494.

————. 1813–1816. *Traité de chimie élémentaire, théorique et pratique.* 1 ed. 4 vols. Paris: Crochard.

————. 1817–1818. *Traité de chimie élémentaire, théorique et pratique.* 2 ed. 4 vols. Paris: Crochard.

Thomson, Thomas. 1804. *A System of Chemistry.* 2 ed. 4 vols. Edinburgh: Bell & Bradfute, & E. Balfour.

Tomic, Sacha. 2003. *Les pratiques et les enjeux de l'analyse chimique des végétaux: Étude d'une culture hybride (1790–1835).* Ph.D. thesis: Université de Paris X—Nanterre.

Trommsdorff, Johann Bartholomäus. 1800–1807. *Systematisches Handbuch der gesammten Chemie zur Erleichterung des Selbststudiums dieser Wissenschaft.* 8 vols. Erfurt: Henning-sche Buchhandlung.

Usselman, Melvyn C., Christina Reinhart, Kelly Foulser, and Alan Rocke. 2005. "Restaging Liebig: A Study in the Replication of Experiments." *Annals of Science* 62(1): 1–55.

Venel, Gabriel François. 1753. "Article: 'Chymie.'" In *Encyclopédie, ou dictionnaire raisonné*, edited by Denis Diderot, vol. III 408–437. Paris: Briasson.

————. 1755. "Essai sur l'analyse des végétaux. Premier mémoire, contenant l'exposition abrégée de mon travail, et des considérations générales sur la distillation analytique des plantes." *Mémoires de mathematique et de physique présentées à l'Académie Royale des Sciences par divers savans* 2: 319–332.

Viel, Claude. 1992. "Guyton-Morveau, père de la nomenclature chimique." In *Lavoisier et la révolution chimique*, edited by Michelle Goupil, 129–170. Paris: SABIX.

Vogel, Rudolph August. 1755. *Institutiones chemiae ad lectionibus academicas accomodatae.* Göttingen: Luzac.

————. 1775. *Lehrsätze der Chemie. Ins deutsche übersetzt, und mit Anmerkungen versehen von Johann Christian Wiegleb.* Weimar: Carl Ludolf Hoffmann.

Weigel, Christian Ehrenfried. 1777. *Grundriß der reinen und angewandten Chemie. Zum Gebrauch academischer Vorlesungen.* Greifswald: Anton Ferdinand Röse.

White, Frederic A. 1961. *American Industrial Research Laboratories.* Washington: Public Affairs Press.

Wiegleb, Johann Christian. 1786. *Handbuch der allgemeinen Chemie.* 2 vols. Berlin and Stettin: Friedrich Nicolai.

Wiesenfeldt, Gerhard. 2002. *Leerer Raum in Minervas Haus: Experimentelle Naturlehre an der Universität Leiden, 1675–1715, History of Science and Scholarship in the Nether-lands.* Amsterdam: Royal Netherlands Academy of Arts and Sciences.

Wilsdorf, Helmut. 1956. *Georg Agricola und seine Zeit.* (= Vol. I of Georg Agricola: *Ausgewählte Werke.* Hans Prescher ed.) Berlin: Deutscher Verlag der Wissenschaften.

Name Index

Subject Index

chemical transformations
  reversible, 38, 47, 56–58, 64–65, 99, 109, 136–146, 204, 230, 296, 301
  see also chemical composition, decomposition and re-composition
chemical workshops, manufactures, and trades, 15, 27–28, 30, 36, 83, 112, 137, 147, 199, 209, 238
  see also apothecary trade, assaying shops, distilleries, and pottery
chemistry
  academic, 3, 36, 83–84, 93, 133, 295–296
  see also chemists, academic
  applied, 3, 27–28, 68, 75, 207–208
  artisanal, 33, 83, 111
  classical, 9, 60, 69, 304
  carbon, 3, 38 62, 68, 74, 287–293
  of gases, 85, 92, 102, 135
  of life, 237, 260–262
  organic, 13, 62, 68–69, 74, 78, 192, 196, 204–205, 251, 264–267, 279–283, 287–289, 293, 296, 303
  pure, 27, 29, 68, 75
  see also alchemy, chymistry, experimental culture of organic chemistry, lectures, and textbooks
chemists, chymists
  academic, 15–16, 18–19, 33–34, 58, 64, 112, 148, 297
  iatrochemists, Paracelsian chymists, 21, 39–45, 49, 53, 55–56, 64–65, 114, 141, 170
  industrial, 37
  self-representation of, 14, 68
  see also apothecaries, assayers, distillers, and mining, officials
chymical species, 49–53
chymistry, 17, 39–40, 49–50, 53–54, 212
  see also alchemy and chemistry
cinis clavelettus, see potash
classification, taxonomy
  artificial / natural, 76–77, 131, 273, 283
  chemical, *see* chemical classification
  explicit / implicit, 49, 77–78, 86, 104, 155–158, 162–163, 179–180
  Linnaean, 63, 79, 131, 260, 275
  pharmaceutical, 205–210

selectivity of, 63–65
  see also mineralogy, classfication and textbooks, organization
classificatory structure, 102, 105–107, 109–110, 128–131, 155, 157, 179, 183, 185–186, 188, 303
  analogous taxa, analogy of taxa, 129, 132, 182–185
  completeness, 99
  dichotomous divisions, exhaustive bifurcations, 50, 273
  encaptic, 50–51, 78, 107, 184
  matrix, 78–79, 85, 107, 187
  rendered by diagrams, 49, 77–78
  rendered by lists, 63, 77–79, 150, 201, 209, 222, 242
  rendered by tables, 75–79, 155–163
  system, 3, 9, 49–51, 53, 63, 68, 78–79, 97, 107, 131–132, 275, 281–282, 287
  taxonomic ranks, 12, 66, 79, 85, 106–107, 191, 201, 268, 278, 300
  taxonomic tree, 78, 183–184
  see also alphabetical order
clissus, 51–52
coal, charcoal, 36, 103, 136, 167, 210, 248–249, 251, 256
  see also carbon
coal tar, 290, 292–293
coffee, 209
coloring parts of plants, *see* vegetable materials, coloring parts of plants
coloring principle, 200, 229
combustion, 8, 36, 104, 117, 167, 181, 183, 185, 250,
combustion method 117, 298–299, 302
Commission d'Agriculture et des Arts, 94
Commission des Poids et Mesures, 94
component, *see* chemical component
composition, *see* chemical composition
compound, *see* chemical compound
confusion, 42, 47
copal officin, see gums, *copal officin*
copper, 15–17, 138, 143–147, 149
corpuscles, 8, 26, 41, 44–49, 61–62
  preserved in chemical transformations, 46–48
  see also atoms

sels salés, 56, 173–175
  theories of, 8, 15, 56, 65
  see also Glauber's salt, *materière saline,*
    and *Sel febrifuge de Sylvius*
saltpeter, *Salt-petre,* 94, 174, 245
  see also niter
sandaracha officin, see gum-resins, *sandaracha officin*
saps, 212–214, 240, 261–263
  watery, 213
saturation, 97, 103, 172–173, 189, 192, 267,
  302
scammony, 199, 239, 258–259
Scheidekunst, see metallurgy, separation
sel, see salt
selection, selectivity, *see* chemical substances, selection of, and classification, selectivity of
Sel febrifuge de Sylvius, 91
semi-metals, *see* metals, half-metals
separation, 40–44, 48, 55, 114–117, 138–
  142, 218
  of gold and silver, 16, 136–137, 140, 145
  see also metallurgy, separation, and met-
    als, separation of
silver, 136–140, 143–146, 149, 166
simples, *simplicia, see* remedies, simple
smelting, *see* metallurgy, smelting
soap, 16, 28, 209, 226, 259, 263
  manufacture of, 28, 259
Society of Arcueil, 276, 285
soda, 16, 94, 142, 176
solutions, 42, 145–147
  acid, 110, 112, 128, 146–147, 152–153
  metallic, 138
  saline, 209
  watery, 56–57, 65–66, 70, 115–116, 147
spectroscopic analysis, 60
  see also chemical analysis
spices, 209
spirits, 39, 43–44, 114, 212, 215
  acid, 113, 210, 212, 226
  ardent, 16
  aromatic, 199–200, 226
  combustible, 209
  composite, 200
  composite aromatic, 200

presiding, 215
saline, 209
of wine, *see* wine, spirit of
Spiritus fumans Libavii, 143–144, 152
spiritus rector, 215, 239, 241
starch, 199–200, 241, 246, 248, 256, 259,
  263, 265, 269–270, 276, 278–279, 288–
  289
stibnite, 138
strychnine, 216, 291
styrax calamita officin, see resins, *styrax
  calamita officin*
suber, 209, 262–263
sublimate, 142–145
substances
  animal, *see* animal materials
  chemical, *see* chemical substances
  exalted, ennobled, 39–40, 49–55, 64, 84
  imperceptible dimension of, 19–20, 22,
    24, 37–58, 59–61, 211, 297–299
  laboratory, *see* chemical substances, pure
  mineral, *see* minerals
  organic, *see* organic materials
  processed, *see* chemical substances, pro-
    cessed
  spiritualization of, 40–43, 54
  transmuted, 52, 55
  vegetable, *see* vegetable materials,
  see also natural philosophy and princi-
    ples
Substances metalliques, see metals, *Substances metalliques*
substantial form, 42, 53
substitution, *see* chemical reaction, substitution
substitution products, 68, 291
Sucre de Saturne, 91
sugar, *sacharum,* 120, 207, 209, 222, 226–
  227, 245–246, 248, 261, 263, 278, 280–
  281
  beet sugar, 28, 200
  cane sugar, 200, 202, 264, 279
  extraction and purification of, 28–29
  of milk, 276, 279
sulfur, 56, 121, 151, 164–165, 167–169, 180
  compounds of, 65–66, 99, 103, 106, 121,
    132, 181